Voyage To Jupiter

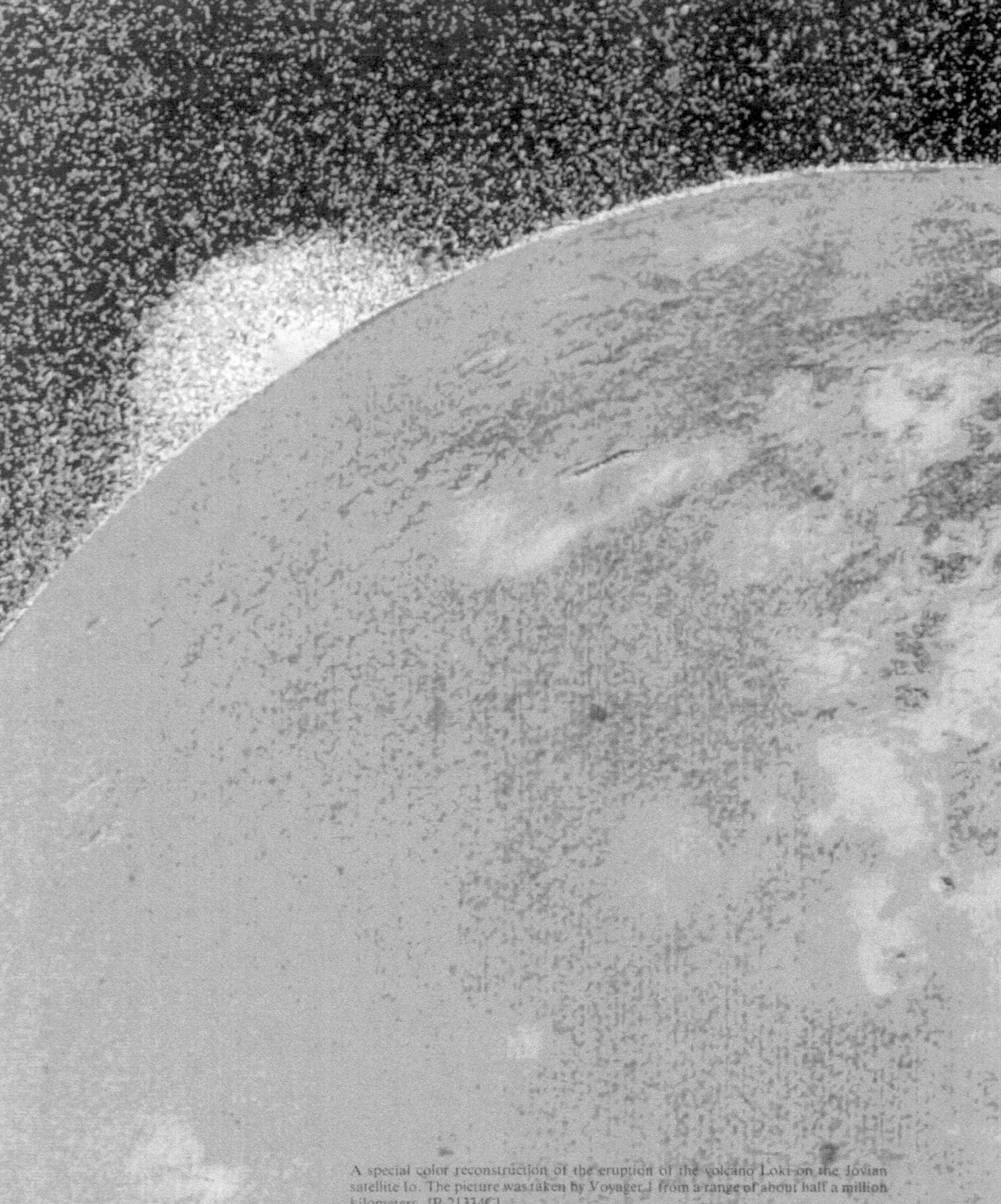

A special color reconstruction of the eruption of the volcano Loki on the Jovian satellite Io. The picture was taken by Voyager I from a range of about half a million kilometers. [P-21334C]

NASA SP-439

Voyage To Jupiter

David Morrison
and Jane Samz

NASA Scientific and Technical Information Branch   1980
National Aeronautics and Space Administration
Washington, DC

For sale by the Superintendent of Documents
U.S. Government Printing Office, Washington, D.C. 20402
Library of Congress Catalog Card Number 80-600126

FOREWORD

Few missions of planetary exploration have provided such rewards of insight and surprise as the Voyager flybys of Jupiter. Those who were fortunate enough to be with the science teams during those weeks will long remember the experience; it was like being in the crow's nest of a ship during landfall and passage through an archipelago of strange islands. We had known that Jupiter would be remarkable, for man had been studying it for centuries, but we were far from prepared for the torrent of new information that the Voyagers poured back to Earth.

Some of the spirit of excitement and connection is captured in this volume. Its senior author was a member of the Imaging Team. It is not common that a person can both "do science" at the leading edge and also present so vivid an inside picture of a remarkable moment in the history of space exploration.

April 30, 1980

Thomas A. Mutch
Associate Administrator
Office of Space Science

INTRODUCTION

The two Voyager encounters with Jupiter were periods unparalleled in degree and diversity of discovery. We had, of course, expected a number of discoveries because we had never before been able to study in detail the atmospheric motions on a planet that is a giant spinning sphere of hydrogen and helium, nor had we ever observed planet-sized objects such as the Jovian satellites Ganymede and Callisto, which are half water-ice. We had never been so close to a Moon-sized satellite such as Io, which was known to be dispersing sodium throughout its Jovian neighborhood and was thought to be generating a one-million-ampere electrical current that in some way results in billions of watts of radio emission from Jupiter.

The closer Voyager came to Jupiter the more apparent it became that the scientific richness of the Jovian system was going to greatly exceed even our most optimistic expectations. The growing realization among Voyager scientists of the wealth of discovery is apparent in their comments, discussions, and reports as recounted by the authors in their descriptions of the two encounters.

Although many of the discoveries occurred in the few weeks around each encounter, they were, of course, the result of more than those few weeks of effort. In fact, planning started a decade earlier, and the Voyager team of engineers and scientists had been designing, building, and planning for the encounters for seven years. The Pioneer spacecraft made the first reconnaissance of Jupiter in 1973-1974, providing key scientific results on which Voyager could build, and discoveries from continuing ground-based observations suggested specific Voyager studies. Voyager is itself just the second phase of exploration of the Jovian system. It will be followed by the Galileo program, which will directly probe Jupiter's atmosphere and provide long-term observations of the Jovian system from an orbiting spacecraft. In the meantime, the Voyager spacecraft will continue their journey to Saturn, and possibly Uranus and Neptune, planets even more remote from Earth and about which we know even less than we knew of Jupiter before 1979.

As is clearly illustrated in this recounting of the voyage to Jupiter, scientific endeavors are human endeavors; just as Galileo could not have foreseen the advancement in our knowledge initiated by his discoveries of the four Jovian moons in 1610, neither can we fully comprehend the scientific heritage that our exploration of space is providing future generations.

April 1980

Edward C. Stone
Voyager Project Scientist

ACKNOWLEDGMENT

The authors are grateful to the many members of the Voyager Project team who not only made this historic mission of exploration possible but also took time from their busy schedules to offer us assistance, information, and encouragement in the preparation of this book. Among many too numerous to name individually, we particularly thank E. C. Stone, A. L. Lane, C. H. Stembridge, R. A. Mills, E. Montoya, M. A. Mitz, and B. A. Smith, L. Soderblom, and their colleagues on the Voyager Imaging Team. We are grateful to F. E. Bristow, D. L. Bane, L. J. Pieri and especially J. J. Van der Woude for their assistance in obtaining optimum versions of the photographs printed here. C. B. Pilcher and I. de Pater kindly made available their groundbased pictures of the Jovian magnetosphere. Many colleagues have read and provided helpful comments on parts of the manuscript, among them G. A. Briggs, S. A. Collins, S. Cruikshank, J. Doughty, A. L. Guin, A. L. Lane, R. A. Mills, E. Montoya, E. C. Stone, J. L. Ward, and especially C. R. Chapman.

CONTENTS

Jupiter is the largest planet in the solar system—a gaseous world as large as 1300 Earths, marked by alternating bands of colored clouds and a dazzling complexity of storm systems. The Voyager mission gave us our first close look at this spectacular planet. [P-21085]

CHAPTER 1

THE JOVIAN SYSTEM

Introduction

In the Sun's necklace of planets, one gem far outshines the rest: Jupiter. Larger than all the other planets and satellites combined, Jupiter is a true giant. If intelligent beings exist on planets circling nearby stars, it is probable that Jupiter is the only member of our planetary system they can detect. They can see the Sun wobble in its motion with a twelve-year period as Jupiter circles it, pulling first one way, then the other with the powerful tug of its gravity. If astronomers on some distant worlds put telescopes in orbit above their atmospheres, they might even be able to detect the sunlight reflected from Jupiter. But all the other planets—including tiny inconspicuous Earth—would be hopelessly lost in the glare of our star, the Sun.

Jupiter is outstanding among planets not only for its size, but also for its system of orbiting bodies. With fifteen known satellites, and probably several more too small to have been detected, it forms a sort of miniature solar system. If we could understand how the Jovian system formed and evolved, we could unlock vital clues to the beginning and ultimate fate of the entire solar system.

Ancient peoples all over the world recognized Jupiter as one of the brightest wandering lights in their skies. Only Venus is brighter, but Venus, always a morning or evening star, never rules over the dark midnight skies as Jupiter often does. In Greek and Roman mythology the planet was identified with the most powerful of the gods and lord of the heavens—the Greek Zeus; the Roman Jupiter.

As befits the king of the heavens, the planet Jupiter moves at a slow and stately pace. Twelve years are required for Jupiter to complete one orbit around the Sun. For about six months of each year, Jupiter shines down on us from the night sky, more brightly and steadily than any star. During the late 1970s it was a winter object, but in 1980 it will dominate the spring skies, becoming a summer "star" about 1982.

Early Discoveries

Even seen through a small telescope or pair of binoculars, Jupiter looks like a real world, displaying a faintly banded disk quite unlike the tiny, brilliant image of a star. It also reveals the brightest members of its satellite family as star-like points spread out along a straight line extended east-west through the planet. There are four of these planet-sized moons; with their orbits seen edge-on from Earth, they seem to move constantly back and forth, changing their configuration hourly.

In January 1610, in his first attempt to apply the newly invented telescope to astronomy, Galileo discovered the four large satellites of Jupiter. He correctly interpreted their motion as being that of objects circling Jupiter—establishing the first clear proof of celestial motion around a center other than the Earth. The discovery of these satellites played an important role in supporting the Copernican revolution that formed the basis for modern astronomy.

A few decades later the satellites of Jupiter were used to make the first measurement of the speed of light. Observers following their motions had learned that the satellite clock seemed to run slow when Jupiter was far from Earth and to speed up when the two planets were closer together. In 1675 the Danish astronomer Ole Roemer explained that this change was due to the finite velocity of light: The satellites only seemed to run slow at large distances because the light coming from them took longer to reach Earth. Knowing the

Galileo's notes summarizing his first observations of the Jovian satellites Io, Europa, Ganymede, and Callisto in January 1610 were made on a piece of scratch paper containing the draft of a letter presenting a telescope to the Doge in Venice. These observations were the result of Galileo's first attempt to apply the telescope to astronomical research.

dimensions of the orbits of Earth and Jupiter and the amount of the delay (about fifteen minutes), Roemer was able to calculate one of the most fundamental constants of the physical universe—the speed of light (about 300 000 kilometers per second).

The four great moons of Jupiter are called the Galilean satellites after their discoverer. Their individual names—Io, Europa, Ganymede, and Callisto—were proposed by Simon Marius, a contemporary and rival of Galileo. (Marius claimed to have discovered the satellites a few weeks before Galileo did, but modern scholars tend to discredit his claim.) Io, Europa, Ganymede, and Callisto are names of lovers of the god Jupiter in Greco-Roman mythology. Since Jupiter was not at all shy about taking lovers, there are enough such names for the other eleven Jovian satellites, as well as for those yet to be discovered.

In the century following Galileo's death, improvements in telescopes made it possible to measure the size of Jupiter and to note that it bulged at the equator. The equatorial diameter

is known today to be 142 800 kilometers, while from pole to pole Jupiter measures only 133 500 kilometers. For comparison, the diameter of the Earth is 12 900 kilometers, only about one-tenth as great, and the flattening of Earth is also much smaller (less than one percent). By measuring the orbits of the satellites and applying the laws of planetary motion discovered by Johannes Kepler and Isaac Newton, astronomers were also able to determine the total mass of Jupiter—about 2×10^{24} tons, or 318 times the mass of the Earth.

Once the size and mass are known, it is possible to calculate another fundamental property of a planet—its density. The density, which is the mass divided by the volume, provides important clues to the composition and interior structure of a planetary body. The density of Earth, a body composed primarily of rocky and metallic materials, is 5.6 times the density of water. The mass of Jupiter is 318 times that of Earth; its volume is 1317 times that of Earth. Thus Jupiter's density is substantially lower than Earth's, amounting to 1.34 times the density of water. From this low density, it was evident long ago that Jupiter was not just a big brother of Earth and the other rocky planets in the inner solar system. Rather, Jupiter is the prototype of the giant, gas-rich planets Jupiter, Saturn, Uranus, and Neptune. These giant planets must, from their low density, have a composition fundamentally different from that of Mercury, Venus, Earth, Moon, and Mars.

Jupiter Through the Telescope

Jupiter is a beautiful sight seen with the naked eye on a clear night, but only through a telescope does it begin to reveal its magnificence. The most prominent features are alternating light and dark bands, running parallel to the equator and subtly shaded in tones of blue, yellow, brown, and orange. However, these bands are not the planet's only conspicuous markings. In 1664 the English astronomer Robert Hooke first reported seeing a large oval spot on Jupiter, and additional spots were noted as telescopes improved. As the planet rotates on its axis, such spots are carried across the disk and can be used to measure Jupiter's speed of rotation. The giant planet spins so fast that a Jovian day is less than half as long as a day on Earth, averaging just under ten hours.

During the nineteenth century, observers using increasingly sophisticated telescopes were

2

able to see more complex detail in the band structure, with wisps, streaks, and festoons that varied in intensity and color from year to year. Furthermore, observations revealed the remarkable fact that not all parts of the planet rotate with the same period; near the equator the apparent length of a Jovian day is several minutes shorter than the average day at higher latitudes. It is thus apparent that Jupiter's surface is not solid, and astronomers came to realize that they were looking at a turbulent kaleidoscope of shifting clouds.

Although the face of Jupiter is always changing, some spots and other cloud features survive for years at a time, much longer than do the largest storms on Earth. The record for longevity goes to the Great Red Spot. This gigantic red oval, larger than the planet Earth, was first seen more than three centuries ago. From decade to decade it has changed in size and color, and for nearly fifty years in the late eighteenth century no sightings were reported, but since about 1840 the Great Red Spot has been the most prominent feature on the disk of Jupiter.

It was not until the twentieth century that the composition of the atmosphere of Jupiter could be measured. In 1905 spectra of the planet revealed the presence of gases that absorb strongly at red and infrared wavelengths; thirty years later these were identified as ammonia and methane. These two poisonous gases are the simplest chemical compounds of hydrogen combined with nitrogen and carbon, respectively. In the atmosphere of Earth they are not stable, because oxygen, which is highly active chemically, destroys them. The existence of methane and ammonia on Jupiter demonstrated that free oxygen could not be present and that the atmosphere was dominated by hydrogen—a reducing, rather than oxidizing, condition. Subsequently, hydrogen was identified spectroscopically. Although much more abundant than methane or ammonia, hydrogen is harder to detect.

In the 1940s and 1950s the German-American astronomer Rupert Wildt used all the available data to derive a picture of Jupiter that is still generally accepted. He noted that both the low total density and the observed presence of hydrogen-rich compounds in the atmosphere were consistent with a bulk composition similar to that of the Sun and stars. This "cosmic composition" is dominated by the two simplest elements, hydrogen and helium, which together make up nearly 99 percent of all the material in

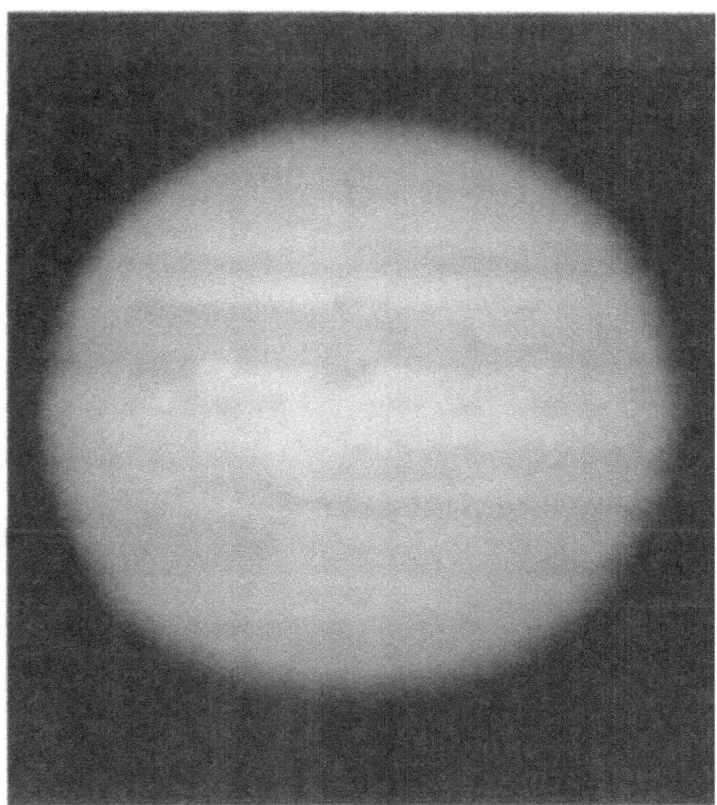

This ground-based photograph of Jupiter showing the Great Red Spot in the southern hemisphere was taken with the Catalina Observatory's 61-inch telescope in December 1966.

the universe. Wildt hypothesized that the giant planets, because of their large size, had succeeded in retaining this primordial composition, whereas the hydrogen and helium had escaped from the smaller inner planets. He also used his knowledge of the properties of hydrogen and helium to calculate what the interior structure of Jupiter might be like, concluding that the planet is mostly liquid or gas. Wildt suggested that there probably was a core of solid material deep in the interior, but that much of Jupiter is fluid—extremely viscous and compressed deep below the visible atmosphere, but still not solid. The atmosphere seen from above is just the thin, topmost layer of an ocean of gases thousands of kilometers thick.

Recent Earth-Based Studies of Jupiter

In the past, a great deal of planetary research was basically descriptive, consisting of visual observations and photography. Beginning in the 1960s, a new generation of planetary

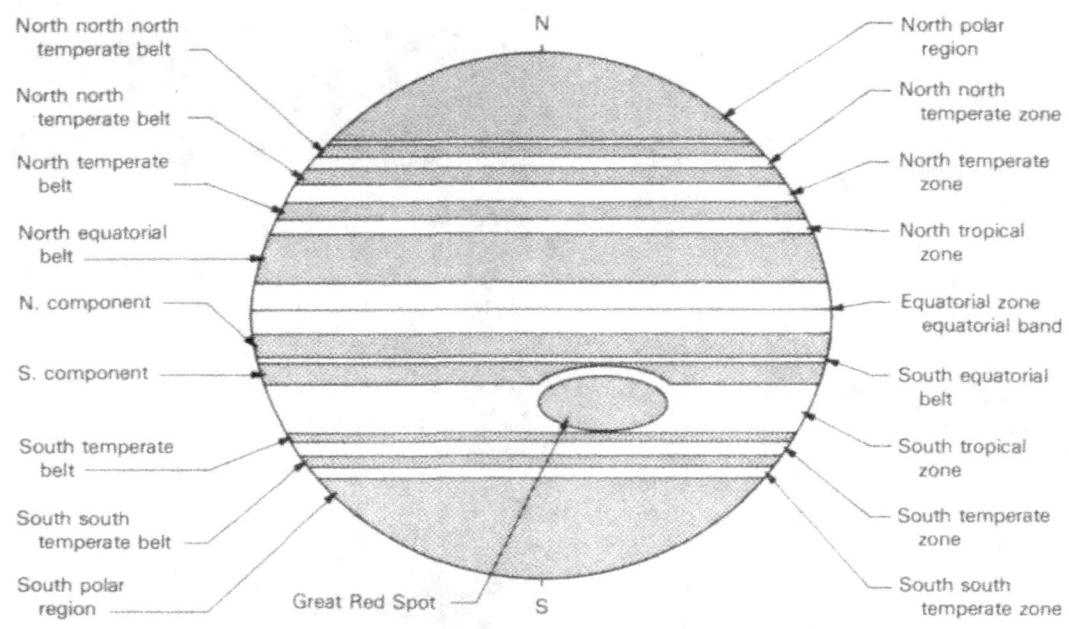

North north north
temperate belt

North north
temperate belt

North temperate
belt

North equatorial
belt

N. component

S. component

South temperate
belt

South south
temperate belt

South polar
region

Great Red Spot

N

S

North polar
region

North north
temperate zone

North temperate
zone

North tropical
zone

Equatorial zone
equatorial band

South equatorial
belt

South tropical
zone

South temperate
zone

South south
temperate zone

The major features of Jupiter are shown in schematic form. The planet is a banded disk of turbulent clouds; all its stripes are parallel to the bulging equator. Large dusky gray regions surround each pole. Darker gray or brown stripes called belts intermingle with lighter, yellow-white stripes called zones. Many of the belts and zones are permanent features that have been named. One feature of particular note is the Great Red Spot, an enigmatic oval larger than the planet Earth, which was first seen more than three centuries ago. During the years the spot has changed in size and color, and it escaped detection entirely for nearly fifty years in the 1700s. However, since the mid-nineteenth century the Great Red Spot has been the most prominent feature on the face of Jupiter. [2935]

scientists began to apply the techniques of modern astrophysics and geophysics to the study of the solar system. Inspired in part by the developing space programs of the United States and the Soviet Union, scientists began to ask more quantitative questions: What are the surfaces and atmospheres made of? What are the temperatures and wind speeds? Exactly what quantities of different elements and isotopes are present? And how can these new data be used to infer the origin and evolution of the planets?

Wildt had already suggested the basic gases in the atmosphere of Jupiter: primarily hydrogen and helium, with much smaller quantities of ammonia and methane. Undetected but possibly also present were nitrogen, neon, argon, and water vapor. The abundance of helium was particularly a problem; although it was presumably the second-ranking gas after hydrogen, it has no spectral features in visible light and its presence remained only a hypothesis, unconfirmed by observation.

Although the presence of a gas can usually be inferred from spectroscopy, solids or liquids cannot normally be detected in this way. Thus

the composition of Jupiter's clouds could not be determined directly. However, the presence of ammonia gas provided an important clue. At the temperatures expected in the upper atmosphere of the planet, ammonia gas must freeze to form tiny crystals of ammonia ice, just as water vapor in the Earth's atmosphere freezes to form cirrus clouds. Most investigators agreed that the high clouds covering much of Jupiter must be ammonia cirrus. But ammonia crystals are white, so the presence of this material provides no explanation for the many colors seen on Jupiter. Additional materials must be present—perhaps colored organic compounds, produced in small amounts by the action of sunlight on the atmosphere.

Because Jupiter is five times farther from the Sun than is the Earth, a given area on Jupiter receives only about four percent as much solar heating as does a comparable area on Earth. Thus Jupiter is colder than Earth; even though it may be warm deep below its blanket of clouds, Jupiter presents a frigid face.

The development of a new science, infrared astronomy, in the 1960s made it possible to measure these low temperatures directly. In

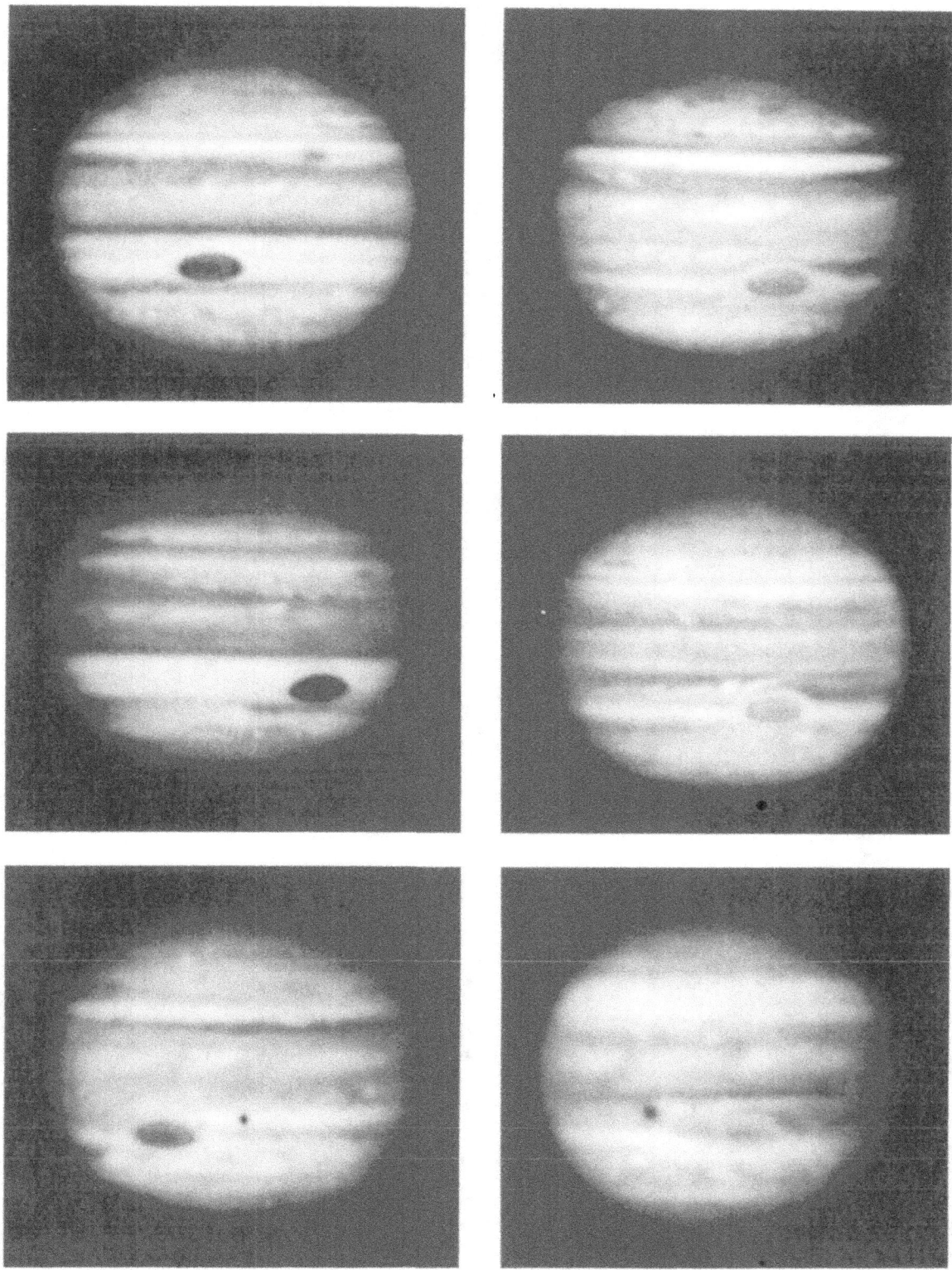

These blue filter photographs of Jupiter were taken at Mauna Kea Observatory, Hawaii. They show changes on Jupiter's surface between 1973 and 1978. The dates of the observations are (top left) July 25, 1973; (top right) October 5, 1974; (middle left) October 2, 1975; (middle right) November 20, 1976; (bottom left) January 28, 1978; (bottom right) December 19, 1978.

the case of a cloudy planet like Jupiter, the infrared emission evident at various wavelengths originates at different depths in the atmosphere. It is a general property of any mixed, convecting atmosphere that the temperature varies with depth; the rate of variation depends only on the composition of the atmosphere, the gravity of the planet, and the presence or absence of condensible materials to form clouds. On Jupiter it is about 1.9° C warmer for each kilometer of descent through the atmosphere. Thus, although the ammonia clouds are very cold, a little above −173° C, if one goes deep enough one can reach temperatures that are quite comfortable. With a variation of 1.9° C per kilometer, terrestrial "room temperature" would be reached about 100 kilometers below the clouds.

To measure the total energy radiated by a planet, it is necessary to utilize infrared radiation at wavelengths more than one hundred times longer than the wavelengths of visible light. Even when detectors were developed that could measure such radiation, it was impossible to observe celestial sources such as Jupiter because of the opacity of the terrestrial atmosphere. Even a tiny amount of water vapor in our own atmosphere can block our view of long-wave infrared. To make the required measurements, it is necessary to carry a telescope to very high altitudes, above all but a fraction of a percent of the offending water vapor.

In the late 1960s a Lear-Jet airplane was equipped with a telescope and made available by NASA to astronomers to carry out long-wave infrared observations from above 99 percent of the terrestrial water vapor. In 1969 Frank Low of the University of Arizona and his colleagues used this system to make a remarkable discovery: Jupiter was radiating more heat than it received from the Sun! Repeated observations demonstrated that between two and three times as much energy emanated from the planet as was absorbed. Thus Jupiter must have an internal heat source; in effect, it shines by its own power as well as by reflected sunlight. Theoretical studies suggest that the heat is primordial, the remnant of an incandescent proto-Jupiter that formed four and one-half billion years ago.

At the same time that the internal heat source on Jupiter was being revealed with long-wave airborne infrared telescopes, a new discovery was being made from short-wave infrared observations. The clouds of Jupiter are too cold to emit any detectable thermal radiation at a wavelength of 5 micrometers (about ten times the wavelength of green light). Nevertheless, images of Jupiter at 5 micrometers revealed a few small spots where large amounts of thermal energy were being emitted. The sources of the energy appeared to be holes or breaks in the clouds, where it was possible to see deeper into hotter regions. The discovery of these hot spots opened the possibility of prob-

Images of Jupiter in visible light (right) and five-micrometer infrared light (left) show the planet's characteristic belts and zones. The infrared image reveals areas that emit large amounts of thermal energy. The source of the energy is thought to be breaks in the Jovian cloud cover, which allow investigators a glimpse of the deep regions of the atmosphere. One of the mysteries of Jupiter concerns its heat balance: The planet appears to radiate more heat than it receives from the Sun. [P-20957]

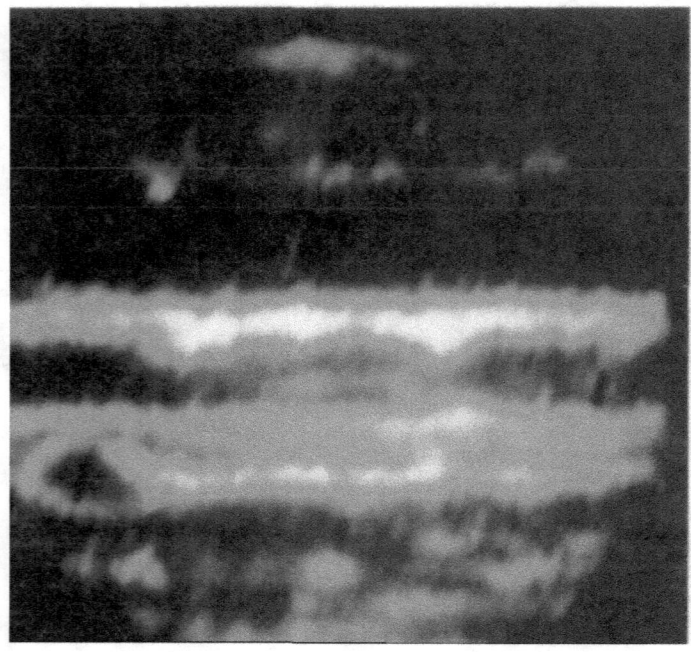 

ing deep regions of the Jovian atmosphere that had previously been beyond the reach of direct investigation.

The Jovian Magnetosphere

The discovery of planetary magnetospheres began in 1959 when the first U.S. Explorer satellite detected the radiation belts around the Earth. Named for James Van Allen of the University of Iowa, whose geiger-counter instrument aboard Explorer 1 first measured them, these belts are regions in which charged atomic particles—primarily electrons and protons—are trapped by the magnetic field of the Earth. They are one manifestation of the terrestrial magnetosphere—a large, dynamic region around the Earth in which the magnetic field of our planet interacts with streams of charged particles emanating from the Sun.

At almost the same time that the terrestrial magnetosphere was being discovered by artificial satellites, astronomers were finding evidence to suggest similar phenomena around Jupiter. Radio astronomy is a branch of science that measures radiation from celestial bodies at radio frequencies, which correspond to wavelengths much longer than those of visible or infrared light. All planets emit weak thermal radio radiation, but in the late 1950s investigators found that Jupiter was a much stronger long-wave radio source than would be expected for a planet with its temperature. This radiation bore the signature of higher-energy processes. Physicists had seen similar emissions produced in synchrotron electron accelerators, huge machines in which electrons are whirled around at nearly the speed of light so that they can be used for experiments in nuclear physics. The Russian theorist I. S. Shklovsky identified the Jovian radio radiation as also resulting from the synchrotron process, due to spiraling electrons trapped in the planet's magnetic field. From the intensity and spectrum of the observed synchrotron radiation, it was clear that both the magnetic field of the planet and the energy of charged particles in its Van Allen belts were much greater than was the case for Earth.

Using radio telescopes of high sensitivity, astronomers determined the approximate strength and orientation of the magnetic field of Jupiter. Although they were able to measure synchrotron radiation only from the innermost parts of the Jovian magnetosphere, they could infer that the total volume occupied by the magnetosphere was enormous. If our eyes were sensitive to magnetospheric emissions, Jupiter would look more than twice the diameter of the full moon in the sky.

All four Galilean satellites orbit within the magnetosphere of Jupiter; in contrast, our Moon lies well outside the terrestrial magnetosphere. Striking evidence of the interaction of the satellites and the magnetosphere was provided when it was found that the innermost large satellite—Io—actually affects the bursts of radio static produced by Jupiter. Only when Io is at certain places in its orbit are these strong bursts detected. Theorists suggested that electric currents flowing between the satellite and the planet might be responsible for this effect.

The Jovian Satellites

For nearly three centuries after their discovery in 1610, the only known moons of Jupiter were the four large Galilean satellites. In 1892 E. E. Barnard, an American astronomer, found a much smaller fifth satellite orbiting very close to the planet, and between 1904 and 1974 eight additional satellites were found far outside the orbits of the Galilean satellites. The outer satellites are quite faint and presumably no more than a few tens of kilometers in diameter, and all have orbits that are much less regular than those of the five inner satellites. Four of them revolve in a retrograde direction, opposite to that of the inner satellites and Jupiter itself.

In 1975 the International Astronomical Union assumed the responsibility for assigning names to the non-Galilean satellites of Jupiter. Following tradition, they named the inner satellite Amalthea for the she-goat that suckled the young god Jupiter. The outer eight were named for lovers of Jupiter: Leda, Himalia, Lysithea, Elara, Ananke, Carme, Pasiphae, and Sinope. For the non-Galilean satellites, the "e" ending is reserved for satellites with retrograde orbits; those with normal orbits have names that end in "a."

Because they are so large, the Galilean satellites have attracted the most attention from astronomers. More than fifty years ago large telescopes were used to estimate their sizes, and a careful series of measurements of their light variation showed that all four always keep the same face pointed toward Jupiter, just as our Moon always turns the same face toward Earth. Also, the subtle gravitational perturbations they exert on each other were used to determine the approximate mass of each.

7

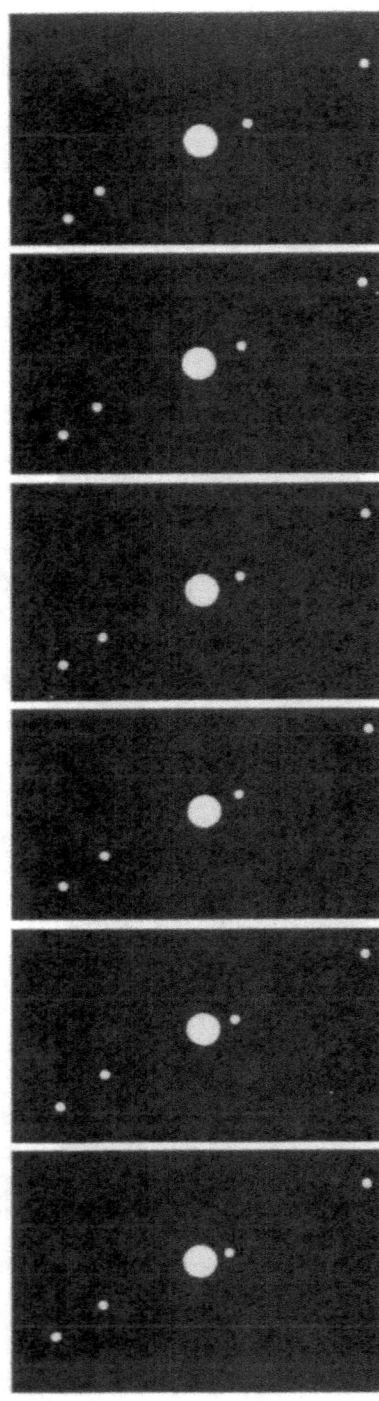

The pattern of the Galilean satellites changes from hour to hour, as seen from Earth. Viewed edge-on, the nearly circular orbits produce an apparent back and forth motion with respect to Jupiter. These images recreate the kinds of observations first made by Galileo in 1610.

Callisto, the outermost Galilean satellite, is larger than the planet Mercury. It also has the lowest reflectivity, or albedo, of the four, suggesting that its surface may be composed of some rather dark, colorless rock. Callisto takes just over two weeks to orbit once around Jupiter.

Ganymede, which requires only seven days for one orbit, is the largest satellite in the Jovian system, being only slightly smaller than the planet Mars. Its albedo is much higher than that of Callisto, or of the rocky planets such as Mercury, Mars, or the Moon. In 1971 astronomers first measured the infrared spectrum of reflected sunlight from Ganymede and found the characteristic absorptions of water ice, indicating that this satellite is partially covered with highly reflective snow or ice.

Europa, which is slightly smaller than the Moon, circles Jupiter in half the time required by Ganymede. Its surface reflects about sixty percent of the incident sunlight, and the infrared spectrum shows prominent absorptions due to water ice; Europa appears to be almost entirely covered with ice. However, its color in the visible and ultraviolet part of the spectrum is not that of ice, so some other material must also be present.

Io, innermost of the Galilean satellites, is the same size as our Moon. It orbits the planet in 42 hours, half the period of Europa. Like Europa, it has a very high reflectivity, but, unlike Europa, it has no spectral absorptions indicative of water ice. Before Voyager, identification of the surface material on Io presented a major problem to planetary astronomers.

When the sizes and masses of these satellites were measured, astronomers could calculate their densities. The inner two, Io and Europa, both have densities about three times that of water—nearly the same as the density of the Moon, or of rocks in the crust of the Earth. Callisto and Ganymede have densities only half as large, far too low to be consistent with a rocky composition. The most plausible alternative to rock is a composition that includes ice as a major component. Calculations showed that if these satellites were composed of rock and ice, approximately equal quantities of each were required to account for the measured density. Thus the two outer Galilean satellites were thought to represent a new kind of solar system object, as large as one of the terrestrial planets, but composed in large part of ice.

In 1973 the attention of astronomers was dramatically drawn to Io when Robert Brown

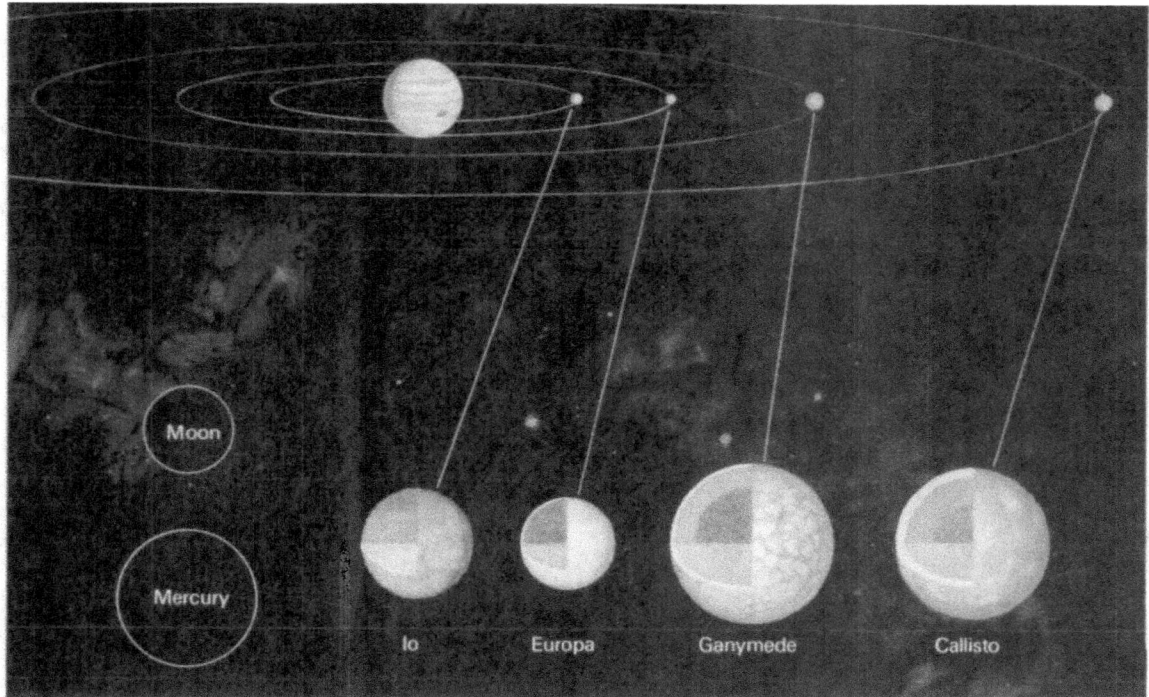

The Galilean satellites in orbit around Jupiter, along with the outer satellites, constitute a miniature solar system. Here they are shown relative to the size of Mercury and that of the Moon. The portrayal of their internal and external composition is based on theoretical models that preceded the Voyager flybys. [PC-17054AC]

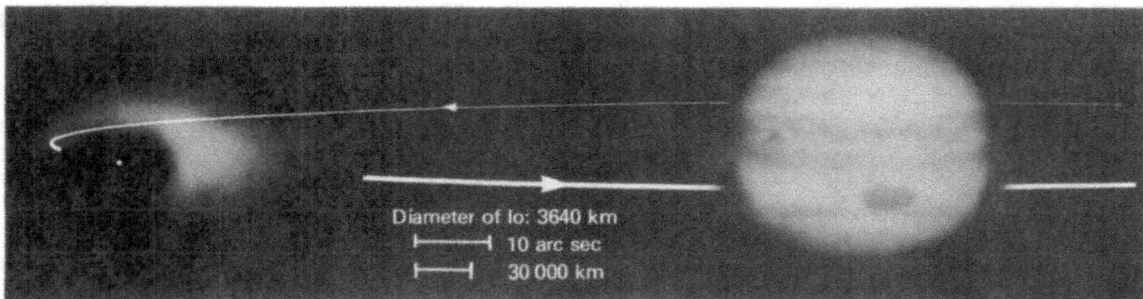

Diameter of Io: 3640 km
10 arc sec
30 000 km

This image of Io's extended sodium cloud was taken February 19, 1977, at the Jet Propulsion Laboratory's Table Mountain Observatory. A picture of Jupiter, drawings of the orbital geometry, and Io's disk (the small circle on the left) are included for perspective. The sodium cloud image has been processed for removal of sky background, instrumental effects, and the like. This photograph demonstrates that the cloud is highly elongated and that more sodium precedes Io in its orbit than trails it. [P-20047]

of Harvard University detected the faint yellow glow of sodium from the region of space surrounding it. It seemed that this satellite had an atmosphere, composed of the metal sodium! Continued observations showed, however, that this was not an atmosphere in the usual sense of the word. The gas atoms were not bound gravitationally to Io, but continuously escaped from it to form a gigantic cloud enveloping the orbit of the satellite. Fraser Fanale and Dennis Matson of the Caltech Jet Propulsion Laboratory suggested that bombardment of Io by high-energy particles from the Jovian Van Allen belts was knocking off atoms of sodium by a process called sputtering, releasing these atoms and allowing them to expand outward to form the observed sodium cloud. No one anticipated then that powerful volcanic eruptions on Io might also be contributing to this remarkable gas cloud.

This picture of the satellites was developed just as the first space probe reached the Jovian system. In the next chapter we describe the Pioneer program by which scientists reached out across nearly a million kilometers of space to explore Jupiter, its magnetosphere, and its system of satellites.

CHAPTER 2

PIONEERS TO JUPITER

Reaching for the Outer Planets

Since the beginning of the Space Age, scientists had dreamed of sending probes to Jupiter and its family of satellites. Initially, robot spacecraft were limited to studying the Earth and its Moon. In 1962, however, the first true interplanetary explorer, Mariner 2, succeeded in escaping the Earth-Moon system and crossing 100 million kilometers of space to encounter Venus, studying Earth's sister planet at close range using half a dozen scientific instruments. By the mid 1960s a U.S. planetary spacecraft had also flown to Mars, there had been a second flyby of Venus, and an ambitious program was under way for two more flybys of Mars in 1969, followed by a Mars orbiter in 1971. Based on this success with the inner planets, NASA scientists and engineers began to plan seriously to meet the challenge of the outer solar system.

Not only Jupiter, but Saturn and even Uranus and Neptune, were considered as possible targets. However, the distances between the outer planets are so vast that many years of flight would be required for a spacecraft to reach them, even using the most powerful rocket boosters then contemplated. If a cautious exploration program were followed, investigating one planet at a time before designing the next mission, it would be well into the twenty-first century before even a first reconnaissance of the solar system could be achieved. A way to bridge the space between planets in a more efficient, economical manner was needed.

In the late 1960s celestial mechanicians—scientists who study the motions of planets and spacecraft—began to solve problems posed by the immensity of the outer solar system. If a spacecraft is aimed to fly close to a planet in just the right way, it can be accelerated by the gravity of the planet to higher speeds than could ever be obtained by direct launch from Earth. If a second, more distant planet is in the correct align-

ment, the gravity boost given by the first encounter can speed the craft on to the second. Jupiter, with its huge size and strong gravitational pull, could be used as the fulcrum for a series of missions to Saturn, Uranus, Neptune, and even distant Pluto. In addition, the early 1980s would offer an exceptional opportunity, one repeated only about once every two centuries. At that time, all four giant planets would be in approximate alignment, so that gravity-assist maneuvers could be done sequentially. A single spacecraft, after being boosted from Jupiter to Saturn, could use the acceleration of Saturn to continue to Uranus, and in turn could be accelerated all the way out to Neptune. Such an ambitious, multiplanet mission was named the Grand Tour.

The first essential step in the Grand Tour was a flyby of Jupiter. However, this planet is ten times farther away from Earth than Venus or Mars. In addition, there were two potentially lethal hazards that had not been faced before in interplanetary flights: the asteroid belt and the Jovian magnetosphere.

The first danger was presented by the many thousands of asteroids that occupy a belt between the orbits of Mars and Jupiter. The largest asteroid, Ceres, was discovered in 1801 and was initially thought to be the "missing planet" sometimes hypothesized as lying between Jupiter and Mars. However, Ceres is only 1000 kilometers in diameter, too small to deserve the title of planet. Hundreds more of these minor planets were discovered during the nineteenth century, and by the 1960s more than 3000 had well-determined orbits. Most were only a few tens of kilometers in diameter, and astronomers estimated that 50 000 existed that were 1 kilometer or more in diameter. Any spacecraft to Jupiter would have to cross this congested region of space.

Even 50 000 minor bodies spread through the volume of space occupied by the asteroid

Pioneer 10 was launched on March 2, 1972, at 8:49 p.m. from Cape Canaveral, Florida. A powerful Atlas-Centaur rocket served as the launch vehicle, which propelled the space probe to its goal nearly a billion kilometers away. The beauty of the night launch was enhanced by the rumbling thunder and flashing lightning of a nearby storm.

belt would present little direct danger, although a chance collision with an uncatalogued object was always possible. Much more serious was the possibility that these larger objects were accompanied by large amounts of debris, from the size of boulders down to microscopic dust, that were undetectable from Earth. Collisions with pebble-sized stones could easily destroy a spacecraft. The only way to evaluate this danger was to go there and find out how much small debris was present.

A second danger was posed by Jupiter itself. In order to use the gravity boost of Jupiter to speed on to another planet, a spacecraft would have to fly rather close to the giant. But this would mean passing right through the regions of energetic charged particles surrounding the planet. Some estimates of the number and energy of these particles indicated that the delicate electronic brains of a spacecraft would be damaged before it could penetrate this region. Again, only by going there could the danger be evaluated properly.

The Pioneer Jupiter Mission

In 1969 the U.S. Congress approved the Pioneer Jupiter Mission to provide a reconnaissance of interplanetary space between Earth and Jupiter and a first close look at the giant planet itself. The Project was assigned by NASA to the Ames Research Center in Mountain View, California. The primary objectives were defined by NASA:

Explore the interplanetary medium beyond the orbit of Mars.

Investigate the nature of the asteroid belt, assessing possible hazards to missions to the outer planets.

Explore the environment of Jupiter, including its inner magnetosphere.

The Pioneer spacecraft was designed for economy and reliability, based on previous experience at Ames with Pioneers 6 through 9, all of which had proven themselves by years of successful measurement of the interplanetary medium near the Earth. Unlike the Mariner class of spacecraft being used to investigate Venus and Mars, the Pioneer craft rotated continuously around an axis pointed toward the Earth. This spinning design was extremely stable, like the wheels of a fast-moving bicycle, and required less elaborate guidance than a nonspinning craft. In addition, the spin pro-

vided an ideal base for measurements of energetic particles and magnetic fields in space, since the motion of the spacecraft itself swept the viewing direction around the sky and allowed data to be acquired rapidly from many different directions. The only major disadvantage of a spinning spacecraft is that it does not allow a stabilized platform on which to mount cameras or other instruments that require exact pointing. Thus the spacecraft design was optimized for measurements of particles and fields in interplanetary space and in the Jovian magnetosphere, but had limited capability for observations of the planet and its satellites.

As finally assembled, the Pioneer Jupiter spacecraft had a mass of 258 kilograms. One hundred forty watts of electrical power at Jupiter were supplied by four radioisotope thermoelectric generators (RTGs), which turned heat from the radioactive decay of plutonium into electricity. The launch vehicle for the flight to Jupiter was an Atlas-Centaur rocket, equipped with an additional solid-propellant third stage. This powerful rocket could accelerate the spacecraft to a speed of 51 500 kilometers per hour, sufficient to escape the Earth and make the billion-kilometer trip to Jupiter in just over two years. The specific scientific investigations to be carried out on Pioneer were selected competitively in 1969 from proposals submitted by scientists from U.S. universities, industry, and NASA laboratories, and also from abroad. Eleven separate instruments would be flown, in addition to two experiments that would make use of the spacecraft itself.

Three complete Pioneer spacecraft, with their payloads of 25 kilograms of scientific instruments, were built: one as a test vehicle and two for launch to Jupiter. One of these—the test vehicle—is now on display at the National Air and Space Museum in Washington. The first opportunity to launch—the opening of the "launch window"—was on February 27, 1972. However, it was not until shortly after dark on March 2 that all systems were ready, and Pioneer 10 began its historic trip to Jupiter.

Pioneer 10 was the first human artifact launched with sufficient energy to escape the solar system entirely. Fittingly, the craft carried a message designed for any possible alien astronauts who might, in the distant future, find the derelict Pioneer in the vastness of interstellar space. A small plaque fastened to the spacecraft told the time and planet from which it had been launched, and carried a symbolic greeting from humanity to the cosmos.

Two identical Pioneer spacecraft were designed and fabricated by TRW Systems Group at their Redondo Beach, California facility. Each weighed only about 260 kilograms, yet carried eleven highly sophisticated instruments capable of operating unattended for many years in space. Data systems on board controlled the instrumentation, received and processed commands, and transmitted information across the vast distance to Earth.

The second Pioneer was to wait more than a year before launch. By following far behind Pioneer 10, its trajectory—in particular how deeply it penetrated the radiation belts during Jupiter flyby—could be modified, depending on the fate of the first spacecraft. At dusk on April 5, 1973, Pioneer 11 blasted from the launch pad at Cape Canaveral and followed its predecessor on the long, lonely journey into the outer solar system.

Flight to Jupiter

Within a few hours of launch, each Pioneer spacecraft shed the shroud that had protected it and unfurled booms supporting the science instruments and the RTG power generators. After each craft had been carefully tracked and precise orbits calculated, small onboard rockets were commanded to fire to correct its trajectory for exactly the desired flyby at Jupiter. Pioneer 10 was targeted to fly by the planet at a mini-

mum distance of 3 Jupiter radii (R_J) from the center, or 2 R_J (about 140 000 kilometers) above the clouds. This close passage, inside the orbit of Io, allowed the craft to pass behind both Io and Jupiter as seen from Earth, so that its radio beam could probe both the planet and its innermost large satellite. Pioneer 11 was intended to fly even closer to Jupiter, but the exact targeting options were held open until after the Pioneer 10 encounter.

On Pioneer 10, all instruments appeared to be working well as the craft passed the orbit of Mars in June 1972, just 97 days after launch. At this point, as it headed into unexplored space, it truly became a pioneer. In mid-July it began to enter the asteroid belt, and scientists and engineers anxiously watched for signs of increasing particulate matter.

Pioneer 10 carried two instruments designed to measure small particles in space. One, with an effective area of about 0.6 square meters, measured the direct impact of dust grains as small as one-billionth of a gram. The other looked for larger, more distant grains by measuring sunlight reflected from them. To the surprise of many, there was little increase in the rate of dust impacts recorded as the craft penetrated more and more deeply into the belt. At about 400 million kilometers from the Sun, near the middle of the belt, there appeared to be an increase in the number of larger particles detected optically, but not to a level that posed any hazard. In February 1973 the spacecraft emerged unscathed from the asteroid belt, having demonstrated that the much-feared concentration of small debris in the belt did not exist. The pathway was open to the outer solar system!

On November 26, 1973, the long-awaited encounter with Jupiter began. On that date, at a distance of 6.4 million kilometers from the planet, instruments on board Pioneer 10 detected a sudden change in the interplanetary medium as the spacecraft crossed the point— the bow shock—at which the magnetic presence of Jupiter first becomes evident. At the bow shock, the energetic particles of the solar wind are suddenly slowed as they approach Jupiter. At noon the next day, Pioneer 10 entered the Jovian magnetosphere at a distance of 96 R_J from the planet.

As the spacecraft hurtled inward toward regions of increasing magnetic field strength and charged plasma particles, the instruments designed to look at Jupiter began to play their role. A simple line-scan camera that could build

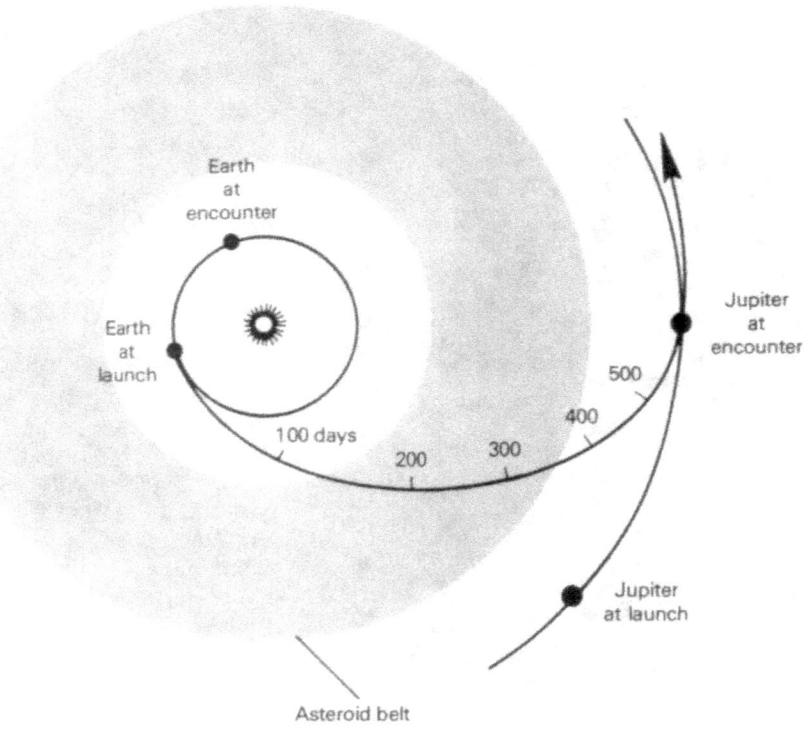

The asteroid belt is a region between Mars and Jupiter that is populated by thousands of minor planets, most only a few kilometers in diameter. Before 1970, some theorists suggested that large quantities of abrasive dust might damage spacecraft passing through the asteroid belt. Pioneer 10 proved that this danger was not present, thus opening the way to the outer solar system.

up an image from many individual brightness scans (like a newspaper picture transmitted by wire) obtained its first pictures of the planet, and ultraviolet and infrared photometers prepared to observe it also. By December 2, when the spacecraft had crossed the orbit of Callisto, the outermost of the large Galilean satellites, the line-scan images were nearly equal in quality to the best telescopic photos taken previously, and as each hour passed they improved in resolution. Near closest approach, Pioneer 10 transmitted partial frames of Jupiter that represented a threefold improvement over any Earth-based pictures ever taken.

Tension increased as the spacecraft plunged deeper into the radiation belts of Jupiter. Would it survive the blast of x-rays and gamma-rays induced in every part of the craft by the electrons and ions trapped by the magnetic field of Jupiter? Several of the instruments measuring the charged particles climbed to full scale and saturated. Others neared their limits but, as anxious scientists watched the data being sent back, the levels flattened off. Meanwhile, the

spacecraft itself began to feel the effects of the radiation, and occasional spurious commands were generated. Several planned high-resolution images of Jupiter and its satellites were lost because of these false signals. But again the system stabilized, and no more problems occurred as, just past noon on December 3, 1973, Pioneer 10 reached its closest point to Jupiter, 130 000 kilometers above the Jovian cloud tops. Pioneer had passed its most demanding test with flying colors, and at a news conference at Ames, NASA Planetary Program Director Robert Kraemer pronounced the mission "100 percent successful." He added, "We sent Pioneer off to tweak a dragon's tail, and it did that and more. It gave it a really good yank, and it managed to survive." The Project Science Chief pronounced it "the most exciting day of my life," and most of the hundreds of scientists and engineers who participated in the encounter probably agreed with him.

Pioneer 11 continued to follow steadily, emerging from the asteroid belt in March 1974. Based on the performance and findings of

Pioneer 10, it was decided to send Pioneer 11 still closer to Jupiter, but on a more inclined trajectory. On April 19 thrusters on the spacecraft fired to move the Pioneer 11 aimpoint just 34 000 kilometers above the clouds of Jupiter.

In using the Pioneer 10 data to assess the hazard to Pioneer 11, scientists had to consider three aspects of the charged particle environment. First was the energy distribution of the particles: The most energetic presented the most danger. Second was the flux, the rate at which particles struck the craft. Third was the total radiation dose. One can make an analogy with a boxing match. The energy distribution tells you how hard the blows of your opponent are. The flux is a measure of how many times a minute he hits you, and the total dose measures how many blows land. The spacecraft reacts just like a boxer; the crucial question is how much total dose it absorbs. Enough radiation blows, and the system is knocked out. The trajectory chosen for Pioneer 11 resulted in higher flux, since the craft probed more deeply into Jupiter's inner magnetosphere than had Pioneer 10. But by moving at a high angle across the equatorial regions where the flux is highest, the total dose could be kept below that experienced by Pioneer 10.

Pioneer 11 entered the Jovian magnetosphere on November 26, 1974, just a year after its predecessor. Closest approach took place on December 2. As with Pioneer 10, the radiation dose taxed the spacecraft to its limit. Again spurious commands were issued, this time affecting the infrared radiometer more than the imaging system. The craft flew at high latitude over the north polar region of Jupiter, an area never seen from Earth, and returned several excellent high-resolution pictures. Once more the little Pioneers had succeeded against the odds in opening the way to the giant planets.

Following their encounters with Jupiter, both Pioneer spacecraft returned to their normal routine of measuring the interplanetary medium. Pioneer 10 had gained speed from the gravity field of Jupiter and became the first craft to achieve the velocity needed to escape from the solar system. Pioneer 11, however, had used the pull of Jupiter to bend its trajectory inward, aiming it across the solar system

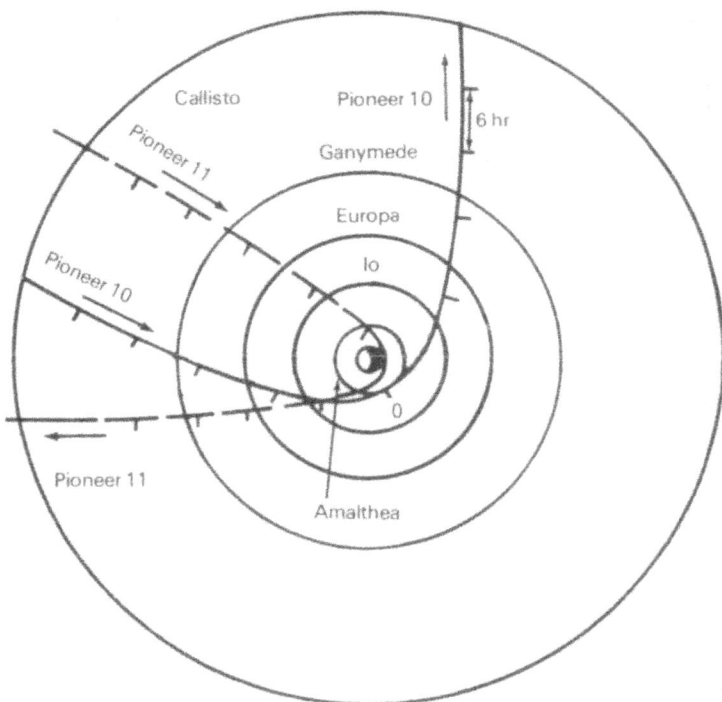

Pioneer 10 and 11 encounters with Jupiter are shown as viewed from the celestial North Pole. Pioneer 10 swung around the giant planet in the counterclockwise direction, while Pioneer 11 followed a clockwise approach. In this view, Jupiter rotates counterclockwise.

toward Saturn. Following the successes at Jupiter, NASA announced that Pioneer 11 would be targeted for a close flyby of Saturn five years later, which was successfully carried out in September 1979. In early 1980, far beyond their design lifetimes, both spacecraft were still performing beautifully.

Jupiter Results

The scientific results of the Pioneer flybys of Jupiter were many and varied. As is always the case, some old questions were answered and new problems were raised by the spacecraft data. Highlights of these results are summarized below.

Photographs of Jupiter. The line-scan imaging systems of Pioneer returned some remarkable pictures of the planet during the two encounters, showing individual features as small as 500 kilometers across. In addition, the Pioneers were able to look at Jupiter from angles never observable from Earth.

One of the discoveries made from these pictures was the great variety of cloud structures near the boundaries between the light zones and dark belts. Many individual cloud patterns suggested rising and falling air. The convoluted swirls evident in these regions appeared to be the result of dynamic motions; unfortunately, with only the few "snapshots" obtained during the hours of the flyby, the actual motions of these clouds could not be followed.

Thermal Emission. In the year the Pioneer Project was begun, astronomers on Earth had measured that Jupiter emitted more heat than it absorbed from the Sun. From the Earth these measurements could be made only of the sunlit part of the planet; neither the night side nor the

In this view of Jupiter, the Great Red Spot is prominent and the shadow of Io traverses the planetary disk. The gross morphology of the belts and zones, with structures showing turbulence and convective cells in the middle latitudes, is clearly seen. The small white spots surrounded by dark rings, seen mainly in the southern hemisphere, indicate regions of intense vertical convective activity, somewhat similar to cumulonimbus or thunderclouds.

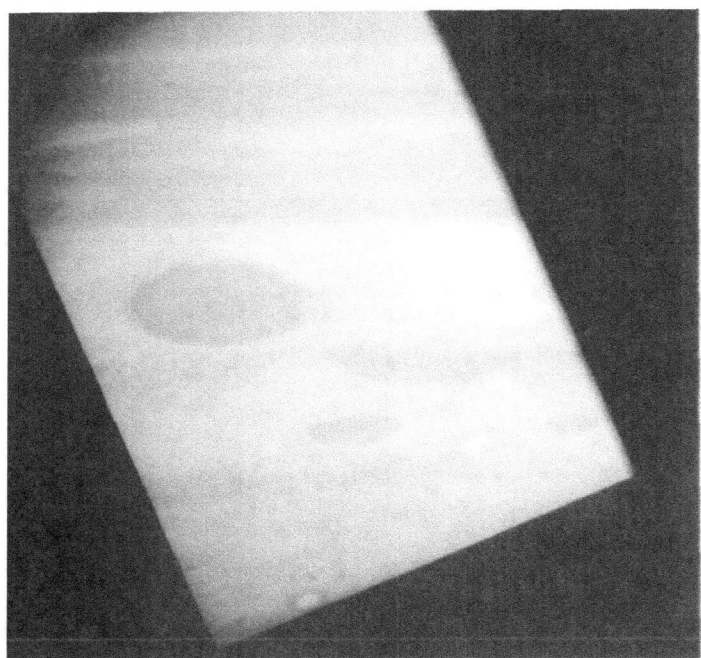

One of the best Pioneer images of Jupiter was obtained at a range of 545 000 kilometers by Pioneer 11. Structure within the Great Red Spot and the surrounding belts and zones can be seen. There was much less turbulent cloud activity around the spot at the time of the Pioneer flybys than was seen five years later by the Voyager cameras.

poles could be seen. One of the main objectives of the Pioneer flybys was to determine the heat budget accurately from temperature measurements at many points on both the sunlit and the night sides.

The Pioneer data confirmed the presence of a heat source in Jupiter and supplied a quantitative estimate of its magnitude. The global effective temperature was found to be $-148°$ C, to a precision of ± 3 degrees. This temperature implies that Jupiter radiates 1.9 times as much heat as it receives from the Sun. The corresponding internal heat source is 10^{17} watts. Surprisingly, the poles were as warm as the equator; apparently, the atmosphere is very efficient at transferring solar heat absorbed near the equator up to high latitudes, or perhaps the internal component of the heat comes preferentially from the polar regions.

Helium in the Atmosphere. The Pioneer infrared experiment made the first measurement of the amount of helium on Jupiter. The ratio of the number of helium atoms to the number of hydrogen atoms was found to be $He/H_2 = 0.14 \pm 0.08$. This is consistent with the known solar ratio of $He/H_2 = 0.11$. Measurements of helium in the upper atmosphere were also made by the ultraviolet experiment.

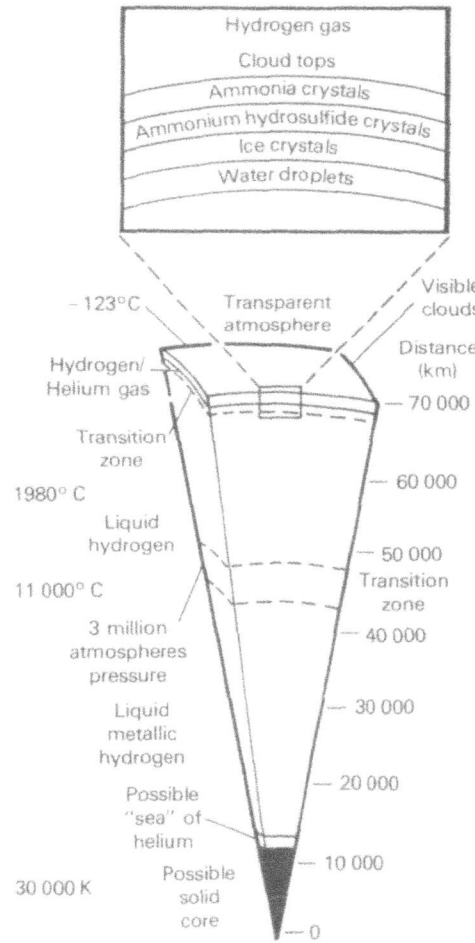

Pioneer 10 confirmed theoretical models of Jupiter that suggest the planet is nearly all liquid, with a very small core and an extremely deep atmosphere. The liquid interior seethes with internal heat energy, which is transferred from deep within the planet to its outer regions. The temperature at the center may be 30 000 K. Since the temperature at the cloud tops is around $-123°$ C, there is a large range of temperatures within the planet.

17

PIONEER SCIENCE INVESTIGATIONS
Project Scientist: J. H. Wolfe, NASA Ames

Investigation	Principal Investigator	Primary Objectives
Magnetic fields	E. J. Smith, JPL	Measurement of the magnetic field of Jupiter and determination of the structure of the magnetosphere.
Magnetic fields (Pioneer 11 only)	N. F. Ness, NASA Goddard	Measurement of the magnetic field of Jupiter and determination of the structure of the magnetosphere.
Plasma analyzer	J. H. Wolfe, NASA Ames	Measurement of low-energy electrons and ions, determination of the structure of the magnetosphere.
Charged particle composition	J. A. Simpson, U. Chicago	Determination of the number, energy, and composition of energetic charged particles in the Jovian magnetosphere.
Cosmic ray energy spectra	F. B. McDonald, NASA Goddard	Measurement of number and energy of very high energy charged particles in space.
Jovian charged particles	J. A. Van Allen, U. Iowa	Measurement of number and energy distribution of energetic charged particles and determination of magnetospheric structure.
Jovian trapped radiation	R. Walker Fillius, UC San Diego	Measurement of number and energy distribution of energetic charged particles and determination of magnetospheric structure.
Asteroid-meteoroid astronomy	R. K. Soberman, General Electric	Observation of solid particles (dust and larger) in the vicinity of the spacecraft.
Meteoroid detection	W. H. Kinard, NASA Langley	Detection of very small solid particles that strike the spacecraft.
Celestial mechanics	J. D. Anderson, JPL	Measurement of the masses of Jupiter and the Galilean satellites with high precision.
Ultraviolet photometry	D. L. Judge, U. Southern California	Measurement of ultraviolet emissions of the Jovian atmosphere and from circumsatellite gas clouds.
Imaging photopolarimetry	T. Gehrels, U. Arizona	Reconnaissance imaging of Jupiter and its satellites; study of atmospheric dynamics.
Jovian infrared thermal structure	G. Münch, Caltech	Measurement of Jovian temperature and heat budget; determination of helium to hydrogen ratio.
S-Band occultation	A. J. Kliore, JPL	Probes of structure of Jovian atmosphere and ionosphere.

Atmospheric Structure. Several Pioneer investigations yielded information on the variation of atmospheric temperature and pressure in the regions above the ammonia clouds. Near the equator, at a level where the atmospheric pressure is the same as that on the surface of the Earth (1 bar), the temperature is −108° C. About 150 kilometers higher, where the pressure drops to 0.1 bar, is the minimum atmospheric temperature of about −165° C. Above this point the temperature rises again, reaching about −123° C near a pressure level of 0.03 bar. Presumably this temperature rise is due to absorption of sunlight by a thin haze of dust particles in the upper atmosphere of Jupiter.

Internal Structure. The measurements of the amount of helium, of the gravitational field, and of the size of the internal heat source on Jupiter greatly clarified scientists' understanding of the deep interior of the planet. Calculations showed that the core of Jupiter must be so hot that hydrogen cannot become solid, but must remain a fluid throughout the interior. Even at great depths, therefore, Jupiter does not have a solid surface. The theory that the

Great Red Spot was the result of interactions with a surface feature below the clouds thus became untenable. Whatever its exact nature, the Red Spot must be a strictly atmospheric phenomenon.

Magnetic Field. Pioneer data showed that the magnetic field of Jupiter has a dipolar nature, like that of the Earth, but 2000 times stronger. The calculated surface fields measured about 4 gauss, compared to a field of about 0.5 gauss on the Earth. The axis of the magnetic field was tilted 11 degrees with respect to the rotation axis, and it was offset by about 10 000 kilometers ($0.1 \ R_J$) from the center of the planet.

Satellite Atmospheres. Two experiments yielded exciting new information on possible atmospheres of the Galilean satellites. First was an occulation, in which the Pioneer 10 spacecraft was targeted to pass behind Io as seen from Earth. At the moments just before the spacecraft disappeared, and just after reemergence from behind the satellite, the radio signal was influenced by a thin layer of ionized gas, in which the electrons had been stripped

Pioneers 10 and 11 did not obtain very detailed pictures of the satellites of Jupiter. The best view was of Ganymede (left), which showed a surface of contrasting light and dark spots of unknown nature. A less detailed image of Europa (right) clearly reveals the illuminated crescent but supplies little information about the surface of this ice-covered satellite.

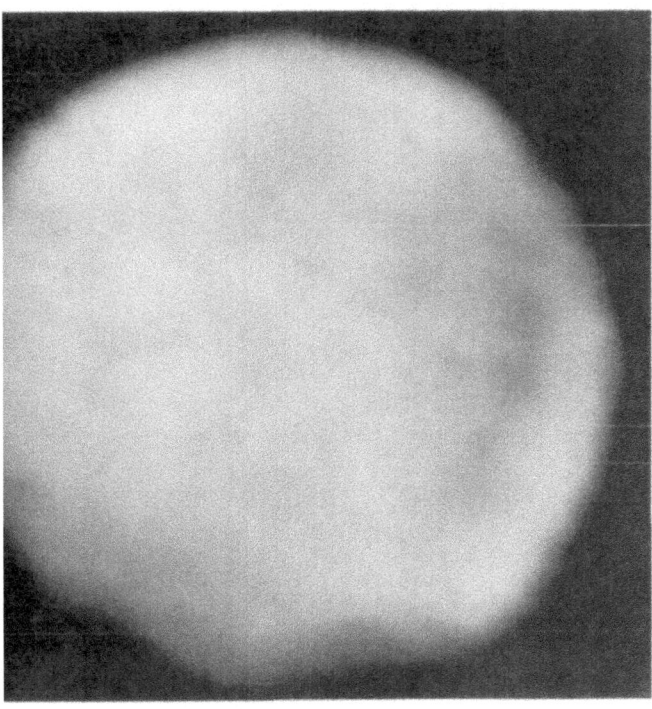 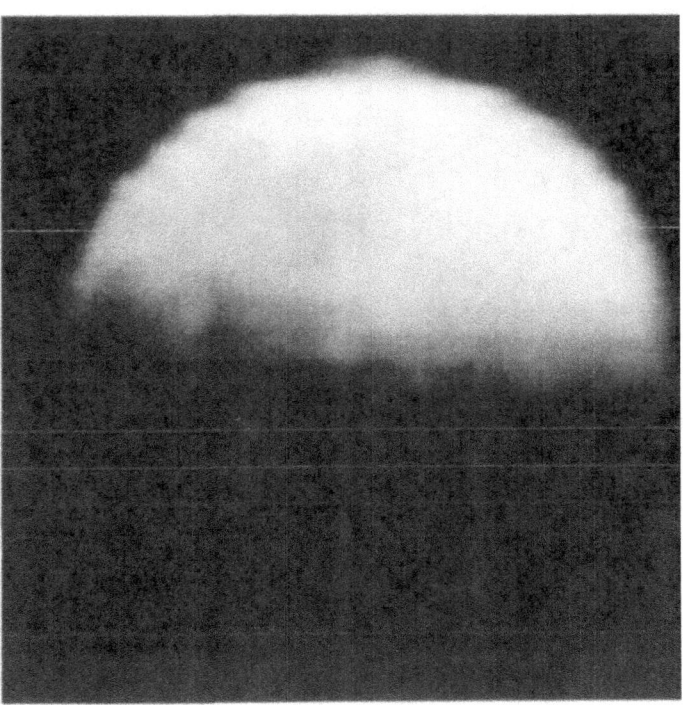

from the atoms by absorbed sunlight or by other processes. The ionosphere thus discovered had a peak density of about 60 000 electrons per cubic centimeter. In addition, a very extended far-ultraviolet glow, probably due to atomic hydrogen, was found near the orbit of Io by the ultraviolet photometer.

Masses of Jupiter and Its Satellites. Precise radio tracking of the Pioneers as they coasted past Jupiter and its satellites revealed that Jupiter is about one percent heavier than had been anticipated, and several satellites were found to have masses that differ by more than ten percent from values determined previously. These improvements in knowledge of the masses were required to achieve the close satellite flybys being planned for later missions.

The Inner Magnetosphere. A great deal of the scientific emphasis of the Pioneer missions was directed at characterizing the particles and fields in the inner magnetosphere, the region in which charged particles are trapped in stable orbits. The Pioneers found that it extended to about 25 R_J, well beyond the orbit of Callisto. Within this region, instruments on the spacecraft recorded the numbers and energy of electrons, protons, and ions. The electrons reached a maximum concentration near 3 R_J, and their numbers remained almost constant from there in toward the planet. The maximum concentration of protons observed by Pioneer 10 was at 3.4 R_J, a little inside the orbit of Io. Pioneer 11 penetrated deeper and found another maximum, about twenty times higher, at 1.9 R_J; at this distance, 10 million energetic protons hit each

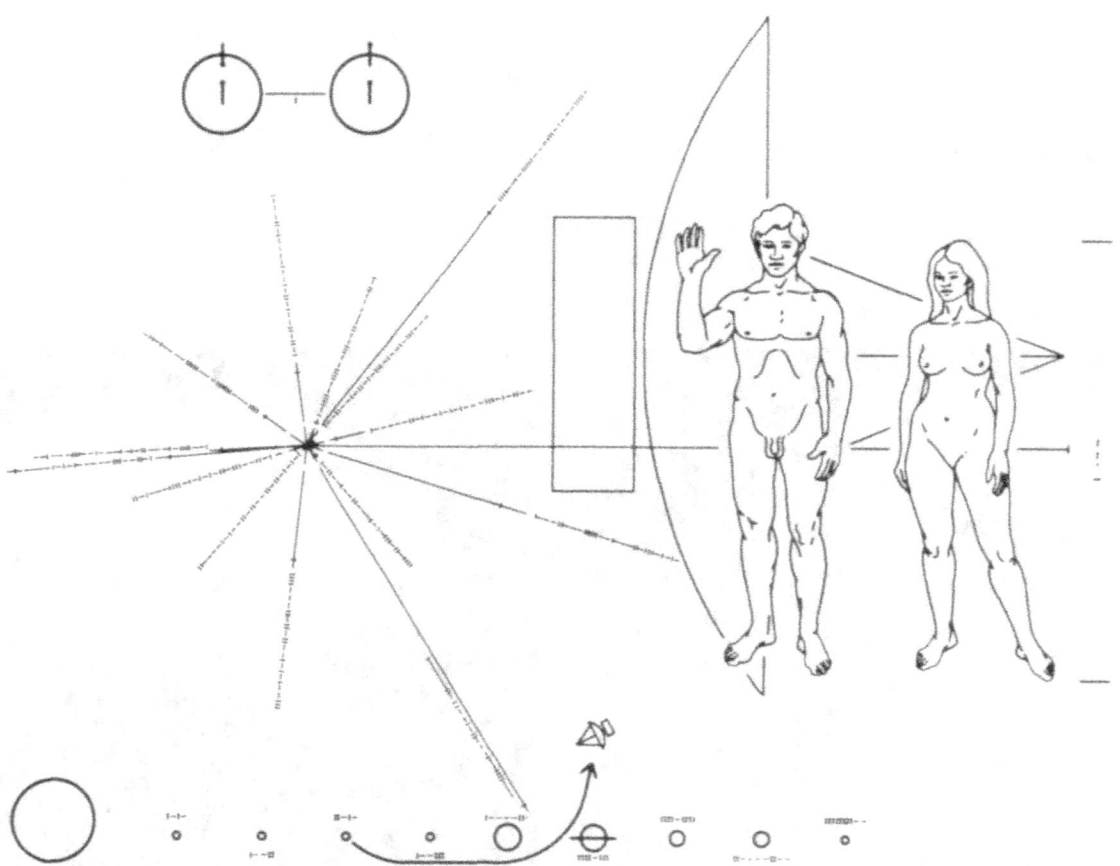

The Pioneer spacecraft carried this plaque on the journey beyond the solar system, bearing data that tell where and when the human species lived and that convey details of our biological form. When Pioneer 10 flew by Jupiter it acquired sufficient kinetic energy to carry it completely out of the solar system. Some time between one and ten billion years from now, the probe may pass through the planetary system of a remote stellar neighbor, one of whose planets may have evolved intelligent life. If the spacecraft is detected and then inspected, Pioneer's message will reach across the eons to communicate its greeting.

square centimeter of the spacecraft every second. It was believed that the gap between these two peaks was due to tiny Amalthea, the innermost satellite, which orbits at 2.5 R_J. Apparently this satellite sweeps up the particles as it circles Jupiter. Another large dip in the proton distribution was attributed to sweeping by Io, with smaller effects seen near the orbits of Europa and Ganymede. There was an additional small effect at 1.8 R_J, later found by Voyager to be due to Jupiter's ring and its fourteenth satellite.

The Outer Magnetosphere. From about 25 R_J outward to its boundary near 100 R_J, the Jovian magnetosphere is a complex and dynamic place. Beyond about 60 R_J, both Pioneers found the boundary to be highly unstable, apparently blown in and out by variations in the pressure of the solar wind, which consists of charged particles flowing outward from the Sun. In this region concentrations of Jovian particles are sometimes seen that rival the inner magnetosphere in intensity. Between 60 R_J and 25 R_J, a region sometimes called the middle magnetosphere, the particle motions are more ordered, and for the most part electrons and protons are carried along with the planet's rotation by its magnetic field. Near the equatorial plane, the flow of these particles produces an electric current circling the planet, and this current in turn generates its own magnetic field, which approaches in strength that of Jupiter itself. Occasionally the outer magnetosphere collapses down to about 60 R_J, and energetic particles are squirted from the middle region into space; these bursts of Jovian particles can sometimes be detected as far away as Earth.

At the same time that Pioneer scientists were analyzing their results and developing new concepts of the Jupiter system, a new team of investigators had been selected for the next mission to Jupiter. From about 1975 on, attention shifted from Pioneer to its successor—Voyager.

CHAPTER 3

THE VOYAGER MISSION

Genesis of Voyager

Voyager had its origin in the Outer Planets Grand Tour, a plan to send spacecraft to all the planets of the outer solar system. In 1969, the same year in which the Pioneer Project received Congressional approval, NASA began to design the Grand Tour. At the same time, the Space Science Board of the National Academy of Sciences completed a study called "The Outer Solar System," chaired by James Van Allen of the University of Iowa, which recommended that the United States undertake an exploration program:

1. To conduct exploratory investigations of the appearance, size, mass, magnetic properties, and dynamics of each of the outer planets and their major satellites;

2. To determine the chemical and isotopic composition of the atmospheres of the outer planets;

3. To determine whether biologically important organic substances exist in these atmospheres and to characterize the lower atmospheric environments in terms of biologically significant parameters;

4. To describe the motions of the atmospheres of the major planets and to characterize their temperature-density-composition structure;

5. To make a detailed study for each of the outer planets of the external magnetic field and respective particle population, associated radio emissions, and magnetospheric particle-wave interactions;

6. To determine the mode of interaction of the solar wind with the outer planets, including the interaction of the satellites with the planets' magnetospheres;

7. To investigate the properties of the solar wind and the interplanetary magnetic field at great distances from the Sun at both low and high solar latitudes, and to search for the outer boundary of the solar wind flow;

8. To attempt to obtain the composition, energy spectra, and fluxes of cosmic rays in interstellar space, free of the modulating effects of the solar wind.

The report also noted that "exceptionally favorable astronomical opportunities occur in the late 1970s for multiplanet missions," and that "professional resources for full utilization of the outer-solar-system mission opportunities in the 1970s and 1980s are amply available within the scientific community, and there is a widespread eagerness to participate in such missions."

An additional Academy study, chaired by Francis S. Johnson of the University of Texas at Dallas and published in 1971, was even more specific: "An extensive study of the outer solar system is recognized by us to be one of the major objectives of space science in this decade. This endeavor is made particularly exciting by the rare opportunity to explore several planets and satellites in one mission using long-lived spacecraft and existing propulsion systems. We recommend that [Mariner-class] spacecraft be developed and used in Grand Tour missions for the exploration of the outer planets in a series of four launches in the late 1970s."

Thus the stage was set to initiate the Outer Planets Grand Tours. NASA's timetable called for dual launches to Jupiter, Saturn, and Pluto in 1976 and 1977, and dual launches to Jupiter, Uranus, and Neptune in 1979, at a total cost over the decade of the 1970s of about $750 million.

A necessary step was to obtain from the scientific community the best possible set of instruments to fly on the spacecraft. Following its initial internal studies, NASA turned for its detailed scientific planning to an open competi-

The Voyager spacecraft are among the most sophisticated, automatic, and independent robots ever sent to explore the planets. Each craft has a mass of one ton and is dominated by the 3.7-meter-diameter white antenna used for radio communication with Earth. Here Voyager undergoes final tests in a space simulator chamber. [373-7162AC]

tion in which any scientist or scientific organization was invited to propose an investigation. In October 1970 NASA issued an "Invitation for Participation in the Mission Development for Grand Tour Missions to the Outer Solar System," and a year later it had selected about a dozen teams of scientists to formulate specific objectives for these missions. At the same time, an advanced spacecraft engineering design was carried out by the Caltech Jet Propulsion Laboratory (JPL), and studies were also supported by industrial contractors. In fiscal year 1972, plans called for an appropriation by Congress of $30 million to fund these developments, leading toward a first launch in 1976.

Even as the scientific and technical problems of the Grand Tour were being solved, however, political and budgetary difficulties intervened. The Grand Tour was an ambitious and expensive concept, designed in the enthusiasm of the Apollo years. In the altered national climate that followed the first manned lunar landings, the United States began to pull back from major commitments in space. The later Apollo landings were canceled, and in fiscal year 1972 only $10 million of the $30 million needed to complete Grand Tour designs was ap-

propriated. It suddenly became necessary to restructure the exploration of the outer planets to conform to more modest space budgets.

Redesign of the Mission

The new mission concept that replaced the Grand Tour dropped the objectives of exploring the outer three planets—Uranus, Neptune, and Pluto. In this way the lifetime of the mission was greatly shortened, placing less stringent demands on the reliability of the millions of components that go into a spacecraft. Limiting the mission to Jupiter and Saturn also relieved problems associated with spacecraft power, and with communicating effectively over distances of more than 2 billion kilometers. The total cost of the new mission was estimated at $250 million, only a third of that previously planned for the Grand Tour. Because it was based on the proven Mariner spacecraft design, the new mission was initially named Mariner Jupiter Saturn, or MJS; in 1977 the name was changed to Voyager. In January 1972 the President's proposed fiscal year 1973 budget included $10 million specifically designated for Voyager; after authorization and

The original plan for the Outer Planets Grand Tour envisaged dual launches to Jupiter, Saturn, and Pluto in the mid-1970s, and dual launches to Jupiter, Uranus, and Neptune in 1979. However, political and budgetary constraints altered the plan, and the Voyager mission to Jupiter and Saturn, with an optional encounter with Uranus, was formulated to replace it. Here the original Grand Tour trajectories from Earth to the outer planets are shown. [P-10612 AC]

appropriation by Congress, the official beginning of Voyager was set for July 1, 1972.

With approval of the new mission apparently assured in the Congress, NASA issued an "Announcement of Flight Opportunity" to select the scientific instruments to be carried on Voyager. Seventy-seven proposals were received; 31 from groups of scientists with designs for instruments, and 46 from individuals desiring to participate in NASA-formed teams. Of these 77 proposals, 24 were from NASA laboratories, 48 were from scientists in various U.S. universities and industry, and 5 were from foreign sources. After extensive review, 28 proposals were accepted: 9 for instruments and 19 for individual participation. The newly selected Principal Investigators and Team Leaders met for the first time at JPL just before Christmas, 1972. To coordinate all the science activity of the Voyager mission, NASA and JPL selected Edward Stone of Caltech, a distinguished expert on magnetospheric physics, to serve as Project Scientist.

The team assembled in 1972 by JPL and its industrial contractors included more than a thousand highly trained engineers, scientists, and technical managers who assumed responsibility for the awesome task of building the most sophisticated unmanned spacecraft ever designed and launching it across the farthest reaches of the solar system. At the head of the organization was the Project Manager, Harris (Bud) Schurmeier. Later, Schurmeier was succeeded by John Casani, Robert Parks, and Ray Heacock. This team had only four years to turn the paper concepts into hardware, ready to deliver to Kennedy Space Center for launch in the summer of 1977.

The Objectives of Voyager

Voyager is one of the most ambitious planetary space missions ever undertaken. Voyager 1, which encountered Jupiter on March 5, 1979, was to investigate Jupiter, its large satellites Io, Ganymede, and Callisto, and tiny Amalthea; Saturn, its rings, and several of its satellites—including Titan, the largest satellite in the solar system. Voyager 2, which arrived at Jupiter on July 9, 1979, was to examine Jupiter, Europa, Ganymede, Callisto, and Saturn and several of its satellites, after which it was to be hurled on toward an encounter with the Uranian system in 1986. Both spacecraft were also designed to study the interplanetary medium and its interactions with the solar wind.

Scientific objectives as well as the orbital positions of the planets and satellites influenced the choice of spacecraft trajectories, which were designed to provide close flybys of all four of Jupiter's Galilean satellites and six of Saturn's satellites—featuring a very close approach to Titan—and occultations of the Sun and Earth by Jupiter, Saturn, Titan, and the rings of Saturn. However, all these objectives could not be accomplished by a single spacecraft. No single path through the Jupiter system, for instance, can provide close flybys of all four Galilean satellites. Specifically, a trajectory for Voyager 1 that included a close encounter with Io precluded a close encounter with Europa and did not allow the spacecraft the option of being targeted from Saturn to Uranus without having to travel through the rings of Saturn. On the other hand, a trajectory that would send Voyager 2 on to Uranus precluded not only a close encounter with Io, but also a close encounter with Titan and with Saturn's rings. The alignment of the planets and their satellites was such that encounter with Jupiter on or before April 4, 1979 was necessary for optimum investigation of Io, but encounter with Jupiter after June 15, 1979 was mandatory for a trajectory that would maintain the option to go on to Uranus. The latter trajectory also allowed Voyager 2 to maintain a healthier distance from Jupiter, avoiding the full dose of radiation that would be experienced by Voyager 1.

Although the two Voyager trajectories do have certain fundamental differences, they also have a great deal in common: Both spacecraft

Project Manager
John Casani

were designed to study the interplanetary medium, and both would investigate Jupiter and Saturn and have close flybys with several of their major satellites. Also, if it were decided not to send Voyager 2 on to Uranus, this spacecraft could be retargeted to a Saturn trajectory similar to that of Voyager 1, providing a close flyby of Titan and a closer look at the rings. The flight paths of Voyager 1 and Voyager 2 complement each other—allowing the planets and some of the satellites to be viewed from a number of angles and over a longer period of time that would be possible with only one spaceprobe—yet the trajectories were also designed to be more or less capable of duplicating each other's scientific investigations. This redundancy helps to ensure, so far as is possible, the success of the mission.

The redundancy built into the mission is evident, not only in the design of the trajectories, but in the spacecraft themselves. Voyagers 1 and 2 are identical spacecraft. In addition, many crucial elements are duplicated on each: For example, the computer command subsystem (CCS), the flight data subsystem (FDS), and the attitude and articulation control subsystem (AACS)—which function as onboard control systems—each have multiple reprogrammable digital computers. In addition, the CCS, which decodes commands from ground control and can instruct other subsystems from its own memory, contains duplicates for all its functional units. The AACS, which controls the spacecraft's stabilization and orientation, has duplicate star trackers and Sun sensors. The communications system contains two radio receivers and four transmitters, two each to transmit both S-band (frequency of about 2295 megahertz) and X-band (frequency of about 8418 megahertz).

The Spacecraft

The Voyager spacecraft are more sophisticated, more automatic, and more independent than were the Pioneers. This independence is important because the giant planets are so far away that the correction of a malfunction by engineers on Earth would take hours to perform. Even at "nearby" Jupiter, radio signals take about forty minutes to travel in one direction between the spacecraft and Earth. Saturn is about twice as far away as Jupiter, and Uranus is twice as far as Saturn, slowing communication even more.

Perched atop the Centaur stage of the launch rocket, each Voyager spacecraft has a mass of 2 tons (2066 kilograms), divided about equally between the spacecraft proper and the propulsion module used for final acceleration

About the same size and weight of a subcompact car, the Voyager spacecraft carry instruments for eleven science investigations of the outer planets and their satellites. Power to operate the spacecraft is provided by three radioisotope thermoelectric generators mounted on one boom; other booms hold the science scan platform and the dual magnetometers.

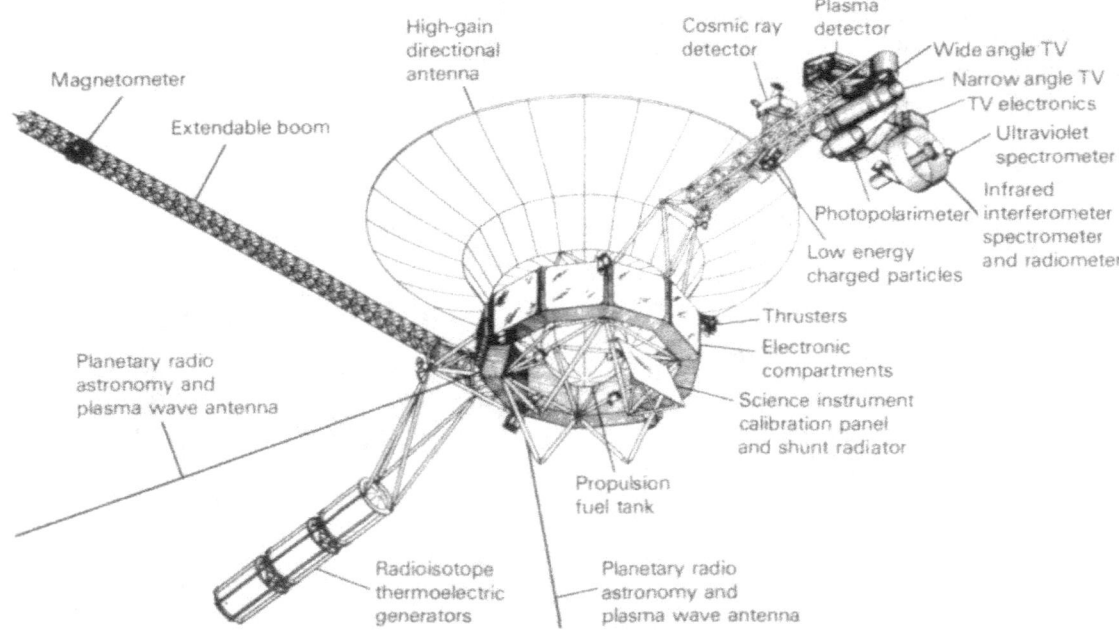

to Jupiter. The Voyager itself, with a mass of 815 kilograms and typical dimensions of about 3 meters, is almost the size and weight of a subcompact car—but enormously more complex. A better analogy might be made with a large electronic computer—but no terrestrial computer was ever asked to supply its own power source and to operate unattended in the vacuum of space for up to a decade.

Each Voyager spacecraft carries instruments for eleven science investigations covering visual, infrared, and ultraviolet regions of the spectrum, and other remote sensing studies of the planets and satellites; studies of radio emissions, magnetic fields, cosmic rays, and lower energy particles; plasma (ionized gases) waves and particles; and studies using the spacecraft radios. Each Voyager has a science boom that holds high and low resolution TV, photopolarimeter, plasma and cosmic ray detectors, infrared spectrometer and radiometer, ultraviolet spectrometer, and low energy charged particle detector; in addition, each spacecraft has a planetary radio astronomy and plasma wave antenna and a long boom carrying a low- and a high-field magnetometer.

Communication with Earth is carried out via a high-gain antenna 3.7 meters in diameter, with a smaller low-gain antenna as backup. The large white dish of the high-gain antenna dominates the appearance of the spacecraft, setting it off from its predecessors, which were able to use much smaller antennas to communicate over the more modest distances that separate the planets of the inner solar system. Although the transmitter power is only 23 watts—about the power of a refrigerator light bulb—this system is designed to transmit data over a billion kilometers at the enormous rate of 115 200 bits per second. Data can also be stored for later transmission to Earth; an onboard digital tape recorder has a capacity of about 500 million (5×10^8) bits, sufficient to store nearly 100 Voyager images.

Project Manager
Ray Heacock

During the early assembly stage, technicians at Caltech's Jet Propulsion Laboratory equip Voyager's extendable boom with low- and high-field magnetometers that measure the intensity and direction of the outer planets' magnetic fields. [373-7179 BC]

Radioisotope thermoelectric generators (RTGs), rather than solar cells, provide electricity for the Voyager spacecraft. The RTGs use radioactive plutonium oxide for this purpose. As the plutonium oxide decays, it gives off heat which is converted to electricity, supplying a total of about 450 watts to the spacecraft at launch. This power slowly declines as the plutonium is used up, with less than 400 watts expected at Saturn flyby five years after launch. Hydrazine fuel is used to make midcourse corrections in trajectory and to control the spacecraft's orientation.

Since the Voyagers must fly through the inner magnetosphere of Jupiter, it was imperative that the hardware systems be able to withstand the radiation from Jovian charged particles. The electronic microcircuits that form the heart and brain of the spacecraft and its scientific instruments are especially susceptible to radiation damage. Three techniques were used to "harden" components against radiation:

1. Special design using radiation-resistant materials;
2. Extensive testing to select those electronic components which come out of the manufacturing process with highest reliability; and

27

VOYAGER'S GREETINGS TO THE UNIVERSE

The Voyager spacecraft will be the third and fourth human artifacts to escape entirely from the solar system. Pioneers 10 and 11, which preceded Voyager in outstripping the gravitational attraction of the Sun, both carried small metal plaques identifying their time and place of origin for the benefit of any other spacefarers that might find them in the distant future. With this example before them, NASA placed a more ambitious message aboard Voyager 1 and 2—a kind of time capsule, intended to communicate a story of our world to extraterrestrials.

The Voyager message is carried by a phonograph record—a 12-inch gold-plated copper disk containing sounds and images selected to portray the diversity of life and culture on Earth. The contents of the record were selected for NASA by a committee chaired by Carl Sagan of Cornell University. Dr. Sagan and his associates assembled 115 images and a variety of natural sounds, such as those made by surf, wind and thunder, birds, whales, and other animals. To this they added musical selections from different cultures and eras, and spoken greetings from Earth-people in sixty languages, and printed messages from President Carter and U.N. Secretary General Waldheim.

Each record is encased in a protective aluminum jacket, together with a cartridge and needle. Instructions, in symbolic language, explain the origin of the spacecraft and indicate how the record is to be played. The 115 images are encoded in analog form. The remainder of the record is in audio, designed to be played at 16⅔ revolutions per second. It contains the spoken greetings, beginning with Akkadian, which was spoken in Sumer about six thousand years ago, and ending with Wu, a modern Chinese dialect. Following the section on the sounds of Earth, there is an eclectic 90-minute selection of music, including both Eastern and Western classics and a variety of ethnic music.

Once the Voyager spacecraft leave the solar system (by 1990, both will be beyond the orbit of Pluto), they will find themselves in empty space. It will be forty thousand years before they come within a light year of a star, called AC + 79 3888, and millions of years before either might make a close approach to any other planetary system. As Carl Sagan has noted, "The spacecraft will be encountered and the record played only if there are advanced spacefaring civilizations in interstellar space. But the launching of this bottle into the cosmic ocean says something very hopeful about life on this planet."

LANGUAGES ON VOYAGER RECORD

Sumerian	Cantonese	Spanish	Hindi	Turkish	Armenian	Swedish	Kannada
Akkadian	Russian	Indonesian	Vietnamese	Welsh	Polish	Ukrainian	Telugu
Hittite	Thai	Kechua	Sinhalese	Italian	Netali	Persian	Oriya
Hebrew	Arabic	Dutch	Greek	Nguni	Mandarin	Serbian	Hungarian
Aramaic	Roumanian	German	Latin	Sotho	Gujorati	Luganada	Czech
English	French	Bengali	Japanese	Wu	Ila (Zambia)	Amoy (Min dialect)	Rajasthani
Portuguese	Burmese	Urdu	Punjabi	Korean	Nyanja	Marathi	

SOUNDS OF EARTH ON VOYAGER RECORD

Whales	Birds	Laughter	Riveter	Tractor	Kiss
Planets (music)	Hyena	Fire	Morse code	Truck	Baby
Volcanoes	Elephant	Tools	Ships	Auto gears	Life signs—
Mud pots	Chimpanzee	Dogs, domestic	Horse and cart	Jet	EEG, EKG
Rain	Wild dog	Herding sheep	Horse and carriage	Lift-off Saturn 5	Pulsar
Surf	Footsteps and	Blacksmith shop	Train whistle	rocket	
Crickets, frogs	heartbeats	Sawing			

VOYAGER RECORD PHOTOGRAPH INDEX

Calibration circle
Solar location map
Mathematical definitions
Physical unit definitions
Solar system parameters
The Sun
Solar spectrum
Mercury
Mars
Jupiter
Earth
Egypt, Red Sea, Sinai Peninsula
    and the Nile
Chemical definitions
DNA structure
DNA structure magnified
Cells and cell division
Anatomy (eight)
Human sex organs
Diagram of conception
Conception
Fertilized ovum
Fetus diagram
Fetus
Diagram of male and female
Birth
Nursing mother
Father and daughter (Malasia)
Group of children
Diagram of family ages
Family portrait
Diagram of continental drift
Structure of Earth
Heron Island (Great Barrier Reef
    of Australia)
Seashore
Snake River and Grand Tetons
Sand dunes

Monument Valley
Forest scene with mushrooms
Leaf
Fallen leaves
Sequoia
Snowflake
Tree with daffodils
Flying insect with flowers
Diagram of vertebrate evolution
Seashell (Xancidae)
Dolphins
School of fish
Tree toad
Crocodile
Eagle
Waterhold
Jane Goodall and chimps
Sketch of Bushmen
Bushmen hunters
Man from Guatamala
Dancer from Bali
Andean girls
Thailand craftsman
Elephant
Old man with beard and glasses
    (Turkey)
Old man with dog and flowers
Mountain climber
Cathy Rigby
Sprinters
Schoolroom
Children with globe
Cotton harvest
Grape picker
Supermarket
Underwater scene with diver
    and fish
Fishing boat with nets

Cooking fish
Chinese dinner party
Demonstration of licking, eating
    and drinking
Great Wall of China
House construction (African)
Construction scene (Amish country)
House (Africa)
House (New England)
Modern house (Cloudcroft, New Mexico)
House interior with artist and fire
Taj Mahal
English city (Oxford)
Boston
UN Building, day
UN Building, night
Sydney Opera House
Artisan with drill
Factory interior
Museum
X-ray of hand
Woman with microscope
Street scene, Asia (Pakistan)
Rush hour traffic, India
Modern highway (Ithaca)
Golden Gate Bridge
Train
Airplane in flight
Airport (Toronto)
Antarctic expedition
Radio telescope (Westerbork,
    Netherlands)
Radio telescope (Arecibo)
Page of book (Newton,
    System of the World)
Astronaut in space
Titan Centaur launch
Sunset with birds
String quartet (Quartetto Italiano)
Violin with music score (Cavatina)

MUSIC ON VOYAGER RECORD

Bach Brandenberg Concerto Number Two, First
    Movement
"Kinds of Flowers" Javanese Court Gamelan
Senegalese Percussion
Pygmy girls initiation song
Australian Horn and Totem song
"El Cascabel" Lorenzo Barcelata
"Johnny B. Goode" Chuck Berry
New Guinea Men's House
"Depicting the Cranes in Their Nest"
Bach Partita Number Three for Violin; Gavotte et
    Rondeaus
Mozart Magic Flute, Queen of the Night (Aria Number 14)
Chakrulo
Peruvian Pan Pipes

Melancholy Blues
Azerbaijan Two Flutes
Stravinsky, Rite of Spring, Conclusion
Bach Prelude and Fugue Number One in C Major from
    the Well Tempered Clavier, Book Two
Beethoven's Fifth Symphony, First Movement
Bulgarian Shepherdess Song "Izlel Delyo Hajdutin"
Navajo Indian Night Chant
The Fairie Round from Pavans, Galliards, Almains
Melanesian Pan Pipes
Peruvian Woman's Wedding Song
"Flowing Streams"—Chinese Ch'in music
"Jaat Kahan Ho"—Indian Raga
"Dark Was the Night"
Beethoven String Quartet Number 13 "Cavatina"

Each Voyager carries a message in the form of a 12-inch gold-plated phonograph record. The record, together with a cartridge and needle, is fastened to the side of the spacecraft in a gold-anodized aluminum case that also illustrates how the record is to be played. [P-19728]

THE BRAINS OF THE VOYAGER SPACECRAFT*

The Voyager spacecraft had greater independence from Earth-based controllers and greater versatility in carrying out complex sequences of scientific measurements than any of their predecessors. These capabilities resulted from three interconnected onboard computer systems: the AACS (attitude and articulation control subsystem); the FDS (flight data subsystem); and the CCS (computer command system). Operating from "loads" of instructions transmitted earlier from Earth, these computers could issue commands to the spacecraft and the science instruments and react automatically to problems or changes in operating conditions.

The complex sequence of scientific observations and the associated engineering functions were executed by the spacecraft under the control of an updatable program stored in the CCS by ground command. At appropriate times, the CCS issued commands to the AACS for movement of the scan platform or spacecraft maneuvers; to the FDS for changes in instrument configuration or telemetry rate; or to numerous other subsystems within the spacecraft for specific actions. The two identical (redundant) 4096-word memories within the CCS contained both fixed routines (about 2800 words) and a variable section (about 1290 words) for changing science sequencing functions. A single 1290-word science sequence load could easily generate 300 000 discrete commands, thus providing significantly more sequencing capability than would be possible through ground commands. A 1290-word sequencing load in the CCS controlled both the science and engineering functions of the spacecraft for a period lasting for 3/4 day at closest approach and for up to 100 days during cruise.

Each 1290-word program (or load) was built from specific science measurement units called links. Some links were used repeatedly in a looping cyclic (like a computer DO loop) to perform the same observation numerous times; other links that involved special measurement geometry or critical timing occurred only once. About 175 science links were defined for the Voyager 1 Jupiter encounter. It took almost two years to convert the desired science objectives and measurements first into links, then into a minute-by-minute timeline for the 98-day encounter period, and finally into the specific computer instructions that could be loaded into the CCS memory for that portion of the encounter time represented by a particular load. The total Voyager 1 Jupiter encounter period used eighteen sequence memory loads, supplemented by about 1000 ground commands to modify the sequences because of changing conditions or calibration requirements.

For the Voyager 2 encounter, concern about the ailing spacecraft receiver limited the number of loads that could be transmitted, particularly while the spacecraft was deep within the Jovian magnetosphere, where radiation effects caused the receiver frequency to drift unpredictably. However, a careful redesign of the planned sequences permitted the accomplishment of very nearly the original set of observations even with these constraints.

*Adapted from a paper by E. C. Stone and A. L. Lane in the Voyager 1 Thirty-Day Report.

Project Manager
Robert Parks

3. Spot shielding of especially sensitive areas with radiation-absorbing materials.

All three approaches were required on the Voyager craft, especially after Pioneer 10 and 11 demonstrated that the radiation at Jupiter was even more intense than had been assumed in early design studies.

The steady streams of engineering and scientific data received on Earth are transmitted from the receiving stations of the Deep Space Network (DSN) to JPL, where the Voyager control functions are centered. There, dozens of technicians check and recheck every subsystem to search for the slightest hint of malfunction. In case of problems, there are thick notebooks of instructions and racks of precoded computer tapes ready to be used to correct any apparent malfunction.

Normally Voyager runs itself. Detailed instructions are programmed into its onboard computers and command systems for dealing with such potential emergencies as a stuck valve in the fuel system, loss of orientation in the star trackers, erratic gyroscope functions, failure of radio communications, or a thousand and one other nightmares. The instructions for operating the scientific instruments are also stored on board, with new blocks of commands sent up once every few days to replace those for tasks already completed. Whether in the calm of cruise mode or the intense excitement of a planetary encounter, the Voyager craft is alone in space, continuously sensing and reacting to its environment, tied by only a tenuous thread of radio communication to the anxious watchers back on Earth.

NASA PLANETARY MISSIONS

Spacecraft	Launch Date	Destination	Encounter Date	Type of Encounter
Mariner 2	8/26/62	Venus	12/14/62	flyby
Mariner 4	11/28/64	Mars	7/14/65	flyby
Mariner 5	6/14/67	Venus	10/19/67	flyby
Mariner 6	2/25/69	Mars	7/31/69	flyby
Mariner 7	3/27/69	Mars	8/05/69	flyby
Mariner 9	5/30/71	Mars	11/13/71	orbiter
Pioneer 10	3/03/72	Jupiter	12/04/73	flyby
Pioneer 11	4/06/73	Jupiter	12/03/74	flyby
		Saturn	9/01/79	flyby
Mariner 10	11/03/73	Venus	2/05/74	flyby
		Mercury	3/29/74	flyby
Viking 1	8/20/75	Mars	6/19/76	orbiter
			7/20/76	lander
Viking 2	9/09/75	Mars	7/07/76	orbiter
			9/03/76	lander
Voyager 1	8/20/77	Jupiter	3/05/79	flyby
Voyager 2	9/05/77	Jupiter	7/09/79	flyby
Pioneer Venus	5/20/78	Venus	12/04/78	orbiter
	8/8/78	Venus	12/09/78	probe

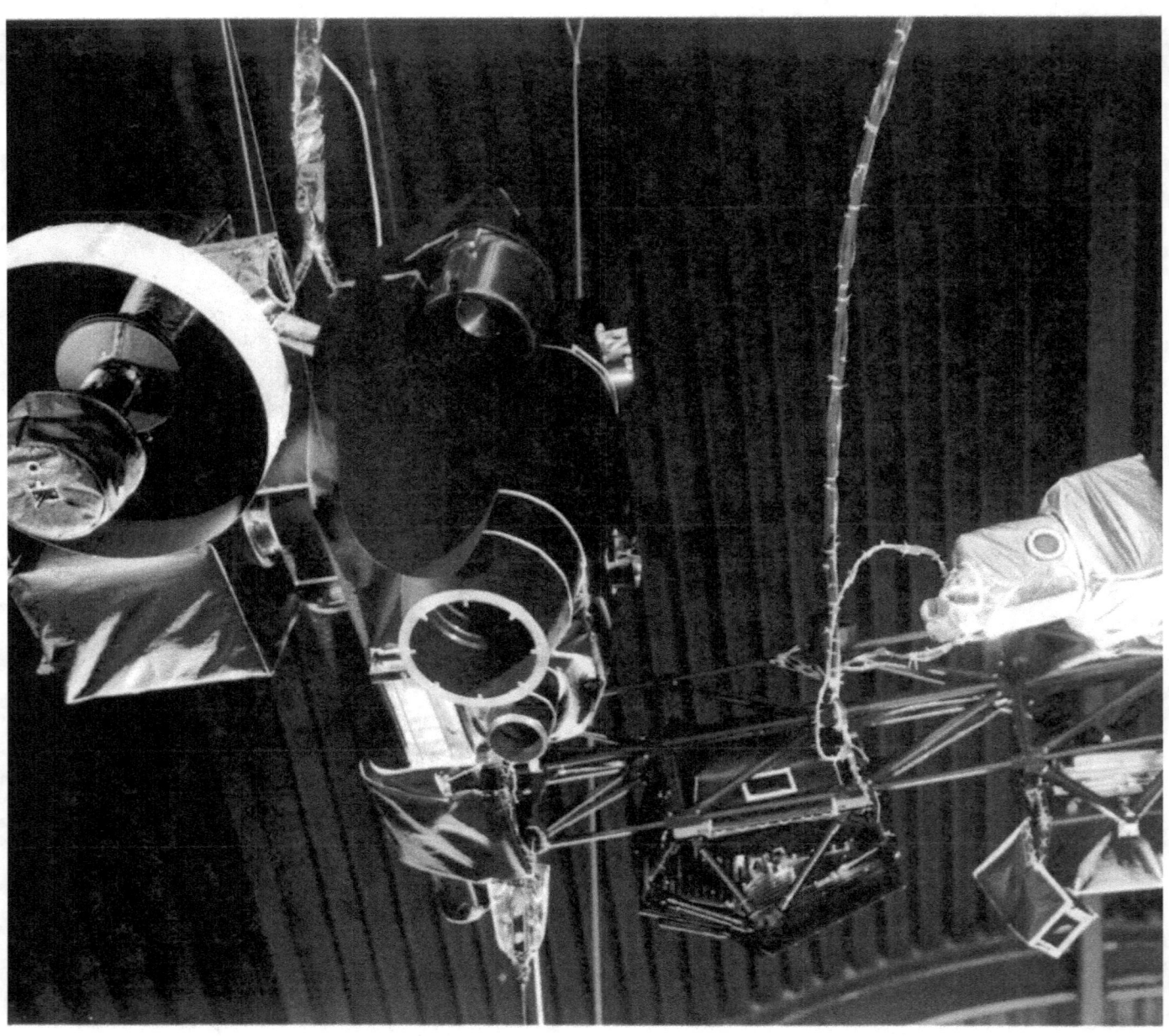

The Voyager scan platform contains sophisticated instruments that gather data for Voyager's remote sensing investigations. Five of the remote-sensing instruments—two TV cameras, the infrared spectrometer, the ultraviolet spectrometer, and the photopolarimeter—are mounted together on the scan platform, which can be pointed to almost any direction in space, allowing exact targeting of the observations. [373-7146BC]

CHAPTER 4

SCIENCE AND SCIENTISTS

Introduction

There are many reasons for sending spacecraft to the planets, but in the final analysis, we send our robot messengers across the vastness of space for the sake of scientific exploration. Science and exploration have always gone hand in hand, whether in the transcontinental journey of Lewis and Clark or the Pacific voyages of Captain James Cook. In this century, as exploration has become more and more dependent on advances in technology, the scientific element has attained increased prominence. The greatest legacy of the NASA Planetary Program is the knowledge it has provided of the other worlds that share our corner of the universe.

Despite the central role of science in motivating missions to the planets, specific scientific considerations are not dominant during most of the development of a mission such as Voyager. The problem of building a spacecraft and getting it to the outer solar system is too demanding. Of the ton of mass in a Voyager spacecraft, the scientific instruments make up only 115 kilograms (about eleven percent). Similarly, the cost of these instruments amounts to only about 10 percent of the cost of the spacecraft and launch vehicle. Unless the launch and operation of the spacecraft are nearly perfect, there can be no scientific return in any case; even the most sophisticated package of scientific instruments will not tell much about Jupiter if, following launch, it rests at the bottom of the Atlantic Ocean. But it is equally true that the ultimate purpose of the mission is scientific discovery, and NASA makes every effort to ensure that the very best instruments are flown and that a broad scientific community is given the opportunity to participate in each mission.

A decade before the 1977 launch, many astronomers and space scientists began their involvement with the Voyager mission through participation in study groups convened by NASA and by the National Academy of Sciences. They came primarily from universities, but also in significant numbers from NASA laboratories, from industry, and from abroad. In 1971 the Outer Planets Grand Tours mission definition group carried out a one-year final study of the mission that was to become Voyager. A competition was held in 1972 to select the Voyager flight instruments and science teams, and a third review stage followed a year later to confirm this selection. Out of this process emerged eleven science investigations, with which more than one hundred scientists were associated. In this chapter we look at the instruments and the persons who designed them for this challenging task.

Direct and Remote Measurements

The diverse measurements made by Voyager of a planet and its environment can be divided into two broad categories, usually called direct or *in situ* measurements and remote sensing measurements. A direct measurement involves the analysis of the immediate environment of the spacecraft; remote measurements can be made by analyzing radiation from distant objects.

The direct measurement instruments on Voyager measure cosmic ray particles, low energy charged particles, magnetic fields, plasma particles, and plasma waves. Their activity began immediately after launch, monitoring the Earth environment and then interplanetary space until the magnetosphere of Jupiter was reached a few days before the actual flyby. Long after the flybys of Jupiter and

Saturn are completed, particles and fields data can continue to be acquired by the Voyagers as they speed outward into previously unexplored regions of space.

The remote sensing investigations are essentially astronomical in nature, measuring the light reflected from or emitted by the planet and its satellites. On Voyager, however, these instruments far outstrip their terrestrial counterparts in capability. Primarily, they derive their advantage from their proximity to what they observe—at its closest, Voyager was more than a thousand times nearer to Jupiter than are Earth-based telescopes; and for the still closer satellite encounters, Voyager was nearly ten thousand times nearer than astronomers on Earth. In addition, Voyager provided perspectives, such as views of the night side of Jupiter,

that are impossible from Earth. Finally, these instruments could exploit the full spectrum of electromagnetic radiation without concern for the opacity of the terrestrial atmosphere, which restricts ground-based astronomers to certain spectral windows and blocks all observation at other wavelengths.

Five of the remote sensing instruments—the two TV cameras, the infrared spectrometer, the ultraviolet spectrometer, and the photopolarimeter—are mounted together on a scan platform. This platform can be pointed to almost any direction in space, allowing exact targeting of the observations.

One remote sensing instrument, the planetary radio astronomy receiver, is not on the scan platform. It measures long-wave radio emission without requiring special pointing.

The fully deployed Voyager spacecraft is capable of a wide variety of direct and remote sensing measurements. The instruments and their objectives were selected many years before the first Jupiter encounter. Because of the exploratory nature of the Voyager mission, every effort was made to fly versatile instruments that could yield valuable results no matter what the nature of the Jovian system. [P-18811AC]

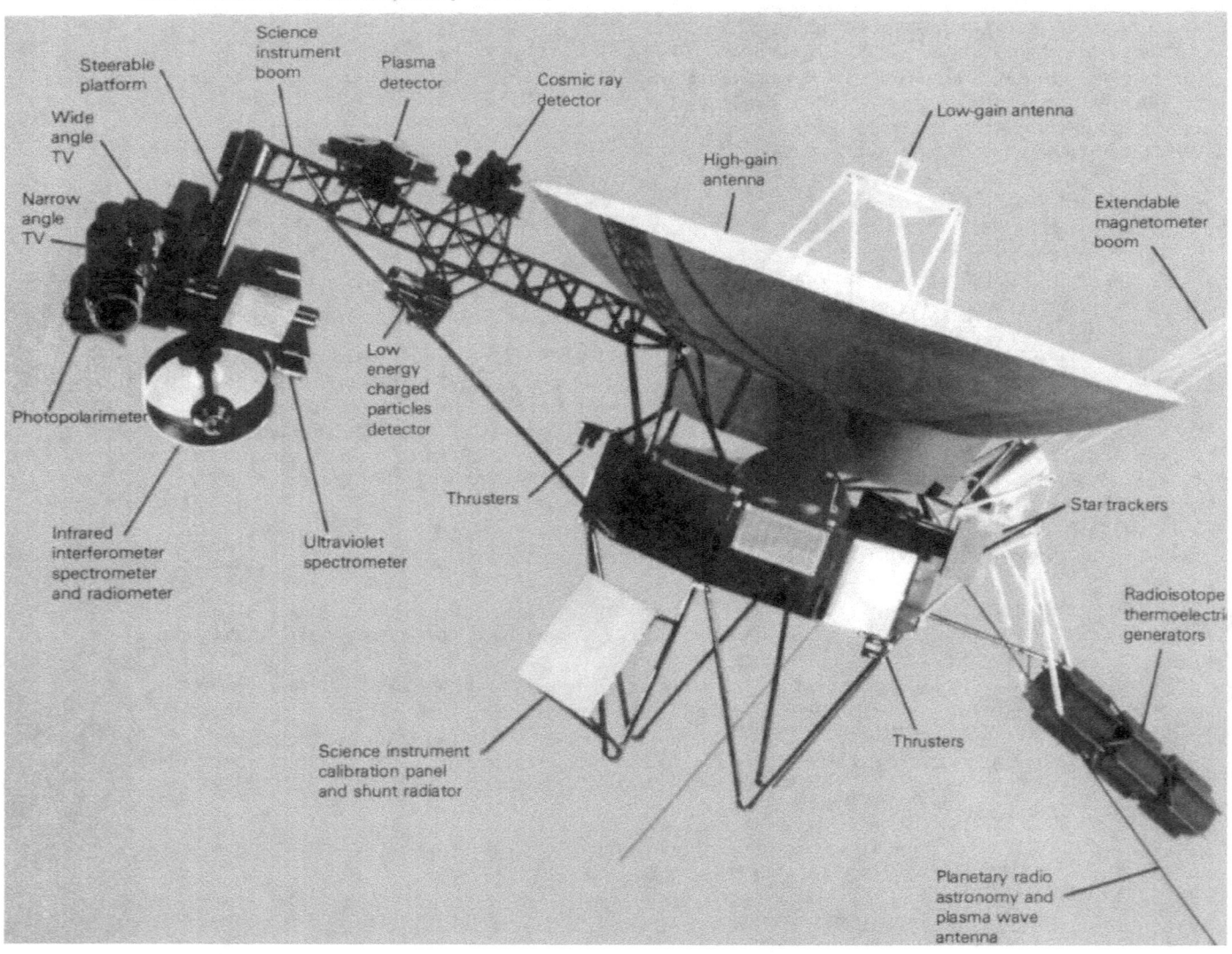

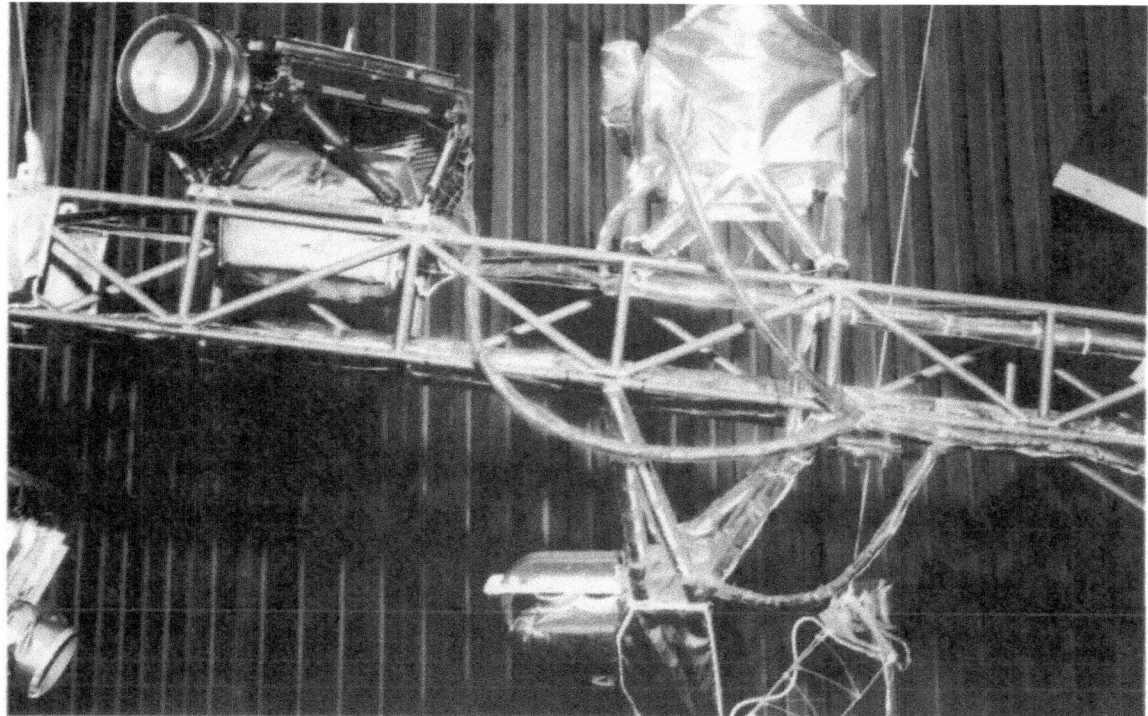

The science instrument boom supports the plasma particle detector, the cosmic ray detector, and the low energy charged particle detector. These instruments began collecting data immediately after launch, monitoring the Earth environment and then interplanetary space until the magnetosphere of Jupiter was reached a few days before the actual flyby. [353-2992BC]

A final Voyager investigation did not fit into this pattern of direct versus remote sensing instruments. In fact, it required no special instrument at all. This investigation deals with radio science, and it utilizes the regular communications link between the spacecraft and Earth to derive the masses of Jupiter and its satellites, to probe the atmosphere of Jupiter, and to study properties of the interplanetary medium.

Imaging

The eyes of Voyager are in its imaging system. Two television cameras, each with a set of color filters, look at the planets and their

The Voyager Imaging Science Team

VOYAGER SCIENCE INVESTIGATIONS

Project Scientist: E. C. Stone, Caltech

Investigation	Principal Investigator or Team Leader	Primary Objectives at Jupiter
Imaging science	B. A. Smith, U. Arizona	High resolution reconnaissance over large phase angles; measurement of atmospheric dynamics; determination of geologic structure of satellites; search for rings and new satellites.
Infrared radiation (IRIS)	R. A. Hanel, NASA Goddard	Determination of atmospheric composition, thermal structure, and dynamics; satellite surface composition and thermal properties.
Ultraviolet spectroscopy	A. L. Broadfoot, Kitt Peak Observatory	Measurement of upper atmospheric composition and structure; auroral processes; distribution of ions and neutral atoms in the Jovian system.
Photopolarimetry	C. F. Lillie/C. W. Hord, U. Colorado	Measurement of atmospheric aerosols; satellite surface texture and sodium cloud.
Planetary radio astronomy	J. W. Warwick, U. Colorado	Determination of polarization and spectra of radio frequency emissions; Io radio modulation process; plasma densities.
Magnetic fields	N. F. Ness, NASA Goddard	Measurement of plasma electron densities; wave-particle interactions; low-frequency wave emissions.
Plasma particles	H. S. Bridge, MIT	Measurement of magnetospheric ion and electron distribution; solar wind interaction with Jupiter; ions from satellites.
Plasma waves	F. L. Scarf, TRW	Measurement of plasma electron densities; wave-particle interactions; low-frequency wave emissions.
Low energy charged particles	S. M. Krimigis, Johns Hopkins U.	Measurement of the distribution, composition, and flow of energetic ions and electrons; satellite-energetic particle interactions.
Cosmic ray particles	R. E. Vogt, Caltech	Measurement of the distribution, composition, and flow of high energy trapped nuclei; energetic electron spectra.
Radio science	V. R. Eshleman, Stanford U.	Measurement of atmospheric and ionospheric structure, constituents, and dynamics; satellite masses.

satellites and transmit thousands of detailed pictures to Earth. The imaging system is probably the most versatile and therefore the most truly exploratory of the Voyager instruments. No matter what is out there, the imaging system will let us see it and hence, we hope, begin to understand its nature.

Unlike other Voyager instruments, the imaging system is not the result of a competition among proposals submitted by groups of

Bradford A. Smith, imaging science Team Leader

scientists. NASA assigned the development of the cameras directly to JPL, to be integrated from the beginning with the design of the Voyager spacecraft and its subsystems. The members of the Voyager Imaging Science Team were selected, as individuals, on the basis of the scientific studies they proposed to carry out. Initially, ten members of the Imaging Science Team were selected, but by the time of the Jupiter encounters, the team had been expanded to 22 scientists.

The Team Leader is Bradford A. Smith, a professor in the Department of Planetary Science at the University of Arizona. Smith was involved in imaging science on several previous missions, including Mariners 6 and 7 and Viking. He was also active in ground-based photography of Jupiter at both New Mexico State University and University of Arizona, and he is a member of the team developing a planetary camera for the Space Telescope, scheduled for operation in Earth orbit in the mid-1980s.

Originally, the Deputy Team Leader was Geoffrey A. Briggs, a young British-born physicist from JPL. However, in 1977 Briggs took a position at NASA Headquarters in Washington and was replaced by geologist

Laurence A. Soderblom of the U.S. Geological Survey in Flagstaff, Arizona. An energetic and articulate scientist, Soderblom, with his interest in satellite geology, complemented Smith, whose personal scientific interests are directed more toward the atmosphere of Jupiter.

The objectives of imaging involved multi-color photography of Jupiter and its satellites. Both wide- and narrow-angle cameras were needed to obtain the highest possible resolution while retaining the capability to study global-scale features on Jupiter and the satellites. In normal photographic terms, both cameras used telephoto lenses. For the wide-angle camera, a focal length of 200 millimeters was selected, giving a field of view of about 3 degrees. This field is similar to that obtained with a 400-millimeter telephoto lens on a 35-millimeter camera. The narrow-angle Voyager camera has a focal length of 1500 millimeters and field of view of 0.4 degrees. The camera optics are combinations of mirrors and lenses, designed for extreme stability of focus and for freedom from distortion.

Each camera has a rotating filter wheel that can be used to select the color of the light that reaches the camera. For the wide-angle camera, these colors are clear, violet, blue, green, orange, and three special bands for selective observation in sodium light (589-nanometer wavelength), and in methane spectral lines at 541 nanometers and 618 nanometers. For the narrow-angle camera, the filters are clear, ultraviolet, violet, blue, green, and orange. To create a color picture, the cameras are commanded to take, in rapid succession, pictures of the same area in blue, green, and orange light. These three pictures can then be reconstructed on Earth into a "true" color image. Other combinations of colors are used to investigate particular scientific problems and to determine the spectrum of sunlight reflected from features on Jupiter and its satellites.

The detector in the cameras is not photographic film but the surface of a selenium-sulfur vidicon television tube, 11 millimeters square. Unlike most commercial TV cameras, these tubes are designed for slow-scan readout, providing 48 seconds to acquire each picture. The shutter speed can be varied from a fraction of a second (for Jupiter and Io) to many minutes (for searches for faint features, such as aurorae on the night side of Jupiter).

Each picture consists of many numbers, each of which represents the brightness of a single picture element, or pixel, on the image. There are 640 000 pixels in each Voyager image, representing a square image of 800 × 800 points. This information content is more than twice that of an ordinary television picture, which has only 520 lines. For each pixel, eight binary numbers are required to specifiy the brightness; the total information in a single image is thus $8 \times 640\,000 = 5\,120\,000$ bits. Even at a transmission of one frame per 48 seconds, the "bit rate" is more than $100\,000$ bits per second. For comparison, the bit rate from the first spacecraft to Mars (Mariner 4) was about 10 bits per second, requiring a week to transmit 21 pictures, with a total information content equivalent to a single Voyager picture. Altogether, Voyager took nearly $20\,000$ pictures at each Jupiter encounter, representing 10^{11}—a hundred billion—bits of information.

Infrared Spectrometer

The infrared investigation on Voyager is based on one of the most sophisticated instruments ever flown to another planet. In the past, most infrared instruments on planetary spacecraft measured at only a few wavelengths, but Voyager carries a true spectrometer, capable of measuring at nearly 2000 separate wavelengths, covering the spectrum from 4 to 50 micrometers.

Twelve scientists, led by Principal Investigator Rudolph Hanel of the NASA Goddard Space Flight Center at Greenbelt, Maryland, proposed this infrared instrument. Hanel is an acknowledged world leader in infrared spectroscopy from space. With his co-workers at Goddard, he has pioneered in adapting the extremely complex art of interferometric spectroscopy to the rigors of space flight. His spectrometers have made many studies of the Earth's atmosphere from meteorological satellites, and a Hanel interferometer flew successfully to Mars on Mariner 9.

The primary goals of the infrared spectrometer investigation are directed toward analysis of the composition and structure of the atmosphere of Jupiter. Among the molecules to be searched for on Jupiter were hydrogen (H_2), helium (He), methane (CH_4), ammonia (NH_3), phosphine (PH_3), water (H_2O), carbon monoxide (CO), simple compounds of silicon and sulfur, and a variety of organic compounds consisting of atoms of carbon and hydrogen (e.g., C_2H_2, C_2H_4, C_2H_6). In addition to indicating the abundance of these constituents of the atmosphere, the infrared spectra also contain information on atmospheric structure, that is, on the variation of temperature and pressure with altitude. The presence of clouds or dust

Rudolph Hanel, infrared spectrometer Principal Investigator

layers can also be inferred from the shapes of spectral lines.

In addition to its spectroscopy of Jupiter, the Voyager instrument could be used as a heat-measuring device to map the temperatures of both the satellites and the atmosphere of Jupiter. Particularly interesting for the satellites are measurements of the surface cooling and heating rates, since these rates reveal the physical compactness of the surface, easily distinguishing between rock and sand or dust.

The infrared instrument is a Michelson interferometer at the focus of a gold-plated telescope of 51-centimeter aperture. The spectrum is not obtained directly, by moving a prism or grating, but indirectly through the interference effects of light of different wavelengths: hence its name, an interferometer. The complex interference pattern generated by the motion of one of the mirrors in the light path is transmitted to Earth, where computer analysis is required to transform it into a recognizable spectrum. The entire instrument, which has a mass of 20 kilograms, is called IRIS, for infrared interferometer spectrometer.

The development of the infrared system for Voyager posed many problems. IRIS was designed to cover the optimum spectral region for studies of the atmospheres of Jupiter and Saturn. However, when it was decided in 1974

38

that an extended Voyager mission might also allow a visit to Uranus, Hanel and his colleagues realized that their instrument had serious deficiencies for investigation of that colder and more distant planet, and they proposed that a modified IRIS (MIRIS) be substituted for the original design. A crash program was authorized to develop MIRIS in parallel with IRIS, and in early 1977, as launch approached, it appeared that the improved instrument would be ready. Problems occurred during testing, however, and for several weeks in June and July the decision hung in the balance. Unfortunately, there simply was not enough time to solve all the problems; both Voyagers were launched carrying the original IRIS, and the final flight qualification of MIRIS came in October, about six weeks after launch. With no other missions planned beyond Saturn, no alternate use for MIRIS has been found; it remains "on the shelf," one of the rare cases where a technological gamble by NASA did not pay off.

Ultraviolet Spectrometer

A second spectroscopic instrument on the Voyager scan platform examines short-wave, or ultraviolet, radiation. The Principal Investigator for the ultraviolet spectrometer (UVS) is A. Lyle Broadfoot of Kitt Peak National Observatory in Tucson, Arizona. Broadfoot's expertise lies in instrumentation for atmospheric

Lyle Broadfoot, ultraviolet spectrometer Principal Investigator

research, and he was previously the Principal Investigator for a similar ultraviolet instrument on the Mariner 10 mission to Venus and Mercury. Associated with Broadfoot in the Voyager investigation are fourteen other scientists from the United States, Canada, and France.

In the upper atmosphere of a planet, the lighter gases tend to diffuse, rising above their heavier neighbors, and unusual chemical reactions take place as the action of sunlight breaks some chemical bonds and stimulates the formation of others. The study of these tenuous regions of planetary atmospheres is called aeronomy. Many of the exotic chemical processes that occur can best be studied by examining radiation of very short wavelength, and it is to these problems that much of the work of the UVS team is directed.

The observations are also sensitive to special processes in the Jovian atmosphere resulting from the magnetosphere. At the very top of the atmosphere, energetic charged particles interacting with the atmospheric molecules produce ultraviolet aurorae that indicate the location and nature of the bombarding electrons and ions. Emissions from atoms in the extended gas clouds around Io and possibly around other satellites were also expected to be observable.

The ultraviolet instrument is a relatively straightforward optical spectrometer, adapted for use in space. A diffraction grating disperses the light into a spectrum, and the other optical elements are gold-coated mirrors. The field of view is rectangular, about 0.1 degrees × 0.9 degrees. An array of sensitive electronic detectors provides simultaneous measurements in 128 wavelength channels over a wavelength range from 50 to 170 nanometers.

Photopolarimeter

The fourth scan-platform instrument is designed to measure the brightness and polarization of light with high precision. Unlike the imaging system, however, it can look at only a single point (one pixel) at a time; thus it sacrifices spatial capability in favor of precision of measurement. The prime objectives of this photopolarimeter are related to study of the clouds of Jupiter.

During the development stage for this instrument, the Principal Investigator was Charles F. Lillie of the Laboratory for Atmospheric and Space Physics of the University of Colorado at Boulder. Later, the Principal Investigator's role was assumed by Lillie's col-

39

Charles W. Hord, photopolarimeter Principal Investigator

took place, leading to the reluctant decision to shut down the Voyager 1 photopolarimeter. On Voyager 2, similar mechanical problems greatly restricted the ability of the instrument to carry out its observations. The photopolarimeter thus was unable to contribute its share to unraveling the mysteries of Jupiter and its satellites.

Planetary Radio Astronomy

The final Voyager remote sensing instrument is designed to measure radio emission from Jupiter and Saturn over a wide range of frequencies. These emissions, which sound like hiss or static if played through an audio receiver, result from interactions of charged particles in the magnetospheres and ionospheres of the giant planets. The planetary radio astronomy (PRA) Principal Investigator is James W. Warwick, an astronomer in the Department of Astro-Geophysics of the University of Colorado at Boulder. Eleven colleagues from the United States and France participate as Co-Investigators. Warwick has been studying Jupiter longer than any other Voyager Principal Investigator. He has monitored its radio emissions since the 1960s, and he played a central role in the discovery that Io influenced these emissions.

Jupiter emits many kinds of radio radiation, ranging from bland thermal emission at short (centimeter) wavelengths, to synchrotron emission from energetic electrons at intermediate (decimeter) wavelengths, to erratic, extremely intense bursts at long (meter and decameter) wavelengths. The origin of these latter, nonthermal emissions constitutes one of the major unsolved problems of the Jovian system,

league, Charles W. Hord, a physicist with considerable experience in space missions, primarily with ultraviolet instruments. Four other scientists are Co-Investigators on the photopolarimeter.

The instrument itself is essentially a small Cassegrain reflecting telescope of 15-centimeter aperture. Two consecutive filter wheels select the color and polarization of the light, which is then detected and measured by a photomultiplier tube. The entire photopolarimeter weighs just 2.5 kilograms.

In spite of their apparent simplicity, however, the Voyager photopolarimeters were plagued with troubles almost from the moment of launch. A succession of mechanical failures on Voyager 1 led to the sticking first of the polarization wheel and then of the filter wheel. Even when repeated commands from Earth managed to free the wheels, their behavior was erratic. Ultimately an electronic failure also

40

James W. Warwick, planetary radio astronomy Principal Investigator

and, more generally, of the physics of plasmas. One of the most important advantages of the Voyager PRA for studying Jovian radio emission lies in its proximity to Jupiter and hence its ability to locate the sources of different kinds of radio bursts.

The PRA instrument consists of an antenna and a radio receiver. The antenna is made up of two thin metal poles, each 10 meters long, extended from the spacecraft after launch at an angle of 90 degrees to each other. These are electrically connected to two receivers of extremely high sensitivity and broad frequency response: from 1.2 kilohertz to 40.5 megahertz. The PRA can operate in a number of modes, depending on the measurements desired. At its lowest level of activity, it monitors intensity in all 198 bands and transmits the data at 266 bits per second. In its highest mode, where searches are made for variations with very short time scales, the data rate goes up to 108 000 bits per second, essentially the same as that required by imaging. In fact, the high-rate PRA data actually use an imaging frame as a display form, and occasionally throughout the mission an unfamiliar looking "image" was transmitted that was actually a block of PRA data—a portrait of electrical events in the Jovian atmosphere and magnetosphere that only Warwick and his colleagues could interpret.

Norman F. Ness, magnetometer Principal Investigator

Magnetometer

The first of the direct sensing instruments to be discussed is the magnetometer, designed to measure the magnetic fields surrounding the spacecraft. Such measurements can be interpreted to yield the intrinsic fields of Jupiter and its satellites and to characterize, in conjunction with data from particle and plasma instruments, the processes taking place in the magnetosphere of the planet. The Principal Investigator for this instrument is Norman F. Ness of the NASA Goddard Space Flight Center. Ness is an intense, competitive scientist

Voyager's 13-meter-long magnetometer boom is shown fully extended. In space, under zero gravity conditions, the triangular epoxy glass mast spirals from its housing and provides a rigid support for two magnetometer instruments—one at the end of the boom and another at about the midpoint. [260-181]

with a great deal of previous experience in spacecraft magnetometers, primarily on board Earth satellites. Ness is also the only Voyager Principal Investigator who has previous experience at Jupiter; he was Principal Investigator on one of the two magnetometer instruments flown on Pioneer 11. For the Voyager investigation, Ness is joined by four colleagues from Goddard and one from Germany.

The magnetometer instrument consists of two systems: a high-field magnetometer and a low-field magnetometer. Each system contains two identical three-axis magnetometers that measure the intensity and direction of the magnetic field. The low-field system requires isolation from magnetic fields induced by electric circuits in the spacecraft itself. To achieve this isolation, it is mounted on the largest component of Voyager—a 13-meter boom, about as long as the width of a typical city house lot. This boom was coiled tightly in a canister during launch; later, when the package was opened, it uncurled and extended automatically.

By using two magnetometers, Ness and his colleagues are able to correct for the residual artificial magnetic fields that reach even 13 meters from the main spacecraft. The dynamic range extends from a maximum field of 20 gauss down to 2×10^{-8} gauss—a factor of one billion. The fields can be measured as frequently as 17 times per second. The total mass, exclusive of the 13-meter boom, is 5.6 kilograms.

Plasma Particles

Plasma is the term given to a "gas" of charged particles; the electrons and protons are separate, yet there are equal numbers of each, producing a zero net charge. If the velocities of the electrons and protons are less than about 0.1 percent of the speed of light, they can be measured by the Voyager plasma instrument; if their energies are higher, they are measured by one of the other two particle instruments—the low energy charged particle (LECP) detector or the cosmic ray detector. The plasma instrument, like the magnetometer, was designed to provide basic data on the particles and fields environment of Voyager.

The Principal Investigator for the plasma instrument is Herbert S. Bridge of the Massachusetts Institute of Technology in Cambridge, Massachusetts. An experienced space physicist, Bridge has flown similar instruments on many

Herbert S. Bridge, plasma particle Principal Investigator

Earth satellites and planetary probes. He also holds the position of Director of the Laboratory for Space Experiments at MIT. On the Voyager experiment, he is joined by one German and ten U.S. Co-Investigators.

The objectives of the plasma investigation are directed toward study of both the interplanetary medium and the Jovian magnetosphere. At Jupiter, Bridge expected to determine the plasma populations and processes in the inner and outer regions of the magnetosphere and in the plasma tail that extends beyond Jupiter, much as a comet tail is blown outward by the solar wind. The densities and temperature of electrons were measured, and their origins determined: Some originate near Jupiter and diffuse outward through the magnetosphere; others derive from the solar wind.

The plasma instrument, with a mass of 9.9 kilograms, was designed to view in two directions: one toward the Earth and Sun, primarily to study the solar wind, and the other sideways, looking toward the direction plasma would flow if it were caught up in the rotating Jovian magnetic field. If it is desired to look in other directions, the entire spacecraft must be tipped, a maneuver that was carried out several times near Jupiter. The detectors directly sense the flow of electrons, protons, and alpha particles (helium nucleii, made up of two protons and two neutrons each). Analysis of the energy spectra can also yield data on positive ions of higher mass.

42

Plasma Waves

The plasma wave investigation on Voyager was a late addition to the scientific payload. It was selected to broaden the capability of the mission to study a wide variety of plasma processes. Because of the electrically charged nature of the plasma, it responds to energy inputs in ways that ordinary gas cannot. One of these modes of response yields plasma waves, which are oscillations in density and electric field that generally cover the audio range of frequencies. Measurement of such waves characterizes the density and temperature of the local plasma surrounding the spacecraft, and it also allows remote sensing detection of distant events from the plasma waves they produce.

The plasma wave Principal Investigator is physicist Frederick L. Scarf of the TRW Defense and Space Systems Group of Redondo Beach, California—he is the only Voyager Principal Investigator to come from industry. Scarf has been associated with many particles and fields investigations in the terrestrial magnetosphere, although he is better known as a theorist than as an experimenter. A member of the Space Science Board of the National

Frederick L. Scarf, plasma wave Principal Investigator

Academy of Sciences, he is familiar in Washington as an eloquent advocate of space physics—the study of plasma-physical processes in the space environment.

The plasma wave instrument shares with the planetary radio astronomy investigation a pair of 10-meter-long antennas. Whereas the PRA uses these as electric antennas to detect radio radiation, the plasma wave system uses them to detect directly the oscillations in the plasma near the spacecraft. Waves are measured over a broad frequency range, from 10 hertz (a bit deeper than the lowest bass note we can hear) to 56 kilohertz (about three times higher than the highest pitch to which the human ear responds). The instrument electronics have a total mass of only 1.4 kilograms.

Low Energy Charged Particles

Charged particles with energies greater than a few thousand electron volts are not easily measured by a plasma instrument such as that designed by Herb Bridge. Instead, these faster moving particles, with speeds up to a few percent the speed of light, are the subject of a pair of Voyager instruments called collectively the LECP, or low energy charged particle instrument. Like the other particles and fields investigation, the LECP is designed to provide basic data on plasma-physical processes in the Jovian magnetosphere and the solar wind, and on their interactions.

The Principal Investigator for the LECP investigation is Stamatios Mike Krimigis, a Greek-born physicist from Johns Hopkins University. Krimigis has participated in a number of satellite studies of the terrestrial magnetosphere, and he now serves as Head of Space Physics and Instrumentation at the Johns Hopkins Applied Physics Laboratory. He is joined in this investigation by one German and five U.S. Co-Investigators.

The LECP instrument consists of two subsystems. The first, called the low energy magnetospheric particle analyzer, is optimized for measurement of particles within the Jovian magnetosphere, with high sensitivity over a broad dynamic range. Measurements of electrons, protons, and other positive ions can be carried out, determining the energy and composition of individual particles. The total energy ranges covered are 10 kiloelectron volts (keV) to 11 million electron volts (MeV) for electrons and 15 keV to 150 MeV for protons and ions.

Stamatios Mike Krimigis, low energy charged particle Principal Investigator

The second LECP subsystem is a low energy charged particle telescope, designed to operate where the density of charged particles is low, such as in interplanetary space or the outer magnetosphere of Jupiter. For protons and positive ions, the energy range is from 50 keV to 40 MeV per nucleon. The energy and species resolution is again sufficient to determine the composition, both chemical and isotopic, of many ions encountered. In order to provide directional discrimination even on a spacecraft of fixed orientation, both LECP subsystems are mounted on a moving platform that steps through eight positions in a time that can be commanded to vary from 48 seconds to 48 minutes. The mass of the instrument and its platform is 6.7 kilograms.

Cosmic Rays

The solar system is constantly bombarded by extremely energetic charged particles. These are called cosmic rays, although they are particles, not photons—"rays" are only produced when the particles strike something, such as the molecules of the Earth's atmosphere, and give up their energy in a flash of x-rays and gamma-rays. One of the Voyager instruments is designed to study these galactic cosmic rays, particularly to look from beyond the orbit of Saturn, where the cosmic ray particles will be less affected by the solar magnetic field and solar wind than they are near Earth.

The cosmic ray Principal Investigator is Rochus E. Vogt of the California Institute of Technology. Vogt has measured cosmic rays from the ground, from balloons, and from spacecraft for many years. During 1977 and 1978 he served as Chief Scientist at JPL, and then assumed the job of directing the physics, mathematics, and astronomy programs at Caltech. Among his six Co-Investigators is Ed Stone, the Voyager Project Scientist.

Because the cosmic ray instrument was not directed principally toward measurements of the Jovian system, it is described only briefly. Like the LECP, it is designed to determine the energy and composition of individual electrons and positive ions. For electrons, the energy range is from 3 to 110 MeV, and for ions from 1 to 500 MeV per nucleon; the corresponding velocities are from about 10 percent to 99 percent of the speed of light. For the positive ions, composition can be determined for elements from hydrogen to iron. At Jupiter, this system could be used to determine the nature of the rare particles accelerated to very high energies in the Jovian magnetosphere.

Radio Science

The final Voyager science investigation is in the field of radio science. No special instrument was required for this study; rather, NASA selected members of a Radio Science Team who proposed investigations that could be carried out using the already existing spacecraft telecommunication system.

The radio science Team Leader is Von R. Eshleman of the Center for Radio Astronomy at Stanford University. Eshleman is a radar physicist who has been interpreting spacecraft radio occultation data since the first such probe was carried out when Mariner 4 passed behind Mars in 1964. The Deputy Team Leader is G. Leonard Tyler, a colleague of Eshleman's at Stanford. There are five other radio team members, four of them from JPL.

The radio science investigations are divided into two groups. The first deals with the atmosphere of Jupiter. During the Voyager flybys, the spacecraft passed behind the planet as seen from Earth, and the radio signal was dimmed by the atmosphere before it was finally extinguished. During an occultation, the propagation of the radio waves is slowed down by passage through the neutral atmosphere and is speeded up by passage through the electrically charged ionosphere. Because of the extreme

Rochus E. Vogt, cosmic ray Principal Investigator

Von R. Eshleman, radio science Team Leader

stability of the ground-based and spacecraft radio transmitters, it is possible to measure these shifts in the signal with high precision. The shifts are proportional to electron density for the ionosphere, and to gas density for the atmosphere. From a careful study of the interactions of the transmitted beam with the Jovian atmosphere, Eshleman and his colleagues can reconstruct a temperature-pressure profile of the ionosphere and the upper atmosphere of Jupiter. The same approach can be used to search for tenuous atmospheres on the satellites.

The second area of study is in the field of celestial mechanics. The frequency stability of the communications system permits measurements of the speed of the spacecraft, relative to Earth, to a precision of one part in several million. By careful tracking, gravitational perturbations on the spacecraft can be detected and used to measure the gravitational fields,

and hence the masses, of Jupiter and its satellites.

These scientific instruments and their objectives were selected many years before the first Jupiter encounter in March 1977. Because Voyager was an exploratory mission, every effort was made to fly versatile instruments that could yield valuable results no matter what the nature of the Jovian system. In addition, the Voyager spacecraft control system permitted the instruments to receive commands from Earth to adjust their sensitivities and observing sequences in response to new information. By the spring of 1977, all the instruments were completed, ready to be installed in the Voyager spacecraft for testing and launch.

CHAPTER 5

THE VOYAGE TO JUPITER—GETTING THERE

Launch

On August 20, 1977, exactly two years after the launch of the Viking spacecraft to Mars, the first of the Voyagers—actually Voyager 2—was boosted into space at 10:29 a.m. EDT, less than five minutes after the launch window opened on the first day of the thirty-day launch period. Sixteen days later, at 8:56 a.m. EDT on Labor Day, September 5, 1977, Voyager 1 was hurled into space on a shorter, faster trajectory than its twin, zipping past the orbit of the Moon only ten hours after launch. Ultimately, Voyager 1 earned its title by overtaking Voyager 2 as both spacecraft journeyed through the asteroid belt, to arrive at Jupiter four months ahead of Voyager 2.

The Voyagers lifted off from Launch Complex 41, Air Force Eastern Test Range, Kennedy Space Center, Cape Canaveral, Florida, atop the giant Titan III-E/Centaur rocket. It was the last time such a launch vehicle was scheduled to be used, as, according to plan, the Space Shuttle would take over in the 1980s. Thus the launching of the two Voyagers signified both an end and a beginning: a once-in-a-lifetime opportunity to explore, in only 8½ years' time, perhaps fifteen major bodies of the outer solar system.

But long before even the first Voyager was to make its closest approach to Jupiter—in fact, even before Voyager 2 was off the launch pad—there were problems to overcome.

In early August 1977, about three weeks before launch, failures in the attitude and articulation control subsystem (AACS) and the flight data subsystem (FDS), two of the spacecraft's three main computer subsystems, prevented the VGR77-2 spacecraft, originally scheduled for launch on August 20, from becoming Voyager 2. Instead, the "spare" spacecraft VGR77-3 was substituted, becoming Voyager 2 upon launch August 20, and VGR77-2, after proper repairs, became Voyager 1. Minor problems continued right up to launch. The low energy charged particle instrument failed and had to be replaced, and as late as T minus five minutes there was a halt in the countdown to check on a stuck valve. Unlike a jinxed dress rehearsal, which is said to "assure" an opening-night success, Voyager 2's prelaunch problems were a portent of difficulties to come.

The Voyager 2 launch was witnessed by thousands as the spacecraft ascended gracefully into the blue Florida sky, accompanied by the deep-throated rumblings of its rocket, echoing for miles across the beaches and scrub forests of Cape Canaveral. The Titan-Centaur performance was nearly flawless, and Voyager 2 quickly achieved an accurate trajectory toward Jupiter. However, even while the engines were still firing, the spacecraft began to experience a baffling series of problems that would absorb the attention of hundreds of persons from Pasadena to Washington, D.C., for the next several weeks until they were brought under control.

During the first minutes of flight, there seemed to be two difficulties with the AACS.

The first picture to capture crescent Earth and crescent Moon in the same frame was taken by Voyager 1, the second-launched spacecraft, on September 18, 1977, at a distance of 12 million kilometers from Earth. On the Earth eastern Asia, the western Pacific, and part of the Arctic can be seen. Since the Moon is much less reflective than the Earth, JPL image processors brightened the lunar image by a factor of three to ensure that both Earth and Moon were visible on this print. [P-19891C]

The Titan/Centaur rocket used to launch Voyager stood as tall as a 15-story building and weighed nearly 700 tons. Here the rocket waits for launch at Kennedy Space Center, with the Voyager spacecraft enclosed in the white protective shroud at the top. [P-19471A]

The first was a problem with one of the three stabilizing gyroscopes, but fortunately, the gyroscope began operating normally without intervention from the ground. The other problem appeared to be with one of the AACS computers; the spacecraft switched to a backup computer during the Titan burn, and initial data transmissions were incomplete. Early analysis seemed to indicate that an event during the launch itself, rather than a faulty spacecraft computer system, was the cause of the data loss. At first, on August 23, officials suspected that perhaps the spacecraft had been bumped by the rocket motor one hour after liftoff and again about seventeen hours later, when telemetry signals indicated that the spacecraft had been jolted. However, by the next day, flight engineers determined that electronic gyrations in the AACS seemed to have caused the difficulty.

Within an hour after launch, Voyager 2's science scan platform boom was to have been fully extended and locked. Instructions to deploy were given, and the boom moved outward; however, there was no signal to indicate that the boom was actually locked in place. Efforts to command the boom to move into the locked position were thwarted by the spacecraft. The first maneuver designed to try to lock the boom was aborted by the computer command subsystem (CCS) when the AACS erroneously indicated that it was in trouble. Three days later another maneuver was scheduled to reprogram the faulty computer in the AACS, to align the Sun sensors, and to try to lock the science boom. To provide a direct check of the boom position, the scan platform was turned so that the TV cameras could see the spacecraft. Careful measurement of these pictures verified that the boom was within ½ degree of full deployment, but still there was no indication that it was locked into place. Ultimately, it was decided that the sensor to signal actuation of the lock was at fault, and that the boom itself was almost certainly fully extended and operational.

Because of the postlaunch problems of Voyager 2, the launch of Voyager 1 was delayed twice—from September 1 to September 3 and then to September 5—in order to inspect Voyager 1's science boom and to try to prevent a repetition of Voyager 2's problems. An extra spring was attached to the science boom to assure its full extension. Finally, as if to make up for the troubles of the first launch, Voyager 1's launch was both "flawless and accurate."

THE TITAN/CENTAUR LAUNCH VEHICLE

The two Voyager spacecraft were carried into space and accelerated toward Jupiter by the Titan III-E Centaur rocket, the largest launch vehicle in the NASA arsenal after the retirement of the Saturn rockets in 1975. The Titan and Centaur vehicles were originally developed separately and have been used with other rocket stages for many NASA launches. They were first combined for the two Viking launches to Mars in 1975, and this powerful four-stage launch vehicle was used again in 1977 for Voyager.

The Titan/Centaur stands nearly 50 meters tall, about the height of a fifteen-story building. Fully fueled, it weighs nearly 700 tons. At takeoff, the thrust of the two solid-propellant Stage-0 motors is about 10.7 million newtons. These motors, which burn for 122 seconds, use powdered aluminum as fuel and ammonium perchlorate as oxidizer. Together, they have a mass of 500 tons.

The first stage of the liquid propellant core of the Titan rocket ignites about 112 seconds after takeoff. The propellant is hydrazine as fuel and nitrogen tetroxide as oxidizer. The first stage is 3 meters in diameter and 20 meters tall. Fueled, it has a mass of 130 tons. The motor provides a thrust of 2.5 million newtons for a duration of 146 seconds.

About 4.3 minutes after takeoff the Titan Stage II liquid propellant motor begins to fire, and the first stage is separated and falls back into the Atlantic. The second stage is 3 meters in diameter and more than 7 meters long, with a fueled mass of 35 tons. The single liquid fuel motor burns for 210 seconds with a thrust of half a million newtons. During the second stage burn, the shroud covering the Voyager spacecraft is jettisoned.

The Centaur and Titan vehicles separate 8 minutes into the flight, and the Centaur main engine begins its burn. The Centaur is nearly 20 meters tall and 3 meters in diameter, with a mass of 17 tons. The motors have a thrust of almost 200 000 newtons, operating on the most powerful chemical fuels known: liquid oxygen and liquid hydrogen. The Centaur burns for only 1 minute and 36 seconds as it attains Earth parking orbit; the engine then shuts down as the vehicle begins a half-hour coasting period that carries it nearly half way around the Earth. During this time, careful tracking of the spacecraft supplies the data needed for Earth-based computers to calculate the proper time to leave parking orbit and start the long trip toward Jupiter.

About 50 minutes after liftoff, from a position high above the Indian Ocean, the second burn of the Centaur main engine begins. Six minutes of additional thrust provides enough energy to break out of Earth's orbit. The Voyager then separates from the Centaur for a final boost toward Jupiter. The solid rocket motor in the spacecraft propulsion module (acting as final stage of this five-stage launch sequence) fires for 45 seconds at a thrust of 68 000 newtons. Just an hour after liftoff, the Voyager spacecraft is on its way, coasting on an orbit toward Jupiter at a speed of more than 10 kilometers per second.

The Voyager spacecraft is dominated by the large 3.7-meter-diameter antenna used for communication with Earth. Here the spacecraft undergoes final tests before launch. The science instrument scan platform is folded against the spacecraft on the right; the three cylinders on the left are the RTG power sources. [260-108BC]

Voyager 2 was the first of the spacecraft to be launched, on August 20, 1977, propelled into space in a Titan/Centaur rocket. [P-19450AC]

All launch and postlaunch events went smoothly. The launch window opened at 8:56 a.m. EDT, and Voyager took off promptly at 8:56:01. The booms and antennas deployed and locked in the first hours after launch; all instruments scheduled to be on were on and working well.

The First Year Is the Roughest

During the autumn of 1977 Voyager 2, and to a lesser extent Voyager 1, continued to plague controllers with erratic actions. Thrusters fired at inappropriate times, data modes shifted, instrument filter and analyzer wheels became stuck, and the various computer control systems occasionally overrode ground commands. Apparently, the spacecraft hardware was working properly, but the computers on board displayed certain traits that seemed almost humanly perverse—and perhaps a little psychotic. In general, these reactions were the result of programming too much sensitivity into the spacecraft systems, resulting in panic overreaction by the onboard computers to minor fluctuations in the environment. Ultimately, part of the programming had to be rewritten on Earth and then transmitted to the Voyagers, to calm them down so that they would ignore minor perturbations, yet still be ready to perform automatic sequences required to protect the spacecraft from major threats. Meanwhile, however, more serious problems were developing.

On February 23, 1978, during a series of movements or slews, Voyager 1's scan platform slowed and stopped before completing the maneuver. This failure caused a great deal of concern, since the scan platform houses the optical instruments that are crucial to the observation of the Jovian system—the ultraviolet spectrometer, the IRIS, the photopolarimeter, and the two TV cameras. At JPL, tests were run on a proof-test model—an exact copy of the Voyager spacecraft—to try to find out why Voyager 1's scan platform had become stuck. On March 17, Voyager 1's scan platform was tested—JPL engineers instructed the platform to move slowly for a short distance, and Voyager responded as ordered. Further tests were conducted on March 23. This time the scan platform was ordered to execute a sequence of four slews, moving away from the part of the sky where the original failure had occurred and ending with the position that it would be most useful to leave the platform in—just in case the platform should become stuck again. On April

4 the scan platform was commanded to perform a sequence of 38 slews, and fifty more slews were performed on April 5. All were successful. Yet engineers were still hesitant to force the platform to move through the region where it had originally stuck, and extensive discussions were held to determine if the Jupiter observations could be carried out without risking a return to the danger area. It was argued, however, that full mobility of the scan platform really was required, and on May 31 commands were sent to maneuver the scan platform through the danger region. It moved normally: The scan platform was operating properly again. After additional slewing tests were run in mid-June, the scan platform was pronounced fit for operation. Engineers suspected that the material caught in the platform gears must have been crushed or moved out of the way by the continued slewing, allowing the platform to move once more.

An even more serious crisis soon endangered the Voyager 2 spacecraft. In late November 1977, the S-band radio receiver began losing amplifier power in its high-gain mode, so the solid-state amplifier was switched to its low-power position. No further problems were noted until April 5, 1978, when Voyager 2's primary radio receiver suddenly failed, and shocked engineers discovered that the backup receiver was also faulty. The trouble was detected after Voyager's computer command subsystem directed the spacecraft to switch from the primary radio receiver to the backup receiver. This command was issued as part of a special protection sequence: If the primary radio receiver receives no commands from Earth for seven days, the backup receiver is switched on instead; if the secondary receiver in turn receives no instructions over a twelve-hour period, the system reverts to the main receiver. When, on April 5, Voyager 2's radio reception was switched from the primary to the secondary receiver, flight engineers found that they were unable to communicate with the spacecraft—the secondary receiver's tracking loop capacitor was malfunctioning. That meant that the secondary receiver could not follow a changing signal frequency sent out from Earth. The frequencies of signals transmitted from Earth are affected by the Doppler effect—just as the siren on a fire engine seems first to rise in pitch as the truck approaches, then falls as the truck speeds away, so the frequency of signals transmitted from Earth fluctuates with the Earth's rotation as the Deep Space Network's radio antennas

move toward or away from the spacecraft. The engineers had to wait until the primary radio receiver was switched back on before they could communicate with the spacecraft. Once the primary receiver was on, Voyager 2 began receiving instructions from Earth, but approximately thirty minutes later, there was an apparent power surge in the receiver. The fuses blew.

Voyager 1 was launched on September 5, 1977. The launch was delayed 5 days to make last-minute adjustments to avoid the postlaunch difficulties experienced by Voyager 2. [P-19480AC]

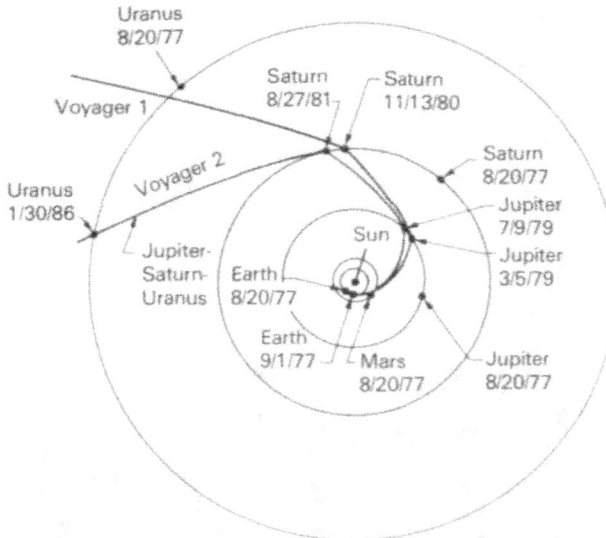

Each Voyager spacecraft follows a billion-kilometer path to Jupiter. Except for minor thruster firings to achieve small trajectory corrections, each Voyager coasts from Earth to Jupiter, guided by the gravitational pull of the Sun. At Jupiter, the powerful tug of the giant planet deflects the spacecraft and speeds them up, imparting an extra kick to send them on their way toward Saturn.

There was no recourse. The main receiver had failed; its loss was permanent. It remained for the engineers to devise a way to communicate with the slightly deaf spacecraft.

Because the switching of the radio receivers was still controlled by the special protection sequence discussed earlier, flight engineers would have to wait for seven days—until April 13—before they could attempt communication with the spacecraft again. During that week special procedures were established and rehearsed so that commands could be sent to Voyager in the short time that the backup receiver would be on. On Thursday, April 13, 1978, the seven days were up and the spacecraft should have shifted from the dead main receiver to the sick backup system. There was just a twelve-hour "window" in which to restore communication. At about 3:30 a.m. PST the Madrid tracking station of the Deep Space Network sent its first order to the spacecraft, approximately 474 million kilometers away. Almost an hour later, word arrived from Voyager that the command had been accepted. (One-way light time for a signal to travel the distance from Earth to Voyager at that time was almost 27 minutes.) Elated flight controllers went

ahead and transmitted nine hours of commands to the spacecraft.

Voyager 2 was successfully commanded again on April 18 and April 26. The April 26 commands included a course change maneuver that was executed properly on May 3. On June 23, Voyager 2 was programmed for a backup automatic mission at Saturn in the event that the secondary radio receiver should also fail. These backup mission instructions would operate all the science experiments, but only a minimum amount of data would be returned, since the scan platform would only be programmed to move through three positions rather than thousands as it would in normal operation. Instructions for a backup minimum automatic encounter at Jupiter were transmitted to Voyager 2 in two segments, the second of these on October 12, 1978.

With the backup instructions recorded on board the spacecraft, Voyager personnel felt their fears partially allayed. If Voyager 2's secondary radio receiver failed, the spacecraft would still obtain some science data at Jupiter and Saturn. But that would mean that there would be no mission beyond Saturn; our first opportunity to explore Uranus, its satellites, its newly discovered ring system, and possibly even to get a look at Neptune, would not come in this century.

Another major concern affecting both Voyager spacecraft was the proper management of hydrazine fuel reserves. Hydrazine is used by the thrusters on the Voyagers for stabilization of the spacecraft and for trajectory correction maneuvers (TCM). Each Voyager was loaded with 105 kilograms of hydrazine budgeted for use on the long flight to Jupiter, Saturn, and beyond. Because of the excellent performance of the launch rockets; both Voyagers required less hydrazine than anticipated for their final boost into proper trajectory toward Jupiter, and at first it looked as though both spacecraft would have plenty of propellant to spare.

Charles E. Kolhase, Manager of Mission Analysis and Engineering for the Voyager Project, later explained the situation: "Voyager 1 should have been launched September 1. Had it been launched on September 1—and I'm glad it wasn't—the maneuver to correct the trajectory for a Titan flyby would have required a change in velocity of 100–110 meters per second—an enormous maneuver—and we would have had a propellant margin for going on to Saturn of

THE DEEP SPACE NETWORK

A vital component of the Voyager Mission is the communications system linking the spacecraft with controllers and scientists on Earth. The ability to communicate with spacecraft over the vast distances to the outer planets, and particularly to return the enormous amounts of data collected by sophisticated cameras and spectrometers, depends in large part on the transmitters and receivers of the Deep Space Network (DSN), operated for NASA by JPL.

The original network of these receiving stations was established in 1958 to provide round-the-world tracking of the first U.S. satellite, Explorer 1. By the late 1970s, the DSN had evolved into a system of large antennas, low-noise receivers, and high-power transmitters at sites strategically located on three continents. From these sites the data are forwarded (often using terrestrial communications satellites) to the mission operations center at JPL.

The three DSN stations are located in the Mohave desert at Goldstone, California; near Madrid, Spain; and near Canberra, Australia. Each location is equipped with two 26-meter steerable antennas and a single giant steerable dish 64 meters in diameter, with approximately the collecting area of a football field. In addition, each is equipped with transmitting, receiving, and data handling equipment. The transmitters in Spain and Australia have 100-kilowatt power, while the 64-meter antenna at Goldstone has a 400-kilowatt transmitter. Most commands to Voyager are sent from Goldstone, but all three stations require the highest quality receivers to permit continuous recording of the data streams pouring in from the spacecraft.

Since the mid-1960s, the DSN's standard frequency has been S-band (2295 megahertz). Voyager introduces a new, higher frequency telemetry link at X-band (8418 megahertz). The X-band signal can carry more information than S-band with similar power transmitters, but it requires more exact antenna performance. In addition, the X-band signal is absorbed by terrestrial clouds and, especially, rain. Fortunately, all three DSN stations are in dry climates, but during encounters the weather forecasts on Earth become items of crucial concern if precious data are not to be lost by storm interference.

As a result of the development of larger antennas and improved electronics, the DSN command capabilities and telemetry data rates have increased dramatically over the years. For example, in 1965 Mariner 4 transmitted from Mars at a rate of only 8⅓ bits of information per second. In 1969, Mariners 6 and 7 transmitted picture data from Mars at 16 200 bits per second. Mariner 10, in 1973, achieved 117 200 bits per second from Mercury. Voyager operates at a similar rate from Jupiter, about six times farther away. Many of these improvements in data transmission result from changes in the DSN rather than in the spacecraft transmitters.

perhaps 4.5 kilograms. But, by launching on the fifth of September we increased our margin to 23 kilograms. Fortunately, for every launch date that went by, that velocity change maneuver was shrinking at a rate of 10 meters per second per day. Now, a 1 meter per second change uses about a pound of hydrazine [about 0.5 kilogram]. So when we launched on the fifth of September, now we suddenly had 40 pounds of hydrazine excess over what we would have had if we had launched on the first of September. As a result, Voyager 1 is in great shape as far as hydrazine is concerned."

Problems with hydrazine management developed, however. Voyager 1's first trajectory correction maneuver achieved only 80 percent of the required speed change. Exhaust plumes from the thrusters apparently struck part of the spacecraft, causing a 20 percent loss in velocity. That being the case, Voyager might require more fuel than had been expected to complete the mission. The extra fuel requirements did not threaten Voyager 1 itself, since it held ample fuel to reach Saturn; the concern was for Voyager 2, where the effective loss of fuel might be enough to jeopardize the Uranus mission.

Because of the plume impingement problem on Voyager 1, Voyager 2's first trajectory correction maneuver was adjusted to allow for the possibility of a 20 percent loss in thrust. The Voyager 2 maneuver was successful, but controllers felt that additional action was required to conserve fuel. One way to save was by reducing requirements on control of the spacecraft orientation. Less control fuel would be needed

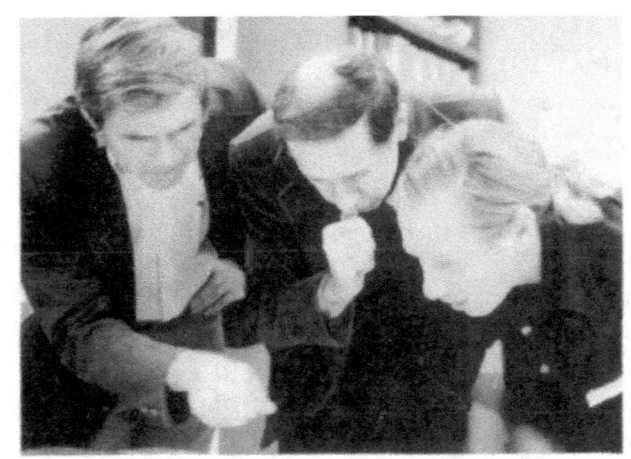

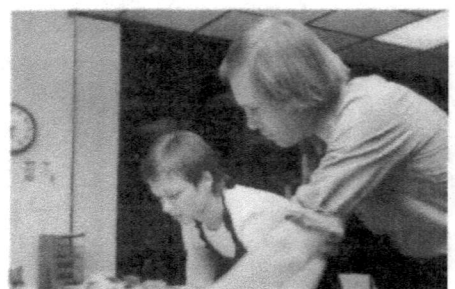

54

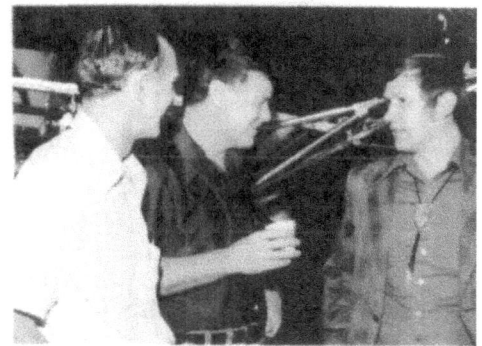

if the already miniscule pressure exerted on the spacecraft by the solar wind could be reduced. Flight engineers at JPL calculated that the pressure would be reduced if the spacecraft were tipped upside down; however, to accomplish this, the spacecraft would have to be steered by a new set of guide stars. By reprogramming the attitude control system it was found possible to substitute the northern star, Deneb, in the constellation of Cygnus, for the original reference star, Canopus, in the southern constellation of Carina. With this change, as well as readjustment of Voyager 2's trajectory near Jupiter, inflight consumption of hydrazine was reduced significantly.

In late August 1978 both Voyagers were reprogrammed to ensure better science results at Jupiter encounter; for example, the reprogramming would prevent imaging (TV) photographs from blurring when the tape recorder was operating. By early November, flight crews had begun training exercises to rehearse for the Voyager 1 flyby of Jupiter on March 5, 1979. A near encounter test was performed on December 12–14, 1978: a complete runthrough of Voyager 1's 39-hour near encounter period, which would take place March 3–5, 1979. Participants included the flight team, the Deep Space Network tracking stations, the scientists, and the spacecraft itself. Results: Voyager and the Voyager team were all ready for the encounter.

Meanwhile, the spacecraft were busy returning scientific data to Earth. Technically, the Voyagers were in the cruise phase of the mission—a period that, for Voyager 2, would last until April 24, 1979, and for Voyager 1, until January 4, 1979, when each spacecraft would enter its respective observatory phase.

Cruise Phase Science

In the first few days after launch, the spacecrafts' instruments were turned on and calibrated; various tests for each instrument would continue to be performed throughout the cruise phase. This period presented a great opportunity for the Voyagers to study the interplanetary magnetic fields, solar flares, and the solar wind. In addition, ultraviolet and infrared radiation studies of the sky were performed. In mid-September 1977 the television cameras on Voyager 1 recorded a number of photographs of the Earth and Moon. A photograph taken September 18 captured both crescent Moon and crescent Earth. It was the first time the two

celestial bodies had ever been photographed together.

In November both Voyagers crossed the orbit of Mars, entering the asteroid belt a month later. On December 15, at a distance of about 170 million kilometers from Earth, Voyager 1 finally speeded past its slower twin. The journey through the asteroid belt was long but uneventful: Voyager 1 emerged safely in September 1978, and Voyager 2 in October. Unlike Pioneers 10 and 11, the Voyagers carried no instruments to look at debris in the asteroid belt.

By April both spacecraft were already halfway to Jupiter and, about two months later, Voyager 1, still approximately 265 million kilometers from Jupiter, began returning photographs of the planet that showed considerable detail, although less than could be obtained with telescopes on Earth. Both the imaging (TV) and the planetary radio astronomy instruments began observing Jupiter, and, by October 2, 1978, officials announced that "the polarization characteristics of Jupiter's radio emissions have been defined. In the high frequencies, there is consistent right-hand circular polarization, while in the low frequencies, there is a consistent left-hand circular polarization. This was an unexpected result." In addition to scientific studies of the interplanetary medium and a first look at Jupiter, the plasma wave instrument, which studies waves of charged particles over a range of frequencies that includes audio frequencies, was able to record the sound of the spacecraft thrusters firing, as hydrazine fuel is decomposed and ejected into space. The sound was described as being "somewhat like a 5-gallon can being hit with a leather-wrapped mallet."

On December 10, 1978, 83 million kilometers from Jupiter, Voyager 1 took photographs that surpassed the best photographs ever taken from ground-based telescopes, and scientists were anxiously awaiting the start of continuous coverage of the rapidly changing cloud forms. These pictures, together with data from several ground-based observatories, were carefully scrutinized by a team of scientists at JPL to select the final targets in the atmosphere of Jupiter to be studied at high resolution during the flyby. The observatory phase of Voyager 1's journey to Jupiter was about to begin.

The Observatory Phase

The observatory phase, originally scheduled to start on December 15, 1978, eighty days

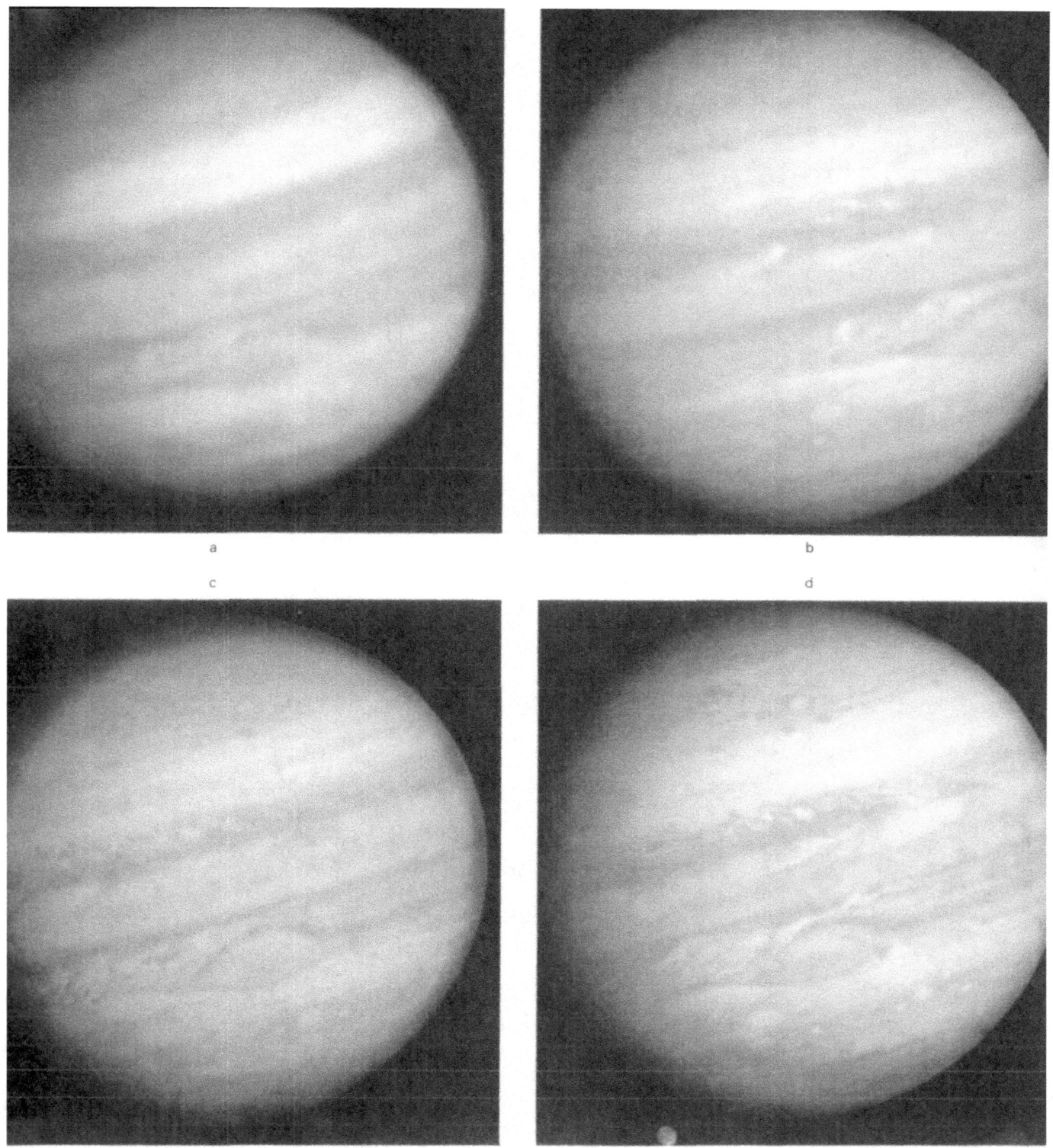

As Voyager 1 approached Jupiter, the resolution of the images steadily improved. In October 1978 (a), at a distance of about 125 million kilometers, the image was less clear than would be obtained with an Earth-based telescope. By December 10 (b) the spacecraft had moved to a distance of 85 million kilometers, and the resolution was about 2000 kilometers, comparable to the best telescopic images. On January 9, 1979 (c), at a distance of 54 million kilometers, the image surpassed all ground-based views and approached the resolution of the Pioneer 10 and 11 photos. In (d), taken January 24 at a distance of 40 million kilometers, the resolution exceeded 1000 kilometers. [P-20790; P-20829C; P-20926C; P-20945C]

before encounter, was postponed until January 4, 1979, to provide the flight team a holiday-season break. For the next two months, Voyager would carry out a long-term scientific study—a "time history"—of Jupiter, its satellites, and its magnetosphere. On January 6, Voyager 1 began photographing Jupiter every two hours—each time taking a series of four photographs through different color filters as part of a long-duration study of large-scale atmospheric processes, so scientists could study the changing cloud patterns on Jupiter. Even the first of these pictures showed that the atmosphere was dynamic "with more convective structure than had previously been thought." Particularly striking were the changes in the planet, especially near the Great Red Spot, that had taken place since the Pioneer flybys in 1973 and 1974. Jupiter was wearing a new face for Voyager.

In mid-January, photos of Jupiter were already being praised for "showing exceptional details of the planet's multicolored bands of clouds." Still 47 million kilometers from the giant, Voyager 1 had by now taken more than 500 photographs of Jupiter. Movements of cloud patterns were becoming more obvious; feathery structures seemed painted across some

of the bands that encircle the planet; swirling features were huddled near the Great Red Spot. The satellites were also beginning to look more like worlds, with a few bright spots visible on Ganymede and dark red poles and a bright equatorial region clearly seen on Io when it passed once each orbit across the turbulent face of Jupiter.

From January 30 to February 3, Voyager sent back one photograph of Jupiter every 96 seconds over a 100-hour period. Using a total of three different color filters, the spacecraft thus produced one color picture every 4¾ minutes, in order to make a "movie" covering ten Jovian "days." To receive these pictures, sent back over the high-rate X-band transmitter, the Deep Space Network's 64-meter antennas provided round-the-clock coverage. Voyager 1 was ready for its far encounter phase.

Far Encounter Phase

By early February Jupiter loomed too large in the narrow-angle camera to be photographed in one piece; 2×2 three-color (violet, orange, green) sets of pictures were taken for the next two weeks; by February 21, Jupiter had grown

The Galilean satellites of Jupiter first began to show as tiny worlds, not mere points of light, as the Voyager 1 observatory phase began. In this view taken January 17, 1979, at a range of 47 million kilometers, the differing sizes and surface reflectivities (albedos) of Ganymede (right center) and Europa (top right) are clearly visible. The view of Jupiter is unusual in that the Great Red Spot is not easily visible, but can just be seen at the right edge of the planet. Most pictures selected for publication include the photogenic Red Spot. [P-20938C]

By February 1, 1979, Voyager 1 was only 30 million kilometers from Jupiter, and the resolution of the imaging system corresponded to about 600 kilometers at the cloud tops of the giant planet. At this time, a great deal of unexpected complexity became apparent around the Red Spot, and movie sequences clearly showed the cloud motions, including the apparent six-day rotation of the Red Spot. [P-20993C]

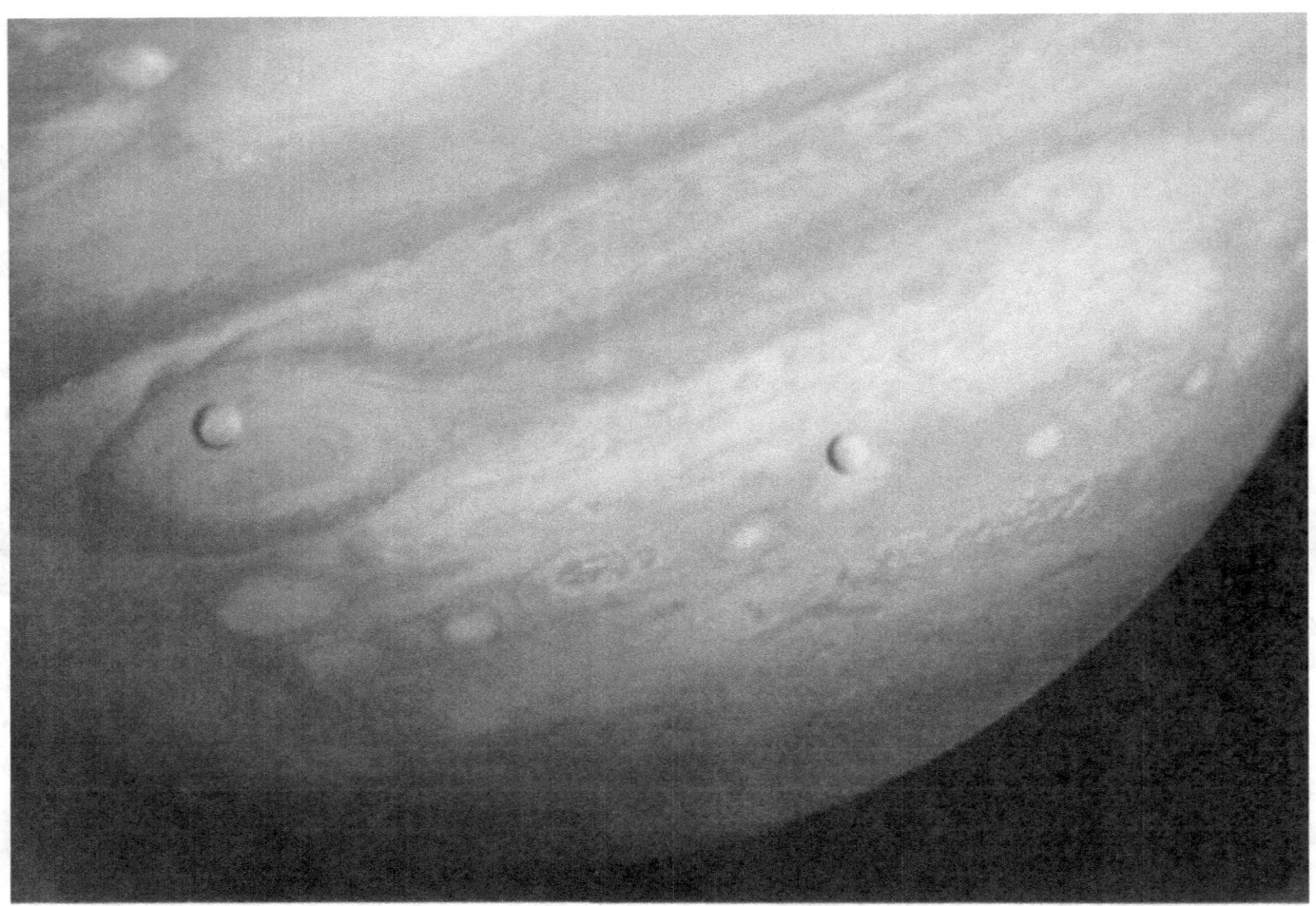

One of the most spectacular planetary photographs ever taken was obtained on February 13 as Voyager 1 continued its approach to Jupiter. By this time, at a range of 20 million kilometers, Jupiter loomed too large to fit within a single narrow-angle imaging frame. Passing in front of the planet are the inner two Galilean satellites. Io, on the left, already shows brightly colored patterns on its surface, while Europa, on the right, is a bland ice-covered world. The scale of all of these objects is huge by terrestrial standards; Io and Europa are each the size of our Moon, and the Red Spot is larger than the Earth. [P-21082C]

too large even for that, and 3 × 3 sets were scheduled to begin. When these sets of pictures are pasted together to form a single picture, the result is called a mosaic.

While the imaging experiments were in the limelight, the other scientific instruments had also begun to concentrate on the Jovian system. The ultraviolet spectrometer had been scanning the region eight times a day; the infrared spectrometer (IRIS) spent 1½ hours a day analyzing infrared emissions from various longitudes in Jupiter's atmosphere; the planetary radio astronomy and plasma wave instruments looked for radio bursts from Jupiter and for plasma disturbances in the region; the photopolarimeter had begun searching for the edge of Io's sodium torus; and a watch was begun for

the bow shock—the outer boundary of the Jovian magnetosphere.

On February 10, Voyager 1 crossed the orbit of Sinope, Jupiter's outermost satellite. Yet the spacecraft even then had a long way to go—still 23 million kilometers from Jupiter, but closing in on the planet at nearly a million kilometers a day. A week later, targeted photographs of Callisto began to provide coverage of the satellite all around its orbit; similar photos of Ganymede began on February 25.

Meanwhile excitement was building. As early as February 8 and 9, delight with the mission and anticipation of the results of the encounter in March were already evident. Garry E. Hunt, from the Laboratory for Planetary Atmospheres, University College,

The Voyager TV cameras do not take color pictures directly as do commercial cameras. Instead, a color image is reconstructed on the ground from three separate monochromatic images, obtained through color filters and transmitted separately to Earth. There are a number of possible filter combinations, but the most nearly "true" color is obtained with originals photographed in blue (480 nanometers), green (565 nanometers), and orange (590 nanometers) light. Before these can be combined, the individual frames must be registered, correcting for any change in spacecraft position or pointing between exposures. Often, only part of a scene is contained in all three original pictures. Shown here is a reconstruction of a plume on Jupiter, photographed on March 1, 1979. The colors used to print the three separate frames can be seen clearly in the nonoverlapping areas. For other pictures in this book, the nonoverlapping partial frames are omitted. [P-21192]

The Voyager Project was operated from the Jet Propulsion Laboratory managed for NASA by the California Institute of Technology. Located in the hills above Pasadena, California, JPL is the main center for U.S. exploration of the solar system. [JB17249BC]

London, a member of the Imaging Team, discussing the appearance of Jupiter's atmosphere as seen by Voyager during the previous month, said "It seems to be far more photogenic now than it did during the Pioneer encounters; I'm more than delighted by it—it's an incredible state of affairs. There are infinitely more details than ever imagined."

The pictures from Voyager are "clearly spectacular," said Lonne Lane, Assistant Project Scientist for Jupiter. "We're getting even better results than we had anticipated. We have seen new phenomena in both optical and radio emissions. We have definitely seen things that are different—in at least in one case, unanticipated—and are begging for answers we haven't got." There was already, still almost a month from encounter, a strong feeling of accomplishment among the scientists and engineers; they had done a difficult task and it has been successful.

By the last week of February 1979, the attention of thousands of individuals was focused on the activities at JPL. Scientists had arrived from universities and laboratories around the country and from abroad, many bringing graduate students or faculty colleagues to assist them. Engineers and technicians from JPL contractors joined NASA officials as the Pasadena motels filled up. Special badges were issued and reserved parking areas set aside for the Voyager influx. Twenty-four hours a day, lights burned in the flight control rooms, the science offices, the computer areas, and the photo processing labs. In order to protect those with critical tasks to perform from the friendly distraction of the new arrivals, special computer-controlled locks were placed on doors, and security officers began to patrol the halls. By the end of the month the press had begun to arrive. Amid accelerating excitement, the Voyager 1 encounter was about to begin.

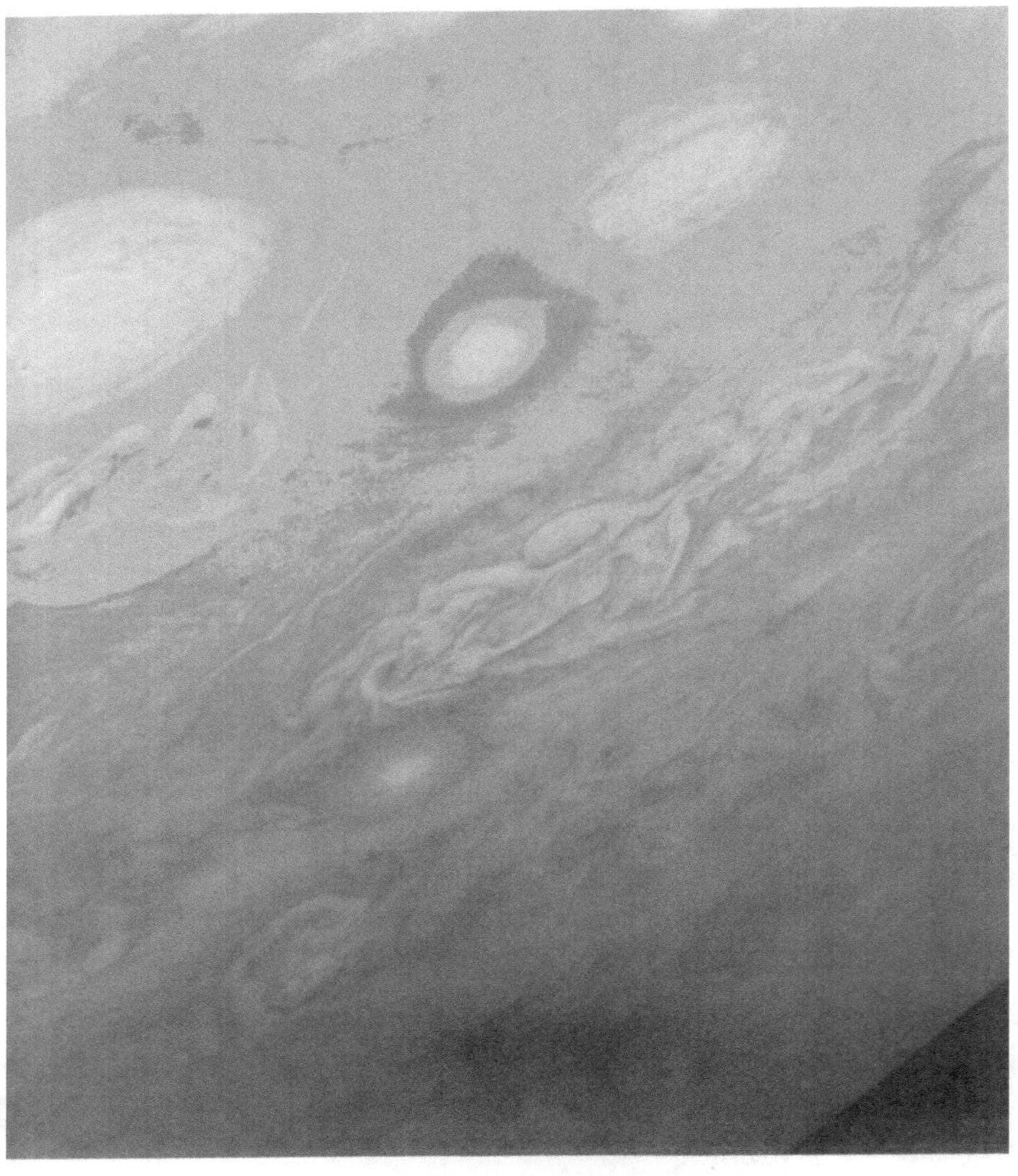

CHAPTER 6

THE FIRST ENCOUNTER

The Giant Is Full of Surprises

The Voyager 1 encounter took place at 4:42 a.m. PST, March 5, 1979. About six hours before, while the spacecraft continued to hurtle on toward Jupiter, overflow crowds had poured out of Beckman Auditorium on the campus of the California Institute of Technology. There had been a symposium entitled Jupiter and the Mind of Man. More than one face that evening had turned to look toward the Pasadena night, for there, glittering against the fabric of the sky, was the "star" of the show—a planet so huge that the Earth would be but a blemish on its Great Red Spot. One might have tried to imagine Voyager 1, large by spacecraft standards, rapidly closing in on Jupiter—a pesky, investigative "mosquito" buzzing around the Jovian system.

Meanwhile, at JPL, pictures taken by Voyager 1 were flashing back every 48 seconds, pictures that were already revealing more and more of Jupiter's atmosphere and would soon disclose four new worlds—the Galilean satellites. The encounter was almost at hand.

4:42 a.m.—Voyager 1 had made its closest approach to Jupiter 37 minutes earlier. Only now were the signals that had been sent out from the spacecraft at 4:05 a.m. reaching the Earth.

But the moment of closest encounter did not have the same impact a landing would have had. Although the champagne would flow later for those who had worked so long and hard on this successful mission, there was none now. In the press room there were just four coffeepots working overtime to help keep the press alert. In the science areas, focus had shifted to the satellite encounters, which would stretch over the next 24 hours; meanwhile, a few persons tried to catch a little sleep at their desks before the first closeups of Io came in. The instant of encounter came . . . and went . . . with no screams, no New Year's noisemakers. All the excitement of the mission lay both behind . . . and ahead.

Thursday, February 22. The encounter activity was kicked off in Washington, D.C., by a press conference at NASA Headquarters. After introductory statements by NASA and JPL officials, some of the scientific results from the observatory and far encounter phases were presented.

Both the plasma wave and the planetary radio astronomy instruments had detected very-low-frequency radio emission that generally increased in intensity whenever either the north or south magnetic pole of Jupiter tipped toward the spacecraft. The plasma wave experiment had also detected radiation that seemed to come from a region either near or beyond the orbit of Io—perhaps even as far as the outer magnetosphere. In addition, Frederick L. Scarf, Principal Investigator for the plasma wave instrument, released a recording of the ion sound waves created by 5 to 10 kilovolt energy protons traveling upstream from Jupiter.

James Warwick, Principal Investigator of the radio astronomy investigation, enthusiastically reported the detection of a striking new low-frequency radiation from Jupiter, at wavelengths of tens of kilometers. Such radio waves cannot penetrate the ionosphere of the Earth, and thus had never been detected before. Because of their huge scale, these bursts probably did not originate at Jupiter itself, but in magnetospheric regions above the planet— perhaps in association with the high-density plasma torus associated with Io. Warwick commented that "only from the magnificent

The smaller-scale clouds on Jupiter tend to be more irregular than the large ovals and plumes. At the lower right, one of the three large white ovals clearly shows internal structure, with the swirling cloud pattern indicating counterclockwise, or anticyclonic, flow. A smaller anticyclonic white feature near the center is surrounded by a dark, cloud-free band where one can see to greater depths in the atmosphere. This photo was taken March 1 from a distance of 4 million kilometers. [P-21183C]

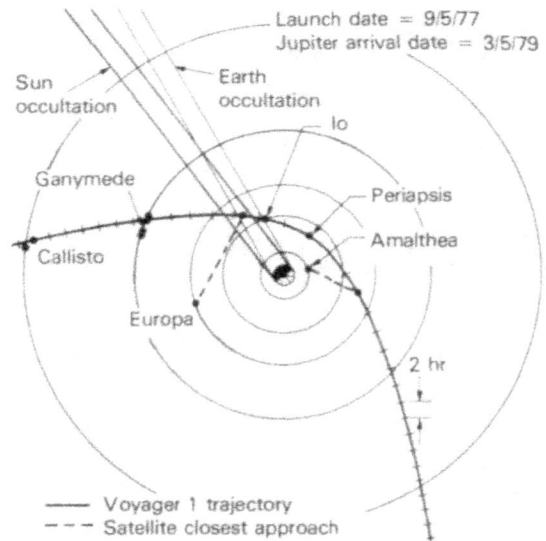

Launch date = 9/5/77
Jupiter arrival date = 3/5/79

Sun occultation

Earth occultation

Io

Ganymede

Periapsis

Amalthea

Callisto

Europa

2 hr

——— Voyager 1 trajectory
– – – Satellite closest approach

The Voyager 1 encounter with Jupiter took place during a little more than 48 hours, from the inbound to the outbound crossing of the orbit of Callisto. This figure shows the spacecraft path as it would be seen from above the north pole of Jupiter. Closest approach to Jupiter was 350 000 kilometers. The close flybys of Io, Ganymede, and Callisto all took place as the spacecraft was outbound.

perspective of the Voyager as it approaches Jupiter have we been able to get this complete picture."

Another unexpected result was announced by Lyle Broadfoot, Principal Investigator for the ultraviolet spectrometer investigation. The scientists had expected to find very weak ultraviolet emissions on the sunlit side of Jupiter, caused by sunlight being scattered from hydrogen and helium in Jupiter's upper atmosphere. "Instead we are seeing a spectacular auroral display. There are two features of the emission—the auroral emission which comes from the planet and a second type of emission which appears to come from a radiating torus or shell around the planet at the orbit of Io. The spectral content of these two radiating sources is distinctly different. What we find is that the auroral emission from Jupiter's atmosphere is so strong that it completely dominates the emission spectrum even on the sunlit side of the atmosphere."

"Not since Mariner 4 carried its TV camera to Mars fifteen years ago have we been less prepared—have we been less certain of what we are about to see over the next two weeks," said Bradford Smith, Imaging Team Leader. He mentioned the time-lapse "rotation movie," in which the colorful planet spun through ten full Jupiter days; tiny images of the satellites passed across Jupiter's face as though being whipped along by the rotation of the giant. A week or so earlier, when this film had been shown for the first time to the full Imaging Team, it provided an occasion for good-humored rivalry between planet people and satellite people, with jokes about the satellites getting in the way of the important studies of Jupiter. For the next few days, the imaging focus remained on Jupiter; it shifted to Io, Ganymede, and Callisto, as each was passed in turn after closest approach to the planet.

Tuesday, February 27. *(Range to Jupiter, 7.1 million kilometers).* At a distance of 660 million kilometers from Earth, within 90 Jovian radii (R_J) of Jupiter, Voyager 1 was prepared to begin the encounter with the planet's magnetosphere. On the previous day the spacecraft had crossed the point, at 100 R_J, at which Pioneers 10 and 11 had found the bow shock, the first indication of the magnetospheric boundary. The start of Voyager's plunge into the Jovian magnetosphere was overdue, and scientists anxiously watched the data from the particles and fields instruments, looking for the first indication of disordered magnetic fields and altered particle densities that would mark the bow shock. Apparently, higher solar wind pressure, associated with increased solar activity since 1974, had compressed the magnetosphere, but no one could predict how strong this compression might be.

For the first time since its discovery, Lyle Broadfoot and his UVS colleagues suggested a probable identification for the unexpected ultraviolet emission near the orbit of Io. The most likely candidate was sulfur atoms with two electrons removed (S III), at an inferred

ENCOUNTER DISTANCES
FOR VOYAGER 1

Object	Range to Center at Closest Approach (kilometers)	Best Image Resolution (km per line pair)
Jupiter	349 000	8
Amalthea	417 000	8
Io	21 000	1
Europa	734 000	33
Ganymede	115 000	2
Callisto	126 000	2

temperature perhaps as high as 200 000 K. An additional indication of sulfur came from Mike Krimigis, who reported that the low-energy charged particles instrument had detected bursts of sulfur ions streaming away from Jupiter that had apparently escaped from the inner magnetosphere. No explanations were offered, however, for the presence of large amounts of this element.

At JPL, a press room had been opened in Von Karman Auditorium to accommodate the hundred or so reporters expected to arrive. To keep all the interested people informed of Voyager progress, frequent television reports were beamed over closed-circuit TV throughout JPL. From an in-lab television studio called the Blue Room, JPL scientist Al Hibbs, who had played a similar role during the Viking Mission to Mars, provided hourly reports and interviewed members of the Voyager teams. As the pressure for constant commentary and instant analysis increased, Garry Hunt of the Imaging Team was also called on to host activities in the Blue Room, where his British accent added an additional touch of class to the operation.

As the encounter progressed, the JPL television reports reached a wider audience. In the Los Angeles area, KCET Public Television began a nightly "Jupiter Watch" program, with Dr. Hibbs as host. During the encounter days, service was extended to interested public television stations throughout the nation. In this way, tens of thousands of persons were able to experience the thrill of discovery, seeing the closeup pictures of Jupiter and its satellites at the same moment as the scientists at JPL saw them, and listening to the excited and frequently awestruck commentary as the first tentative interpretations were attempted. Unfortunately, the commercial television networks did not make use of this opportunity, and the greatest coverage available to most of the country was a 90-second commentary on the nightly network news.

Wednesday, February 28. *(Range to Jupiter, 5.9 million kilometers).* At 6:33 a.m., at a range of 86 R_J, Voyager 1 finally reached Jupiter's bow shock. But by 12:28 p.m. the solar wind had pushed the magnetosphere back toward Jupiter, and Voyager was once more outside, back in the solar wind. Not until March 2, at a distance of less than 45 R_J, would the spacecraft enter the magnetosphere for the final time.

At 11 a.m. the first daily briefing to the press was given. "After nearly two months of atmospheric imaging and perhaps a week or two of satellite viewing, [we're] happily bewildered," said Brad Smith. The Jovian atmosphere is "where our greatest state of confusion seems to exist right at the moment, although over the next several days we may find that some of our smirking geology friends will find themselves in a similar state. I think, for the most part, we have to say that the existing atmospheric circulation models have all been shot to hell by Voyager. Although these models can still explain some of the coarse zonal flow patterns, they fail entirely in explaining the detailed behavior that Voyager is now revealing." It was thought, from Pioneer results, that Jupiter's atmosphere showed primarily horizontal or zonal flow near the equatorial region, but that the zonal flow pattern broke down at high latitudes. But Voyager found that "zonal flow exists poleward as far as we can see."

Smith also showed a time-lapse movie of Jupiter assembled from images obtained during the month of January. Once each rotation, approximately every ten hours, a color picture had been taken. Viewed consecutively, these frames displayed the complex cloud motions on a single hemisphere of Jupiter, as they would be seen from a fixed point above the equator of the planet. The film revealed that clouds move around the Great Red Spot in a period of about six days, at speeds of perhaps 100 meters per second. The Great Red Spot, as well as many of the smaller spots that dot the planet, appeared to be rotating anticyclonically. Anticyclonic motion is characteristic of high-pressure regions, unlike terrestrial storms. Smith noted that "Jupiter is far more complex in its atmospheric motions than we had ever imagined. We are seeing a much more complicated flow of cyclonic and anticyclonic vorticity, circulation. We see currents which flow along and seem to interact with an obstacle and turn around and flow back." There is a Jovian jet stream that is "moving along at well over 100 meters per second. Several of these curious little dark features that appear to be small brown spots near Jupiter's north temperate region have been seen to overtake one another and gobble each other up. And then they occasionally spit out a piece here and there as they move along."

Thursday, March 1. *(Range to Jupiter, 4.8 million kilometers).* At 5 a.m., at a distance of

The southern hemisphere of Jupiter presents a tremendous diversity of atmospheric structure and motion. The Great Red Spot rotates counterclockwise in about six days; above and below it high-speed jet streams flow to the right and the left, while a complex, dynamic cloud pattern develops in its wake. This picture was taken on February 25, when Voyager 1 was 9 million kilometers from the planet. [P-21151C]

71 R$_J$, Voyager crossed the bow shock for the third time, catching up with the contracting magnetosphere of the planet. About noon, at 66 R$_J$, the spacecraft finally reached the boundary of the magnetosphere, called the magnetopause. Herbert Bridge, the plasma instrument Principal Investigator, noted that the solar wind pressure as monitored by Voyager 2, still between the Sun and Jupiter, had been for several days from two to five times greater than its level during the Pioneer 10 and 11 encounters. Presumably, this high pressure was the cause of the compressed state of the magnetosphere. However, in the previous few hours

the solar pressure had dropped, so Bridge anticipated that the Jovian magnetosphere might soon inflate and expand outward.

Fred Scarf intrigued the press with a tape of the sounds made by high-energy protons coming upstream from Jupiter. The plasma wave instrument had recorded the noise of the protons, mixed with the noise of the spacecraft itself, producing sound effects that sounded somewhat like a mixture of singing whales, a Nor'easter, and the Daytona 500.

At the press briefing, interest in the imaging results began to shift from Jupiter toward the satellites. Pictures of each of the four big Galilean moons revealed bright and dark features as small as about 200 kilometers across. Unfortunately, this resolution is not enough to be diagnostic—the spots cannot be interpreted in terms of recognizable geological features, such as mountains or craters. Today one could only speculate, but tomorrow or the next day the answers would begin to come in. Deputy Imaging Team Leader Larry Soderblom conveyed his excitement through a metaphor that would be repeated many times during the next week: "We're beginning a stage in this mission which represents, I think, one of the most exciting points in man's scientific exploration of the solar system—in the next few days, we'll explore four new worlds," seeing in

a few days' time what it took us centuries to learn about other worlds in our solar system. In terms of our experience with the exploration of Mars, "it is about 1700 AD this morning, tomorrow it will be about 1800, and it will be about 1976 [the year of Viking] by Tuesday evening."

Friday, March 2. *(Range to Jupiter, 3.6 million kilometers)*. Early in the morning, a twelvefold increase in solar wind pressure caused another contraction of the magnetosphere, which was behaving like a spring, compressing in response to outside forces. As the magnetopause boundary moved rapidly inward, it crossed the spacecraft at 59 R_J from the planet. An hour later the bow shock also flashed past, and Voyager was once more in the interplanetary medium. By noon reinflation of the magnetosphere began again; Voyager crossed the bow shock for the fifth and final time at 55 R_J, followed by three more magnetopause crossings, as the magnetospheric boundary flopped in and out between 45 R_J and 50 R_J.

Some Project officials began to worry about the contracted state of the magnetosphere. The radiation "hardening" of Voyager was carried out to protect against the energetic particle fluxes observed by Pioneers 10 and 11.

A few days before encounter, the Voyager images of the larger Galilean satellites, Callisto and Ganymede, were beginning to show distinctive surfaces with many bright spots. These two pictures were taken on March 5 at a range of 8 million kilometers; the resolution is about 100 kilometers. Although extremely tantalizing, these images were uninterpretable, because the spots could not be associated with any recognizable geological features, such as mountains or craters. Like a naked-eye view of the Moon, these pictures seemed to reveal more than was actually meaningful. [P-21188C and P-21150C]

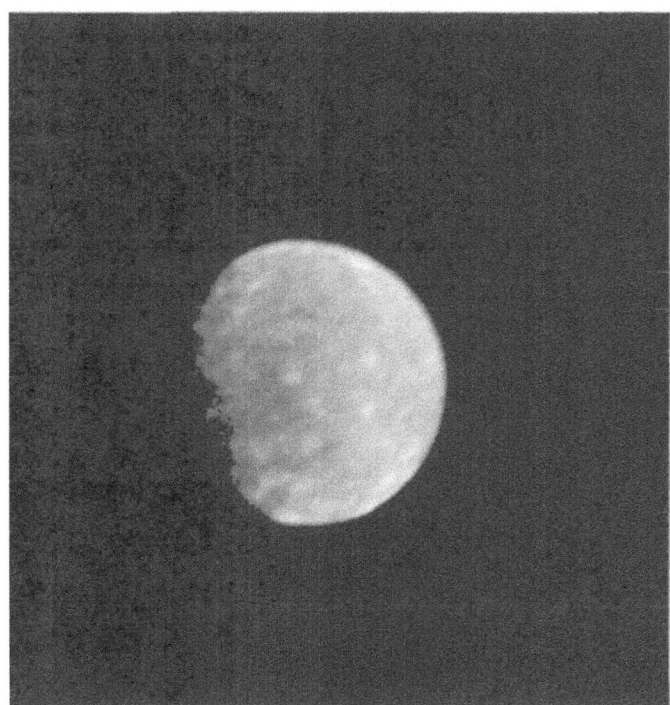 

VOYAGER 1 BOW SHOCK (S) AND
MAGNETOPAUSE (M) CROSSINGS

Boundary	Day	Distance (R_J)
Inbound		
S	2/28	86
S	2/28	82
S	3/01	72
M	3/01	67
M	3/02	59
S	3/02	58
S	3/02	56
M	3/03	47
Outbound		
M	3/13	158
M	3/13	163
M	3/13	165
S	3/16	199
S	3/18	227
S	3/18	227
S	3/19	240
S	3/20	256
S	3/20	258

Under the new conditions, would the particles in the inner magnetosphere be more concentrated and perhaps increase the radiation levels beyond the design limits? Scientists asked how long the compression might last and speculated about how much energy might be pumped into the Jovian radiation belts, but only time could provide the answers.

Accurate recording of the X-band data being sent from Voyager at 115 thousand bits per second required fairly clear weather at the tracking sites. The three Deep Space Network (DSN) stations in California, Spain, and Australia are located at normally dry sites. However, early in the morning, heavy rain at the Australian site interfered with reception of the Voyager signal for fourteen minutes. Fortunately, the loss occurred when the DSN tracking of the spacecraft was being switched from Goldstone, California, to the Australian station. Mission control was able to extend Goldstone coverage for several minutes so that only about three minutes' worth of data was lost completely.

A more positive announcement was made at the press conference that morning by Donald Shemansky of the Ultraviolet Spectroscopy Team: the discovery of a high-energy torus of doubly ionized sulfur (S III) circling Jupiter in the region of Io's orbit. "We were surprised out

of our chairs to see a spectacularly bright emission in the 650–1100 angstrom region, immediately implying that we were looking at a plasma that had to be at a temperature of about 100 000 degrees." The scientists estimated that the density of this torus must be at least 500 ions per cubic centimeter, and that the power needed to keep this plasma at such a high temperature must be in the neighborhood of 500 billion watts. As Al Hibbs mentioned on "Jupiter Watch" that night, 500 billion watts of power is the total amount of installed generating capacity of the United States.

Saturday, March 3. *(Range to Jupiter, 2.5 million kilometers).* Early in the morning the final crossing of the magnetopause took place at 47 R_J, and Voyager finally joined the Jovian system. At the end of the day, the rapidly moving spacecraft crossed the orbit of Callisto, but that satellite itself was on the far side of the planet. Not until the outbound crossing of its orbit on March 6 would closeup views of Callisto be obtained.

During the preceding night, a severe summer thunderstorm in Australia again caused a loss of data. A line of storms over the tracking station blocked the high-rate, X-band science data for three hours and twenty minutes. Attempts were made to save a part of the data by commanding the spacecraft to slow its transmission rate, rather like speaking slowly to a partially deaf listener. But by the time the craft received the signal and responded, the storms had intensified, and no signal could get through. Closeups of the Great Red Spot and an extended series of observations of the glowing sodium cloud around Io were lost.

A black-and-white movie assembled from photos obtained in January and February was shown at the press conference. The movie was taken from observatory phase pictures photographed in blue light. From these pictures, the imaging team "put together this so-called 'blue movie'," said Brad Smith, introducing the film. The film showed changes taking place in Jupiter's atmosphere over a period of about seventy Jovian days. Near the equator there were "bright plumes floating by—at high speed the plumes seem to wave around, something like a flag waving in the breeze." Farther north, along the edge of the north temperate zone, one could see one of the dark ovals—"a rather fuzzy one would move up, catch up with the one just ahead of it, get stuck to the outside and roll around on it for a while, then get ejected a little

At a resolution of about 100 kilometers a planetary surface just begins to reveal its personality. Io was photographed against the disk of Jupiter on February 26, from a distance of about 8 million kilometers. Such early pictures whetted the appetites of the Voyager scientists at JPL as they anxiously speculated about what they would find in closer views of the surface of this remarkable satellite. [P-21185 B/W]

later.'' Dr. Smith also showed new closeup views of the region around the Red Spot, showing not only the anticyclonic features but filamentary "spaghettilike" material which was rotating in a cyclonic direction, indicating a low-pressure region. "The filamentary material still seems to be rather a mystery—very difficult to see the details of the motion." But the photographs were already showing what seemed to be stream lines in the white ovals. "In appearance the white ovals seem to resemble the Red Spot. The stream lines are at least suggestive of divergent flow, that is, material in each of these

spots which is upwelling in these areas and then moving out tends to go around in a counterclockwise anticyclonic motion but may, at the same time, be slowly diverging outward."

The large white ovals are about forty years old. Dr. Smith explained how they formed: Between 1939 and 1940, "where those three white spots exist right now, was a rather bright band similar to the north temperate zone we see on Jupiter right now. In that time period of a year or so, three darkish spots formed. A dark cloud spread out at each one of those three locations and just kept spreading longitudinally until the white material condensed between them." In the end, everything was dark except for the three ovals, which have persisted ever since.

Torrence Johnson of the Imaging Team commented on a photo of Io in which features resembling circular crater-type structures seemed to be visible. "Whatever they are, [those circular features] are approaching the size of the things that we would call basins if they were impact structures on other planets. We don't really know whether they're impact structures. They have some characteristics that look reminiscent of impact structures. They could be endogenic—volcanic in origin—or internally generated in some other way." Johnson also showed another photograph of Io, this one taken against black sky, showing a "strikingly different face" looking, perhaps, like someone's nightmare, glaring back at the intruder from Earth. One huge feature—a "bullseye" or "hoof print" on Io—appeared to be approximately 1000 kilometers long. No one had ever seen such a feature, and the imaging scientists could only speculate about its significance.

A few days earlier, someone had posted in the Imaging Team area a quote from a 1975 review paper on the Jovian satellites by David Morrison and Joseph Burns. The section on Io began, "Io is one of the most intriguing objects in the solar system." This statement seemed more and more appropriate as Voyager images

Small-scale structures in the jet streams of Jupiter's north tropical zone reveal details of atmospheric circulation. The small dark oval near the right edge of the zone may offer a glimpse deep into Jupiter's atmosphere. Between the regularly spaced dark ovals near the bottom of the frame are more small-scale features that are being studied for their roles in Jovian atmospheric activity. The blue-gray regions along the shear line between the equatorial zone and the north equatorial belt also appear to be windows into the deeper regions of the atmosphere. This photo was taken February 19 by Voyager 1 from a distance of 14 million kilometers. [P-21160C]

improved. Johnson referred to this day as equivalent to the "late 1960s" in our study of the Jovian satellites. "We can see much more clearly than ever before, but still not clearly enough to provide understanding of what we are seeing."

Sunday, March 4. *(Range to Jupiter, 1.2 million kilometers).* At 4:37 a.m., the near encounter phase began: Voyager 1 was almost there! In the press room, someone taped a fortune cookie message to the Voyager TV monitor: "There is a prospect of a thrilling time ahead for you."

Pulled by the powerful gravity of Jupiter, the spacecraft was now on a curved path through the inner Jovian system. At 2 p.m. PST, it crossed the orbit of Ganymede, and later in the afternoon it passed within less than 2 million kilometers of Europa, providing Voyager 1's closest look at this satellite. During the afternoon and evening, a number of views were obtained of Amalthea, the small inner satellite, at a range of less than 500 000 kilometers. At 8 p.m. the orbit of Europa was crossed, and increasing attention was drawn to the coming encounter with Io. At about 7 p.m. a full-frame color sequence of Io was received with a resolution of 16 kilometers. During the night, as Imaging Team members scratched their heads trying to prepare a press release caption to interpret the peculiar structures seen, the JPL Image Processing Lab rushed to prepare a color version for release the next day.

The Great Red Spot became more and more spectacular as Voyager 1 approached, with each day revealing new and intricate detail in the clouds. This view was obtained on March 1 at a distance of 5 million kilometers; the smallest features that can be made out are about 100 kilometers across. To the west of the Great Red Spot is a region of great turbulence, and to the south is one of the three white ovals. [P-21182C]

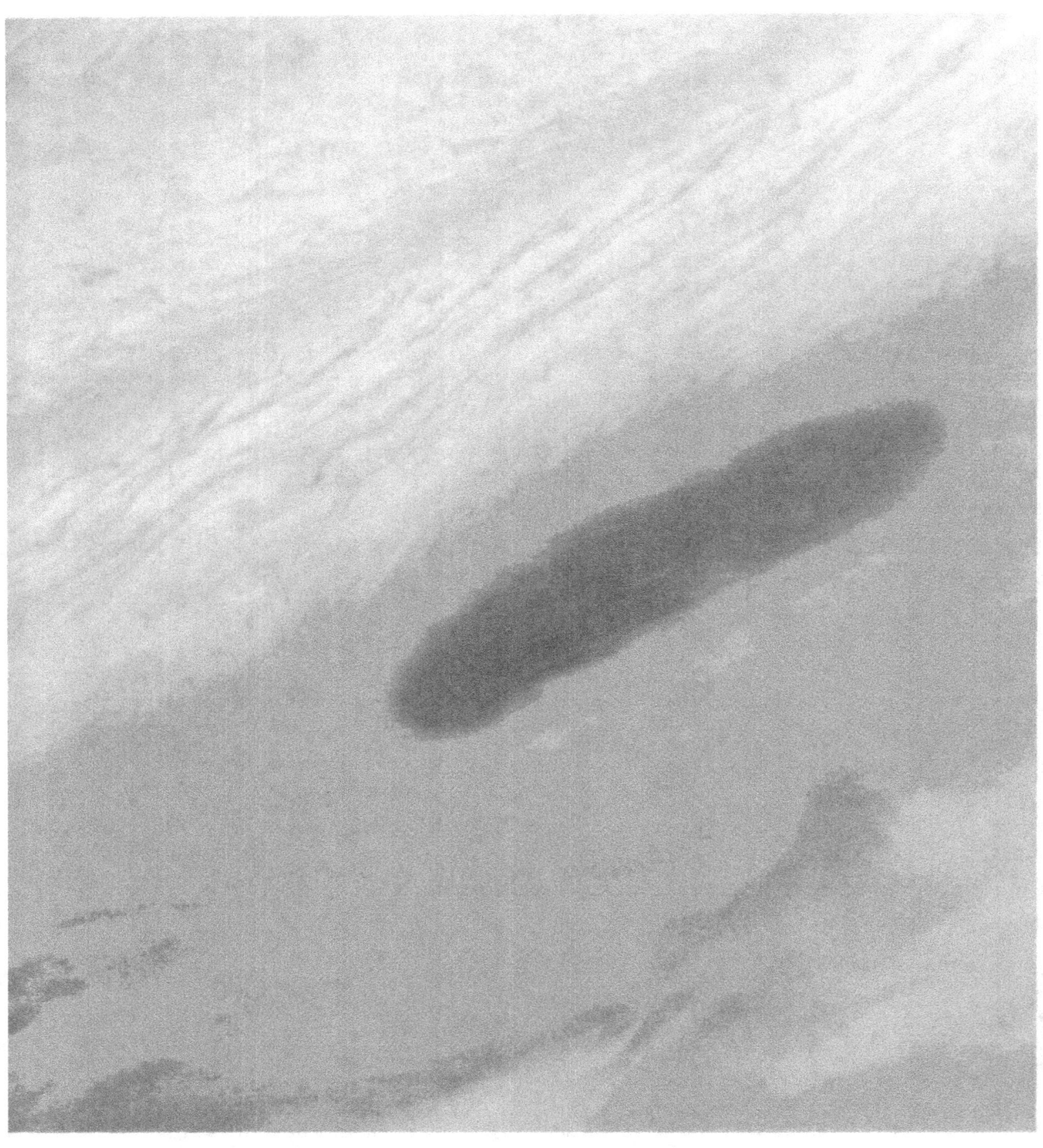

Large brown ovals in the northern hemisphere of Jupiter are apparently regions in which an opening in the upper, ammonia clouds reveals darker regions below. This oval, about the same length as the diameter of the Earth, was at latitude 15°N. Features of this sort are not rare on Jupiter and have an average lifetime of one to two years. Above the feature is the pale orange north temperate belt, bounded on the south by the high-speed north temperate current, with winds of 120 meters per second. The range to Jupiter at the time this photograph was obtained on March 2 was 4 million kilometers, with the smallest resolvable features being 75 kilometers across. [P-21194C]

Once again, tracking station problems interrupted the smooth flow of data from the spacecraft. Failure at Madrid to reestablish the correct receiver frequency after a spacecraft maneuver resulted in the loss of 53 minutes of irreplaceable data; an additional eleven minutes were lost an hour later. As the first problem was being corrected, the tracking antenna became misaligned, resulting in a noisy data return. Meanwhile, only a few of the watchers at JPL mourned the lost data, so exciting were the other new results that kept pouring in.

"For the highlights of this morning Dr. Soderblom will come up and show some beautiful satellite pictures," Brad Smith announced at the daily press briefing. Larry Soderblom's excitement was hard to contain. "Today is probably going to turn out to be one of the most memorable days in our exploration program. For the planetary geologists it's truly Christmas Eve. We see tonight the beginning of the exploration of four new worlds. We're racing through time and space at an incredible rate—the rate at which we are learning things is awe-inspiring in itself." Callisto was still too far away to see well "but the things we're seeing on the closer three satellites have really got us charged in anticipation." Ganymede—"What can I say? Loops and swirls and incredible patches that are difficult now to hazard a guess about." Io—an eerie-looking red, orange, and yellow world—"this one we've got all figured out. [Laughter from the press and applause.] It is covered with thin candy shells of anything from sulfates and sulfur and salts to all kinds of strange things."

The low energy charged particle instrument had discovered high-speed sulfur in the outer Jovian magnetosphere, ten times as far from the planet as the sulfur torus around Io. The high-speed sulfur, which whizzed by Voyager 1 at 8000 kilometers per second, was first detected as the spacecraft crossed the magnetopause and entered the magnetosphere. Apparently this sulfur had picked up speed as it moved outward from Io's torus, but the mechanism for this acceleration was not known. Roby Vogt reported that the cosmic ray instrument was detecting two distinct groups of atoms closer in to the planet. One group was

The best Voyager 1 photos of Europa were obtained on March 4 from a distance of about 2 million kilometers. This view of the hemisphere centered at about 300° longitude has a resolution of about 40 kilometers. Most of the surface is bland and highly reflective, being composed almost entirely of water-ice. No craters resulting from meteoric impacts can be seen. The most striking visual features, which set Europa off from the other satellites, are the dark streaks, as much as several thousand kilometers long, that cross the surface. Just barely visible to Voyager 1, these streaks were dramatically apparent to Voyager 2 in its closer flyby in July. [P-21208C]

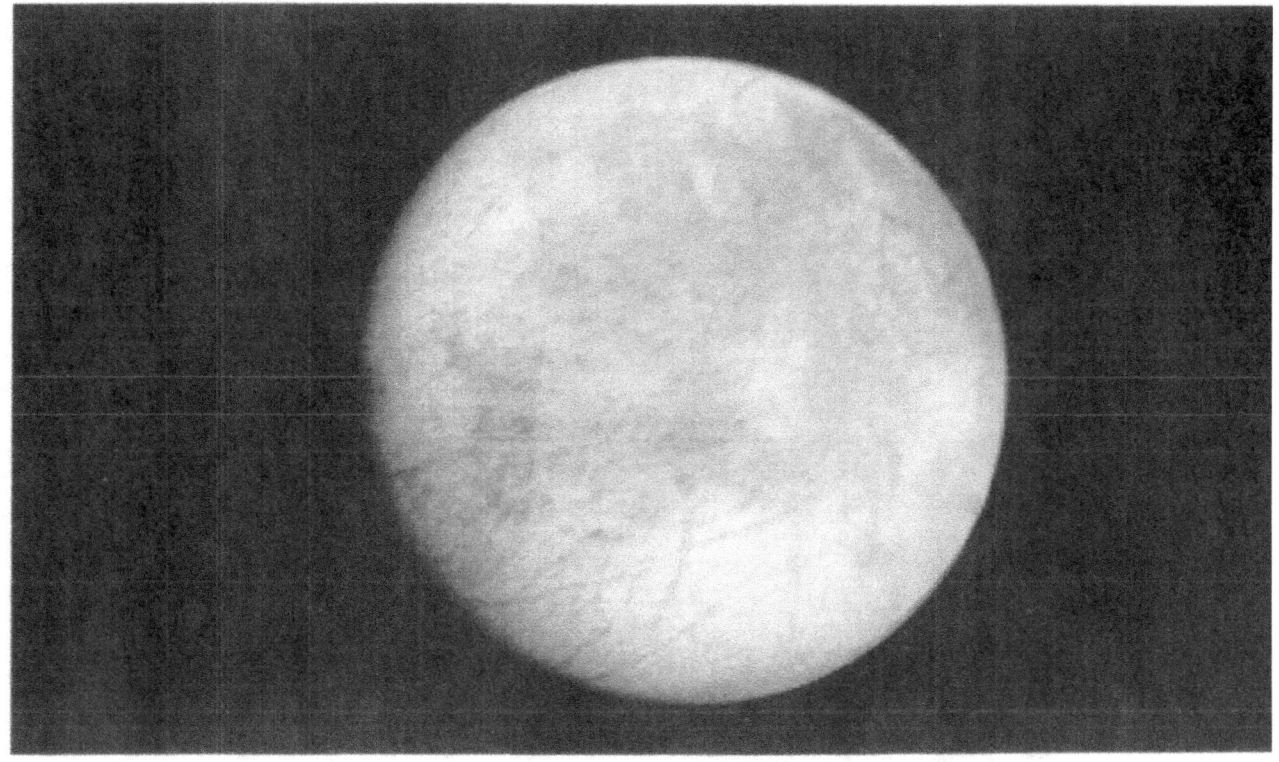

apparently of solar composition (derived from the solar wind), and the other showed enhanced oxygen as well as sulfur. The plasma instrument had also begun to measure sulfur in the same form (S III) as that detected by the UVS in the Io torus. Where, the scientists asked, could the oxygen and sulfur be coming from? Could Io be the source of these atoms?

At one point, as questions from the press became too detailed, asking for "instant analyses" of the data, Dr. Smith was prompted to make the following statement: "I want to say something about what's going on right now in the imaging area because we get a lot of pointed questions. We have a remarkably good system that is getting extremely good photographs. Not only are the cameras working well, but Jupiter and the satellites are cooperating by showing a lot of truly remarkable detail. And it may sound unprofessional, but a lot of the people up in the Imaging Team area are just standing around with their mouths hanging open watching the pictures come in, and you don't like to tear yourself away to go and start looking at numbers on a printout. We will do that, but in the meantime we're just caught up in the excitement of what's going on."

While the press briefing was underway, between 11:38 and 11:50 a.m., a special experiment was being tried on board the Voyager. As the spacecraft passed through the plane of the equator of Jupiter, it aimed its narrow-angle camera to a point in space halfway between the cloud tops and the orbit of Amalthea and took a single, eleven-minute exposure. The purpose: to search for a possible faint ring around Jupiter, which, if present, could best be seen by sighting along the plane of the equator. The image appeared briefly on the TV monitors, clearly showing something—a strange band of light—streaking across the center of the frame. Up in the imaging area, analyses began at once to determine if the strange band were really the sought-for ring, but it would not be until March 7 that the identification would be confirmed and the discovery announced.

Monday, March 5. (*Encounter Day. Minimum range to Jupiter, 780 000 kilometers; speed of spacecraft, almost 100 000 kilometers per hour*). Many celebrities, including the Governor of California, spent the night at JPL to witness the historic occasion. In Washington,

The "bullseye" of Io was first photographed at a range of 2.8 million kilometers on March 3, and this color version was released on March 4. At the time, Imaging Team scientists wrote that "The large heart-shaped feature with a dark spot near its center could be Io's equivalent of an impact basin such as Mare Orientale on the Moon. Its outer dimensions are about 800 by 1000 kilometers. Subsequent high-resolution coverage should reveal whether the small dark spots are impact craters, or perhaps something more exotic such as volcanoes. The reddish color of Io has been attributed to sulfur in the salts which are believed by some to make up the surface of Io." It would be another week before this feature, later named "Pele" for the Hawaiian volcano goddess, would be recognized as an active, erupting volcano. [P-21187C]

D.C., a special TV monitor was set up in the White House for the President and his family.

As late Sunday night eased into the early morning of encounter, closeup images of Jupiter, looking more like abstract art than like planetary science, flashed across the TV screens, and verbal images far less wild then the visuals from Jupiter were heard from commentators and from members of the press. "Are you sure Van Gogh didn't paint that?" "That's not Jupiter; it looks like a closeup of a salad." "They're not showing us Jupiter, that's some medical school anatomy slide."

Shortly before closest approach to Jupiter, Voyager began its intensive observations of Io. Much of this information, taken while the Australian station was tracking the spacecraft, was recorded on Voyager's onboard tape-recorder for playback later that day. But even before the results of that imaging were known, Larry Soderblom was calling Io "one of the most spectacular bodies in the solar system." As more and more vivid photos of Io appeared on the monitors, members of the Imaging Team in the Blue Room buzzed with excitement. "This is incredible." "The element of surprise is coming up in every one of these frames." "I knew it would be wild from what we saw on approach but to anticipate anything like this would have required some sort of heavenly perspective. I think this is incredible."

At 7:35 a.m. Voyager was scheduled to pass through the flux tube of Io, the region in which tremendous electric currents were calculated to be flowing back and forth between the satellite and Jupiter. Norm Ness suggested, after examining magnetometer data, that Voyager skirted the edge of the flux tube, and that the current in the tube was about one million amps. As the flux tube results were received, champagne bottles began to pop in the particles and fields science offices, in celebration of the successful passage through the inner magnetosphere. Meanwhile, at 7:47 a.m., closest approach to Io occurred, at a range of only 22 000 kilometers. Voyager was 25 000 times closer to this satellite than were the watchers on Earth.

At 8:14 a.m., while still within 30 000 kilometers of Io, the spacecraft passed out of sight behind the edge of Jupiter. All scientific data for the next two hours and six minutes were stored on the onboard tape recorder for later transmission to Earth. Meanwhile, the radio communication signal was used to probe the atmosphere of Jupiter, yielding a profile of electron density in the ionosphere and of the gas pressure and temperature in the upper atmosphere. While out of sight from Earth, at 9:07 a.m., Voyager plunged into the shadow of Jupiter. As the Sun set on the spacecraft, the ultraviolet instrument used the absorption of sunlight to determine the composition and temperature of the upper atmosphere. In the darkness, the infrared IRIS measured the night-side temperatures of the planet, and long-exposure images were taken to search for aurora, lightning, and fireballs in the Jovian atmosphere. At 10:20 a.m., Voyager reappeared from behind Jupiter and radio contact was restored; at 11:24 a.m., it emerged from shadow into sunlight, speeding on toward encounter with Ganymede.

At 8 a.m. a special press conference was held to mark the successful Jupiter flyby. Noel Hinners, Associate Administrator for Space Science and the highest ranking NASA official present, congratulated all those who had made the Voyager Mission a success. The encounter was the "culmination of a fantastic amount of dedication and effort. The result is a spectacular feat of technology and a beginning of a new era of science for the solar system. Just watching the data come in has been fantastic. I had a fear that things on the satellites were going to look like the lunar highlands. Nature wins again. If we're going to see exploration of this nature occurring in the 1980s and 1990s we must continue to expound the results of what we're finding here, the role of exploration in the history of our country, what it means to us as a vigorous national society."

As time passed, it became apparent that Voyager 1 had been affected by Jupiter's radiation environment. The basic timing—the main clock on the spacecraft—had slowed down. First it slowed by 6.3 seconds, but by March 6 it was found to have slowed a total of eight seconds. In addition, the two central computers apparently got out of synchronization both with themselves and with the flight data subsystem. On March 6 it was reported that the spacecraft cameras were shuttering one frametime (48 seconds) early; this was partly offset by the eight-second spacecraft "masterclock" slowdown resulting in images (according to our clocks) being photographed forty seconds early. This timing error resulted in the camera taking some pictures while the scan platform was moving, causing some blurred images. A number of the highest resolution images of Io and Ganymede were seriously degraded by this malfunction.

At the regular 11 a.m. press briefing, Brad Smith glowed. "We're all recovering from what I would call the most exciting, the most fascinating, what may ultimately prove to be the most scientifically rewarding mission in the unmanned space program. The Io pictures this morning were truly spectacular and the atmosphere up in the imaging area was punctuated by whoops of joy or amazement or both." The new color photo of Io taken the night before was released, showing strange surface features in tones of yellow, orange, and white. The image defied description; the Imaging Team used terms like "grotesque," "diseased," "gross," "bizarre." Smith introduced the picture with the comment, "Io looks better than a lot of pizzas I've seen." Larry Soderblom added, "Well, you may recall [that we] told you yesterday that when we flew by we'd figure all this out. I hope you didn't believe it."

One thing was certain: There were no impact craters on Io. Unless the satellites of Jupiter had somehow been shielded from the meteoric impacts that cratered objects such as the Moon, Mars, and Mercury, the absence of craters must indicate the presence of erosion or of internal processes that destroy or cover up craters. Io did not look like a dead planet. Imaging Team member Hal Masursky, looking at the "pizza" picture, estimated that the surface of Io must be no more than 100 million years old—that is, some agent must have erased impact craters during the last 100 million years. This interpretation depended on how often catering impacts occur on Io. No one could be sure that there had been any interplanetary debris in the Jovian system to impact the surfaces of the satellites. Perhaps none of them would be cratered. The forthcoming flybys of Ganymede and Callisto would soon provide this information.

The close flyby of Ganymede took place at 6:53 p.m., at a range of 115 000 kilometers. During the preceding four hours, photos re-

When the first color close-up of Io was released, Imaging Team Leader Brad Smith said that he had seen "better looking pizzas"; hence this view, taken March 4 at a range of about 860 000 kilometers, became known as "the pizza picture." The circular feature in the center (the piece of pepperoni) was later revealed to be the active volcano Prometheus, but at the time of its release this lovely but bizarre picture baffled scientists and press alike. [P-21457C]

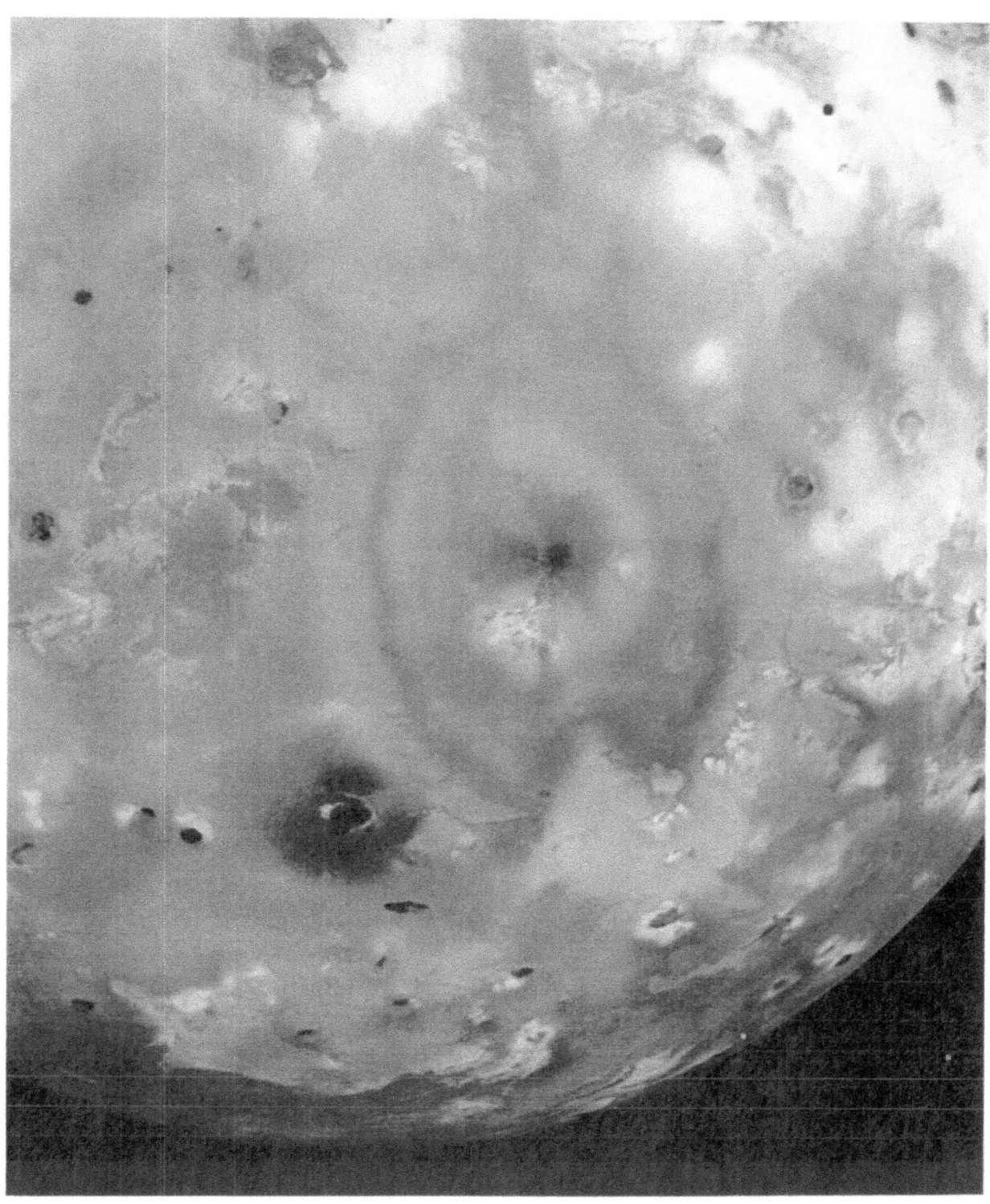

The giant volcanic feature Pele, about 1000 kilometers across, mystified Voyager scientists when this picture of Io was taken on March 5 from a distance of about 400 000 kilometers. The brilliant colors, the strange shapes of the surface deposits, and the absence of impact craters all testified that Io was unlike any world previously encountered in the exploration of the solar system. [P-21226C]

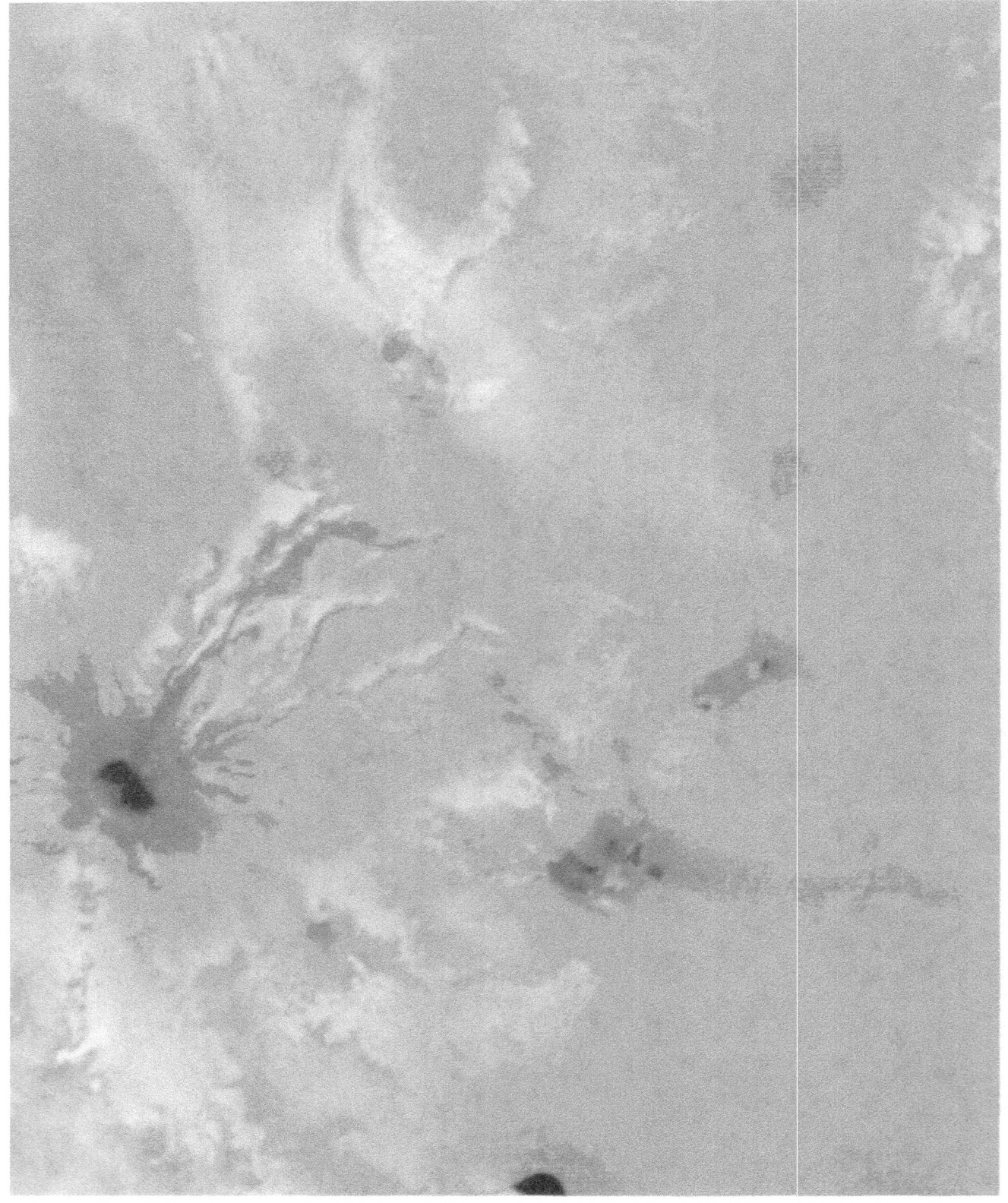

As Voyager 1 approached Io, the images of the surface became more and more spectacular. On the morning of March 5, at a range of 130 000 kilometers, this picture was taken centered near longitude 320° and latitude 10°S. The width of the picture is about 1000 kilometers (the distance from the Mexican border to the northern edge of California). There are no impact craters, signifying a geologically young surface, and the dark center with radiating red flows indicates recent volcanic activity of some sort. [P-21277C]

vealed a surface covered with impact craters. Watching these photos and supplying commentary for the television listeners, David Morrison remarked, "While I'm delighted to see craters, it's just the opposite of what I would have expected. I was telling everyone a few days ago that I thought Io would have plenty of craters and that Ganymede, because of the ice surface, simply would not be able to hold large craters over geological time. So this is fascinating and this is confusing—both what has happened on Io to erase the craters and why Ganymede's surface is strong enough to preserve them."

Just before closest approach, at 6:35 p.m., the ultraviolet instrument watched as a bright star passed behind Ganymede and reemerged ten minutes later. No dimming that could be attributed to an atmosphere was seen; when the data were analyzed later, scientists set an upper limit for any gas on this satellite at one-billionth of the atmospheric pressure at the surface of the Earth.

As encounter day drew to a close, celebrations took place all over JPL. For many, however, the excitement was tempered by exhaustion. After 48 hours of intense activity, sleep was imperative for some. But the close approach to Callisto was still to come, as was an examination of the data already received.

Tuesday, March 6. Voyager was now receding from Jupiter, accelerated on a new trajectory—one that would speed it on toward its November 1980 encounter with Saturn. But first came the encounter with Callisto, with closest approach at 9:50 a.m., at a range of 126 000 kilometers. The satellite was littered with craters and there appeared one huge "bullseye" pattern that might have been the result of an impact. As the photos of Callisto came in, Garry Hunt de-

After its close flyby of Io, Voyager 1 headed for Ganymede, the largest of the Galilean satellites. This global view of Ganymede, taken on March 4 at a range of 2.6 million kilometers, shows features as small as 50 kilometers across. At the time, Voyager scientists speculated that the numerous white spots were impact craters, surrounded in some cases by icy ejecta blankets splashed onto the surrounding surface. However, many narrow white streaks, especially those in the lower left quadrant, promised new and exciting geological features on this satellite. [P-21207C]

As Voyager 1 approached Ganymede on March 5, many strange new surface features became visible. In this frame, taken from a distance of 250 000 kilometers and showing features as small as 5 kilometers across, three distinct types of terrain are seen: polygons of old dark surface, extensive areas of lighter, younger material, and brilliant white ejecta patterns (probably water-ice) around fresh craters. The bright rays in the upper part of the picture are 300-500 kilometers long; at the bottom are several craters with only faint, muted ejecta patterns. [P-21262C]

Voyager 1 found that the surface of Ganymede was geologically very complex. This frame, taken March 5 from a range of 250 000 kilometers, shows a region about 1000 kilometers across centered near longitude 0° and latitude 20°S. The surface displays many craters, some (probably the younger ones) with bright ray systems. Bright grooved bands traverse the surface in various directions. One of these bands, running in a north-south direction in the lower left of the picture, is offset along a white line that may represent a fault. Ganymede is the only Galilean satellite to show indications of such lateral offsets in the crust. [P-21266]

scribed the scene in the imaging area: "The activity around the monitors now is quite incredible with people caught breathless by the pictures coming in. Every satellite we've seen has been a different world." Asked how he felt about the images of Jupiter's atmosphere, he replied, "I'm absolutely delighted with what I've seen, and I'm delighted Voyager 2 is not far behind since I'm convinced that we'll see yet another face of Jupiter by then. The weather will have changed by July."

At the morning press briefing, the wealth of new data began to be revealed. "If Jupiter had ever posed for Monet, it would probably have turned out like this," said Brad Smith as he introduced some enhanced (exaggerated) color images that had been specially processed to show more detail in the Great Red Spot region. Indeed, these photographs did look like paintings with Jupiter displaying some of its best abstract art.

The first picture of Amalthea was shown, revealing an elongated, dark, reddish object about 265 kilometers long. Smith reported: "There is actually some structure that one can see—a crater—a couple of bright features for

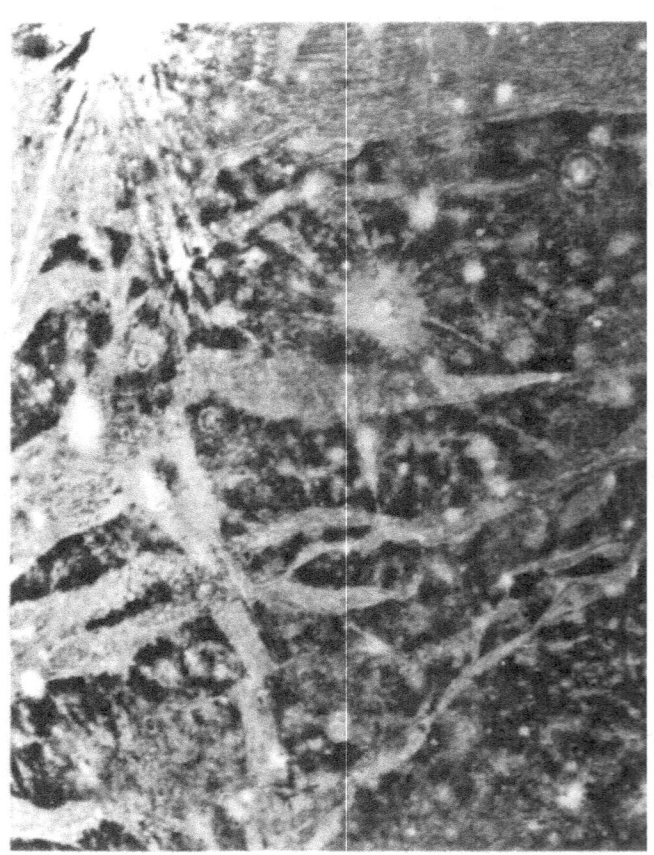

which we have no interpretation and some other evidence of cratering. It doesn't look like much, but after all, Amalthea has never been seen as anything more than a point of light from the Earth, and, in fact, there are very few people that have even seen it as a point of light. I doubt that more than one astronomer out of a hundred has actually seen Amalthea.''

Larry Soderblom introduced new photographs of the satellites. Io still showed no craters, even at high resolution. The craters that "should" have been on Io were on Ganymede and Callisto. Ganymede not only had craters, it had fault lines as well. "There is transverse motion along these faults. Things get offset,

apparently, for hundreds of kilometers. So it's the first time we've seen major kinds of transverse motion on the surface of another planet.'' Ganymede, in effect, had shown evidence of having its own icy version of the San Andreas Fault.

Although Callisto was heavily covered with craters up to 200 kilometers in diameter, Dr. Soderblom commented on the absence of larger impact basins. Perhaps there was one in the huge bullseye feature; however, it was not a "standard" looking basin like those on the Moon. Callisto was "extremely smooth and free of any relief. The structure [the bullseye], if impact caused, shows no relief; the limb does

Amalthea was thought to be the innermost satellite of Jupiter until Voyager 2 discovered tiny Adrastea (1979J1). The Voyager 1 camera revealed Amalthea as an irregular dark reddish object with dimensions of 270 × 160 kilometers. These three images have resolutions of (top right, b) 25 kilometers; (bottom left, c) 13 kilometers; (bottom right, d) 8 kilometers. [260-503]

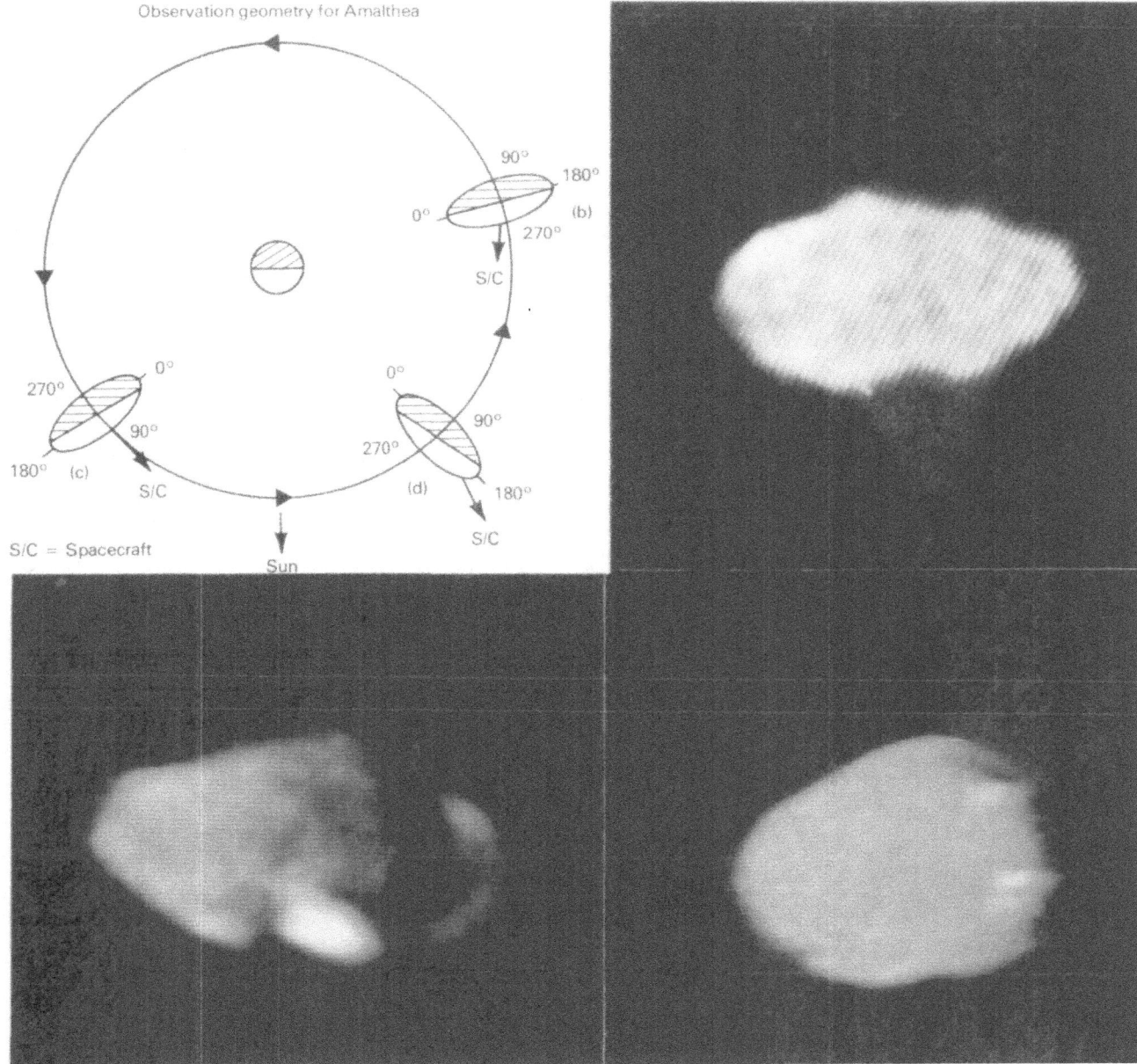

not show any relief; maybe it's possible that Callisto cannot support relief.'' By the next day, the geologists on the Imaging Team were becoming more convinced that the Callisto ''bullseye'' was the frozen remnant of an enormous impact into Callisto's surface. Since Callisto is composed in large part of water and has an icy crust, the team speculated that any raised features created by the impact would eventually ''slide'' back into the surface, ''and the ripple marks from the shock wave caused by the impact were frozen'' into the surface.

Lyle Broadfoot announced that the ultra-violet spectrometer had detected very strong auroral emission in Jupiter's north and south polar regions. The aurora seemed to be caused primarily by the excitation of molecular hydrogen, although some atomic hydrogen was also detected. Auroral emissions from helium atoms were not detected.

The IRIS infrared measurements required more computer processing than other Voyager data, and therefore they were not available until a day or two after the observations were

Callisto was the last of the Galilean satellites to be studied by Voyager 1. In this photo, taken March 5 from a distance of 1.2 million kilometers, with a resolution of about 25 kilometers, the extensive cratering of the surface began to be apparent. Near the upper left edge is the large impact basin Valhalla; the numerous light spots are craters 100 kilometers or more in diameter. This is the same side of Callisto that was photographed at higher resolution during the Voyager 1 flyby of the satellite a day later. [P-21284C]

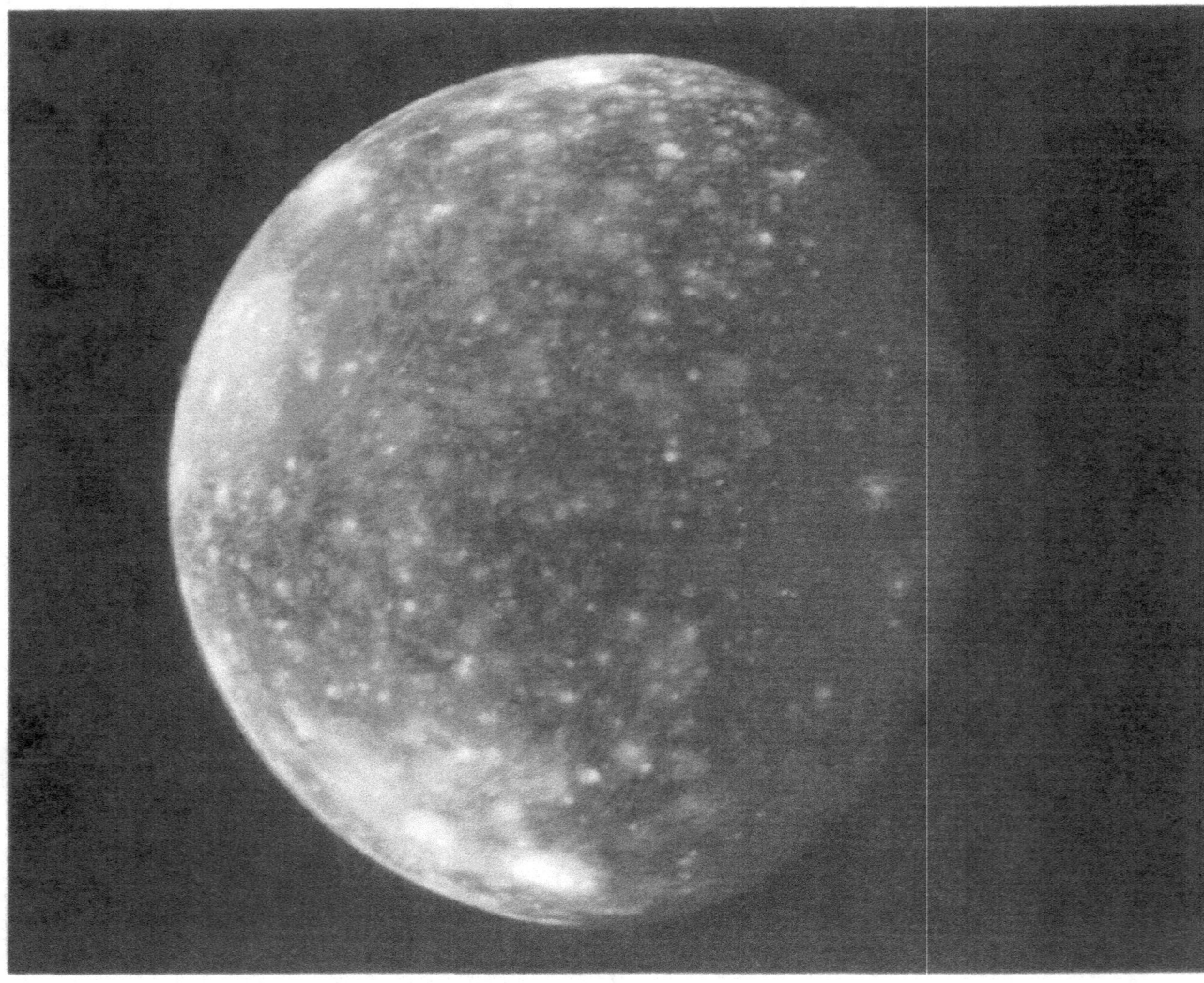

The largest ancient impact basin on Callisto is called Valhalla. The central light area is about 600 kilometers across. Surrounding it is a set of concentric low ridges, looking like frozen ripple marks, extending about 1500 kilometers from the center. This picture was taken by Voyager 1 on March 6 at a range of 350 000 kilometers. [P-21287C]

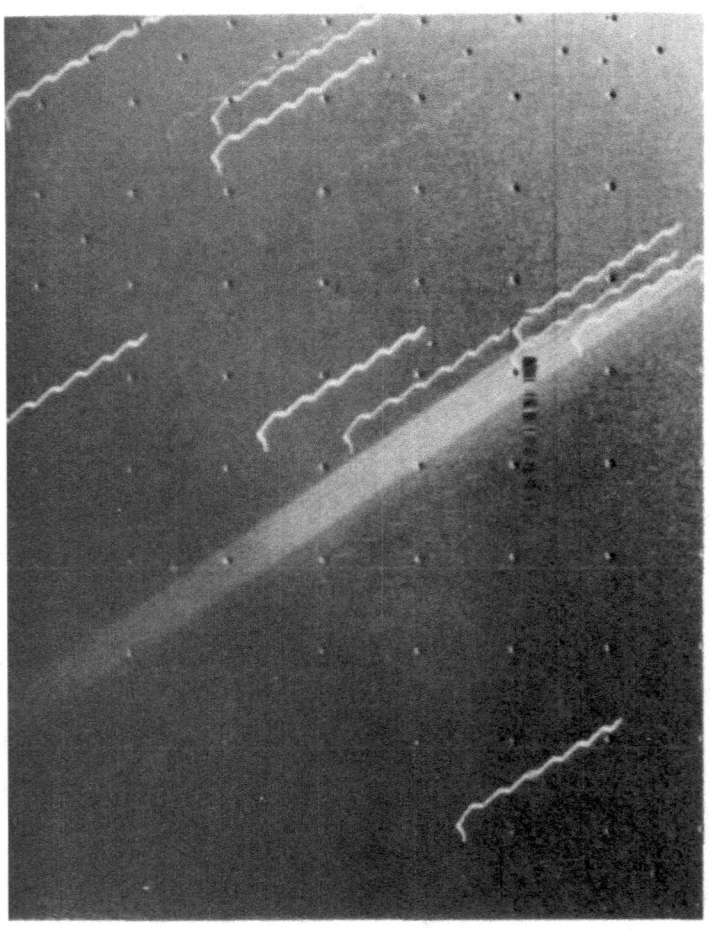

made. However, Rudy Hanel already had two new results to report. First, the Great Red Spot was about 3° C cooler than its surroundings, and this cooling extended many kilometers above the clouds, into the thin upper atmosphere. Second, the thermal emission from Io was peculiar, with an unexpected shape to the spectrum. Tentatively, Hanel suggested there might be some hot spots on the surface of the satellite.

Jupiter has been full of surprises, but the excitement was far from over. Major discoveries were yet to be made.

Wednesday, March 7. At the press briefing, Brad Smith made a spectacular announcement: "This morning I would like to add yet another important discovery to be claimed by this outstanding mission—that of a thin flat ring of particles surrounding Jupiter. Thus Jupiter now joins Saturn and Uranus to become the third planet of our solar system known to possess a planetary ring system and leaves Neptune as the only member of the group of giant planets without a known ring. The discovery of the ring was unexpected in that the current theory which treats long-term stability of planetary rings would not predict the existence of such a ring around Jupiter. The single Voyager

Voyager 1 discovered the rings of Jupiter on March 4 in a single eleven-minute exposure with the narrow-angle camera. Spacecraft motion during the time exposure streaked the picture, as can be seen from the hairpin-like images of the stars. (The star field was unusually rich, since it happened to include the Beehive star cluster in Cancer.) The ring image itself is a multiple exposure, with six separate images side by side. The ring does not extend out of the right side of the picture, indicating that this image captured the outer edge of the ring, about halfway between the cloud tops and the orbit of Amalthea. [P-21258]

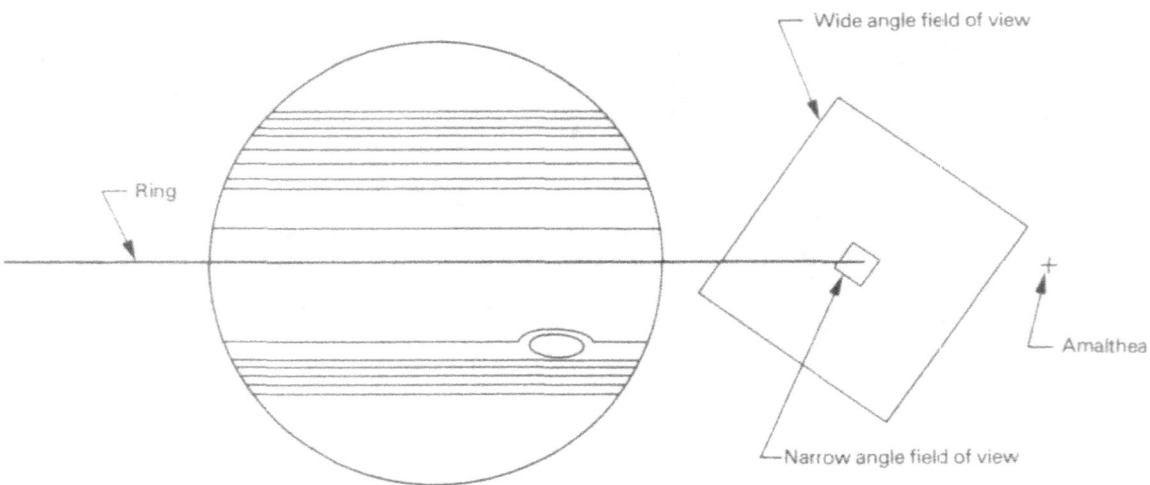

Since the Voyager 1 ring photograph was taken exactly edge-on, it was not possible to determine the width of the ring. In this artist's conception, the ring is drawn as if it were ribbon-like, with very little width, quite unlike the broad flat rings of Saturn. Voyager 2 later showed this to have been a lucky guess. [P-21259]

camera image which recorded the ring was planned by the Voyager Imaging Team several years ago, not really with any great expectation of a positive result, but more for the purpose of providing a degree of completeness to Voyager's survey of the entire Jupiter system. The observation, as planned, involved looking off to the right of the limb of Jupiter in the planet's equatorial plane at the exact moment that the spacecraft would be crossing the equatorial plane. The image actually was taken at 16 hours and 52 minutes before encounter from a distance of about 1.2 million kilometers. Exposure time was 11.2 minutes. We weren't certain of the exact moment that we would cross the equatorial plane, so we planned to open our shutter and leave it open as we went through."

The image showed six exposures of the ring, together with streaked trails of background stars. Smith reported that the thickness of the ring was less than 30 kilometers, and that it extended to a point 128 000 kilometers from the center of Jupiter, or 57 000 kilometers above the clouds. In response to this discovery, scientists were eagerly planning to alter the Voyager 2 encounter sequence to try to obtain additional information on the ring.

The newspapers carried stories of the ring discovery on Thursday, March 8. One story was seen by two University of Hawaii astronomers, Eric Becklin and Gareth Wynn-Williams, who were observing at Mauna Kea Observatory, at an altitude of 4200 meters on the Big Island of Hawaii. Within two days they had succeeded in detecting the rings by their reflected sunlight, at an infrared wavelength of 2.2 micrometers, providing a rapid confirmation of the Voyager discovery.

As the plasma measurements from the Io flyby were analyzed, an additional clue to the origin of sulfur and oxygen was revealed. Herb Bridge reported the detection of sulfur dioxide (SO_2), the simplest molecule composed of these two atoms.

The high-resolution tape recorded pictures of Io baffled imaging scientists. A number of features looked like volcanic flows; together with the absence of impact craters, these features indicated a geologically active planet. A central point of discussion was the recent theoretical work on Io by Stanton Peale of the University of California and two NASA scientists, Pat Cassen and Ray Reynolds. These authors had just published a paper in the March 2 issue of *Science* showing how tidal heating from Jupiter's gravity could melt the interior of Io. They wrote that "widespread and recurrent surface volcanism might occur." It began to look as if the prediction had been correct.

Thursday, March 8. The last Voyager 1 press briefing was held. Each speaker was allotted only a few minutes, prompting Larry Soderblom to preface his remarks by trying to explain how difficult it was to describe four new planets—the Galilean satellites—in the time allowed. "Torrence [Johnson] was sitting with me last night, puzzling. He said, 'You know, Larry, it's sort of like imagining we'd flown into the solar system the day before yesterday, and said, "There's a thing we'll call Mercury, and there's the Moon, and there's Earth, and there's Mars. Now let's explain them in ten minutes".' " There was Callisto, with the highest density of craters of any Galilean satellite—the oldest of the Galilean surfaces—featuring a huge "bullseye" that is "the largest single contiguous feature seen so far in the solar system." There was Ganymede, cratered, but also overrun with fault lines that looked, according to one person, like "tire tracks in the desert," showing a surface that "had laterally slid—faulted and sheared and sheared again—twisted and torn apart." There was Io, the most bizzare—the one that scientists had thought would be most lunarlike—showing a surface that had apparently been "cooked and steamed and fumed out leaving deposits all over the sur-

face much like you might see around a fumarole at Yellowstone National Park. The fact that these things exhibit such youth makes it likely that the planet is still volcanically active." There was Europa, with huge linear features unlike those of the other three Galileans—Europa the mystery satellite, waiting for the probing eyes of Voyager 2 to survey it in early July.

Ed Stone summed it all up: "I think we have had almost a decade's worth of discovery in this two-week period, and I think that all of the people who have been talking to you feel the same saturation of new information which has occurred. And in fact, we will probably be studying it in great detail for at least five years."

Over the next day or so the press packed up and went home. The TV monitors showed the spacecraft's parting glance at a crescent Jupiter, only hinting at the vastness of space Voyager 1 would travel until its encounter with Saturn in the autumn of 1980. Maybe now there would be a relative calm that would allow the scientists to begin analyzing that "decade's worth of data." But things were not to be calm just yet.

Fire and Brimstone

At about 5 a.m. on March 8, Voyager had taken a historic picture. Looking back at a crescent Io from a distance of 4.5 million kilometers, the camera had been used to obtain a long-exposure view for the spacecraft navigation team—one that showed the satellite against the field of background stars. During the day, Linda Morabito, an optical navigation engineer, began to work with this picture on her computer-controlled image display. She noted what appeared to be a crescent cloud, extending beyond the edge of Io. But Io has no atmosphere, so a cloud rising hundreds of kilometers above the surface did not seem to make sense.

The next day, working with her colleagues, Morabito eliminated all possibilities for the new feature on Io except the obvious—a cloud. If it were a cloud, it must be the result of an ongoing volcanic eruption of incredible violence. The picture was shown to members of the Imaging Team, who agreed with the identification. But it was Friday and Brad Smith and Larry Soderblom, along with most of the other team members, had left for the weekend to try to get some rest. The picture would have to wait two more days.

HIGHLIGHTS OF THE VOYAGER 1 SCIENTIFIC FINDINGS*

Atmosphere

Uniform wind speeds for cloud features with widely different size scales, suggesting that mass motion and not wave motion is being observed.

Rapid formation and spreading of bright cloud material, perhaps the result of disturbances that trigger convective activity.

A pattern of east-west winds in the polar regions, previously thought to have been dominated by convective upwelling and downwelling.

Anticyclonic motion of material associated with the Great Red Spot, with a rotational period of about six days.

Interactions of smaller spots with the Great Red Spot and with each other.

Auroral emissions in the polar regions, both in the ultraviolet (which were not present during the 1973 Pioneer encounter) and in the visible.

Cloud-top lightning bolts, similar to terrestrial superbolts, and meteoritic fireballs.

A temperature inversion layer in the stratosphere and a temperature of 160 K at the level at which the atmospheric pressure is 0.01 bar.

Very strong ultraviolet emission from the disk, indicating a thermospheric temperature of more than 1000 K.

A hot (1100 K) upper ionosphere on the dayside that was not observed by Pioneer 10, suggesting there may be large temporal or spatial changes.

An atmospheric composition with volume fraction of helium of 0.11 ± 0.03.

A substantially colder atmosphere above the Great Red Spot than in the surrounding regions.

Satellites and Rings

At least eight currently active volcanoes on Io, probably the result of tidal heating of the interior of the satellite, with plumes extending up to 250 kilometers above the surface.

A large hot spot on Io near the volcano Loki that is about 150 K warmer than the surrounding surface.

Numerous intersecting, linear features on Europa, possibly due to crustal cracking.

Two distinct types of terrain, cratered and grooved, on Ganymede, suggesting that the entire ice-rich crust was once under tension.

An ancient, heavily cratered crust on Callisto, with vestigial rings of enormous impact basins since erased by flow of the ice-laden crust.

The elliptical shape of Amalthea (270×160 kilometers).

A faint ring of material about Jupiter, with an outer edge of 128 000 kilometers from the center of the planet.

Magnetosphere

An electrical current of more than a million amperes flowing in the magnetic flux tube linking Jupiter and Io.

Very strong ultraviolet emissions from ionized sulfur and oxygen in the Io plasma torus, indicating a hot (hundred thousand degree) plasma that evidently was not present at the time of Pioneer 10 encounter.

Plasma electron densities exceeding 4500 per cubic centimeter in some regions of the Io plasma torus.

A cold, corotating plasma inside 6 R_J with ions of sulfur, oxygen, and sulfur dioxide.

High-energy trapped particles inside 6 R_J with significantly enhanced abundances of oxygen, sodium, and sulfur.

Hot plasma near the magnetopause predominantly composed of protons, oxygen, and sulfur.

Jovian radio emission at kilometer wavelengths, which may be generated by plasma oscillations in the Io plasma torus.

Corotating plasma flows unexpectedly far from Jupiter in the dayside outer magnetosphere.

Evidence suggesting a transition from closed magnetic field lines to a Jovian magnetotail at about 25 R_J from Jupiter.

Whistler emission interpreted as lightning whistlers from the Jovian atmosphere.

---

*Adapted from a summary prepared by E. C. Stone and A. L. Lane for the Voyager 1 Thirty-Day Report.

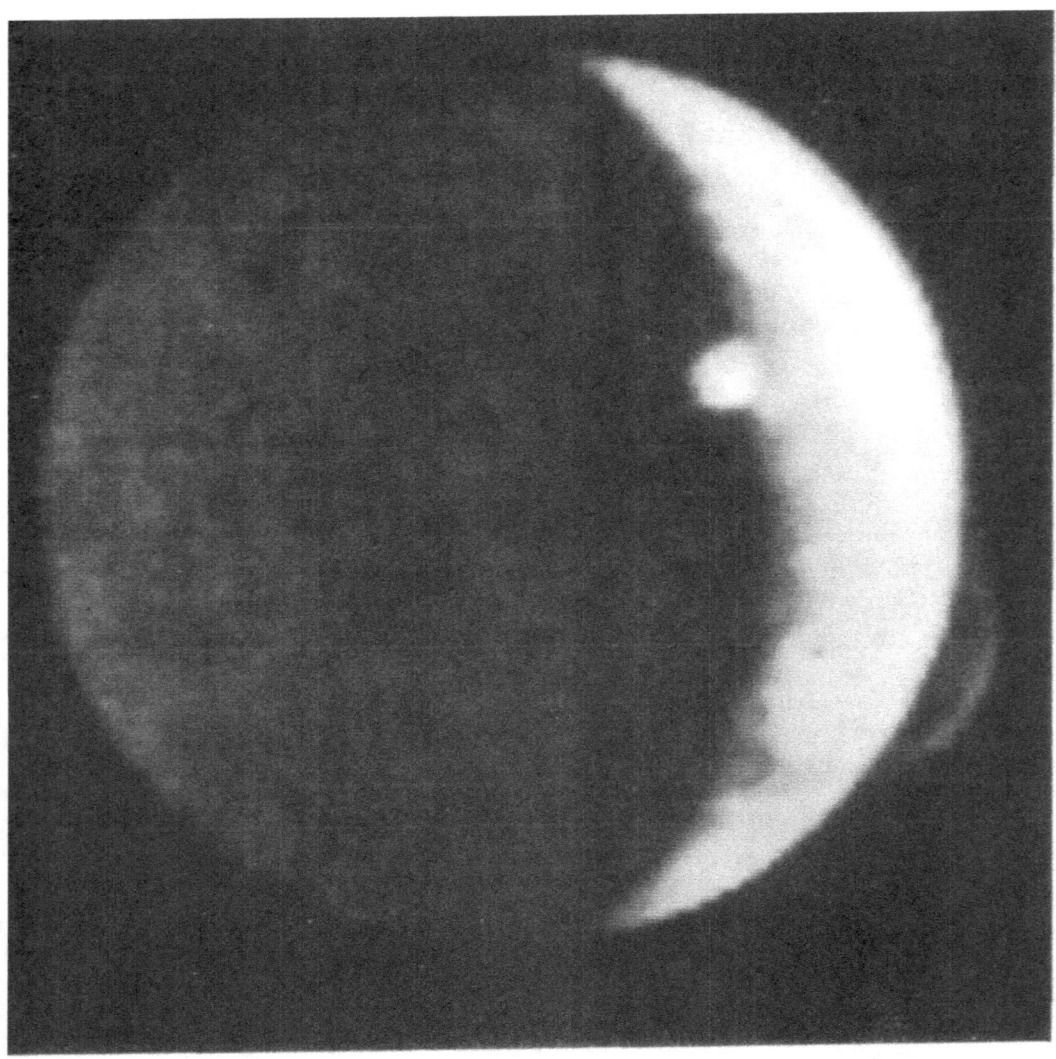

The dramatic discovery of active volcanoes on Io was made by Linda Morabito and her colleagues from this navigation picture, taken March 8 at a range from Io of 4.5 million kilometers. On the bright edge, the immense plume of volcanic ash from Pele (P_1) rises nearly 300 kilometers above the surface. At the terminator, the border between day and night on Io, a second smaller cloud from the volcano Loki (P_2) catches the sunlight. These two eruptions—captured on this single discovery photograph—are much larger than the largest terrestrial volcanic eruption known. [P-21306 B/W]

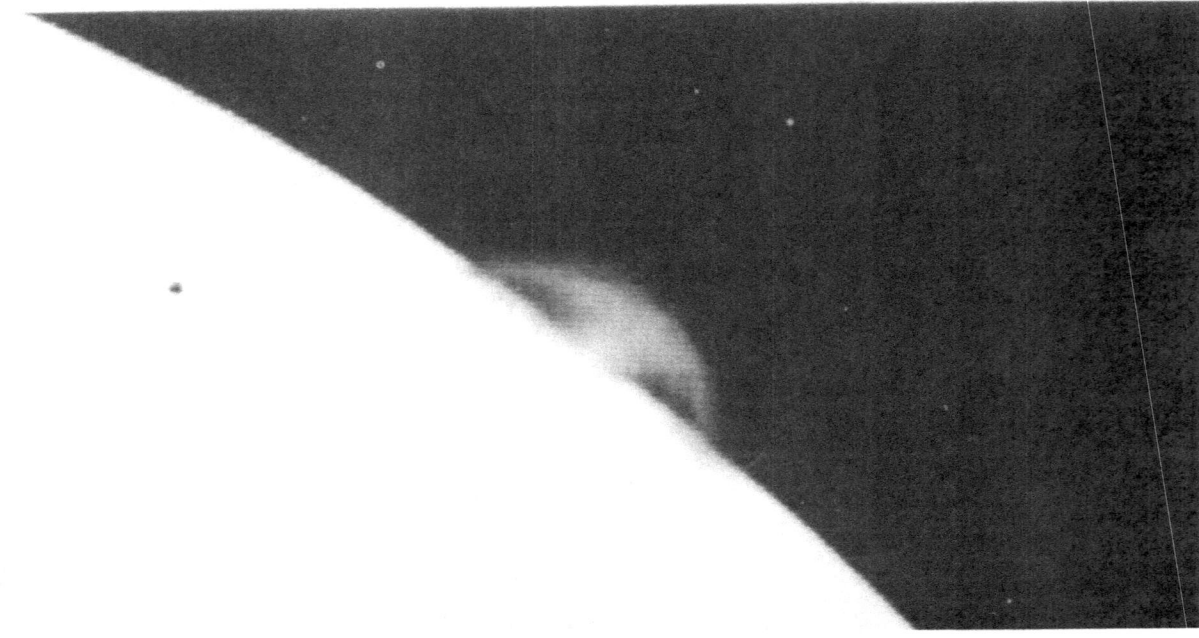

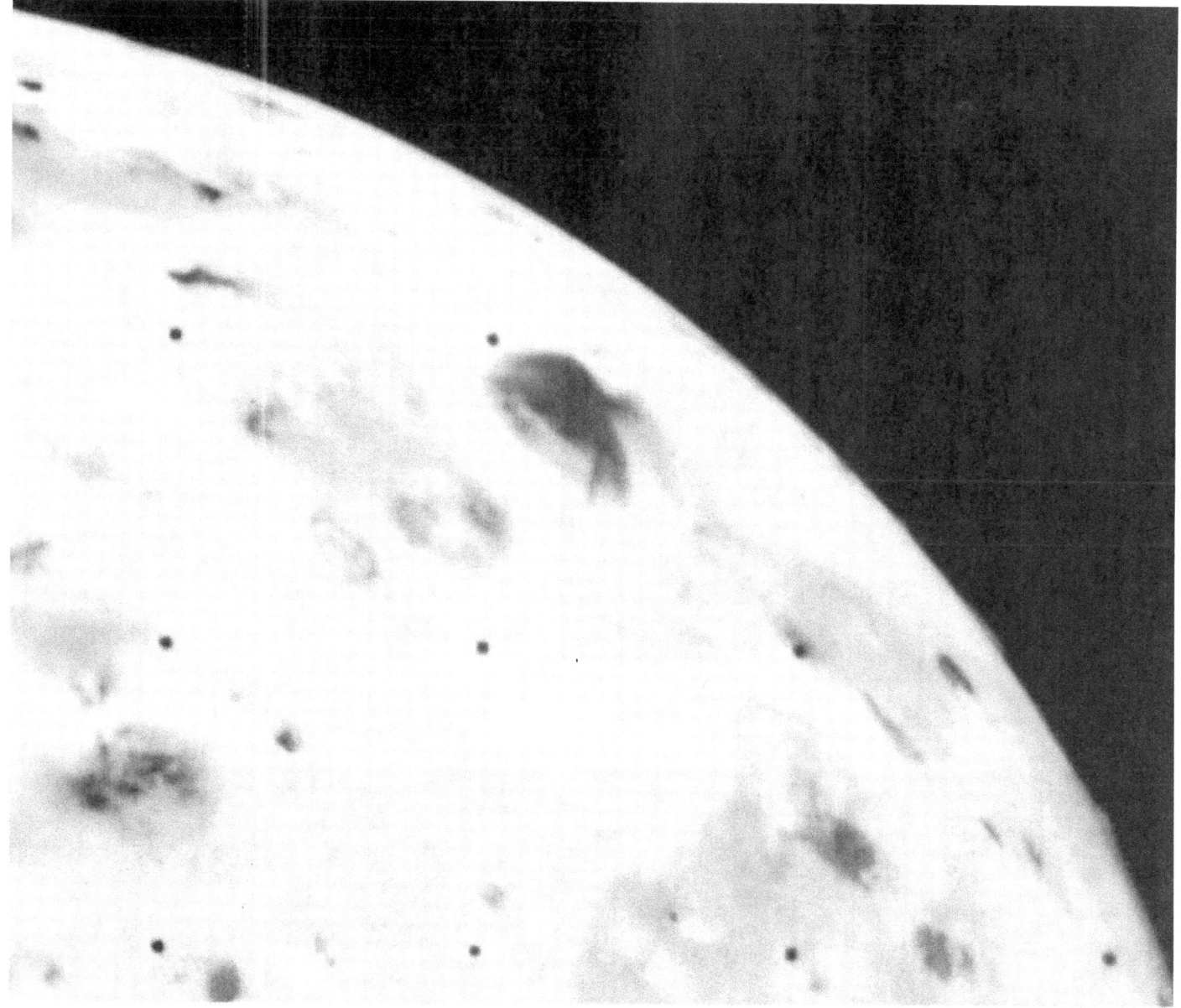

Once the existence of giant volcanic eruptions on Io was recognized, a reexamination of the Voyager 1 encounter pictures revealed many more plumes. These two views of Prometheus (P₃) were found by Joseph Veverka and Robert Strom on March 12 when they reproduced earlier pictures. (Bottom left) The plume is silhouetted against the black space, although it is also possible to see dark "feet" where the falling material reaches the surface. (Above) The complex jets of material are clearly seen as dark streaks against the light background of the surface of Io. The plume itself rose more than 100 kilometers above Io's surface. [P-21295 and P-21294]

Meanwhile, new information about Jupiter was released to the public. A long-exposure (three minutes and twelve seconds) image of the dark side of the planet, taken with the wide-angle camera while in the shadow of the planet, caught Jupiter showing off some Jovian "fire-works." A long, broad, white streak across the picture was a visible aurora, the largest aurora ever seen—almost 29 000 kilometers long. In addition, nineteen smaller bright splotches, looking insignificant by comparison, were in reality "superbolts" of lightning. Since huge electrical discharges such as lightning can, under the right circumstances, power chemical reactions that form complex organic molecules, the discovery of lightning on Jupiter could have profound implications. Was "lightning-inspired" organic synthesis going on in Jupiter's atmosphere? No one knew.

Returning to JPL on Sunday night, Brad Smith got his first look at the Morabito picture of the volcanic cloud. Early Monday morning,

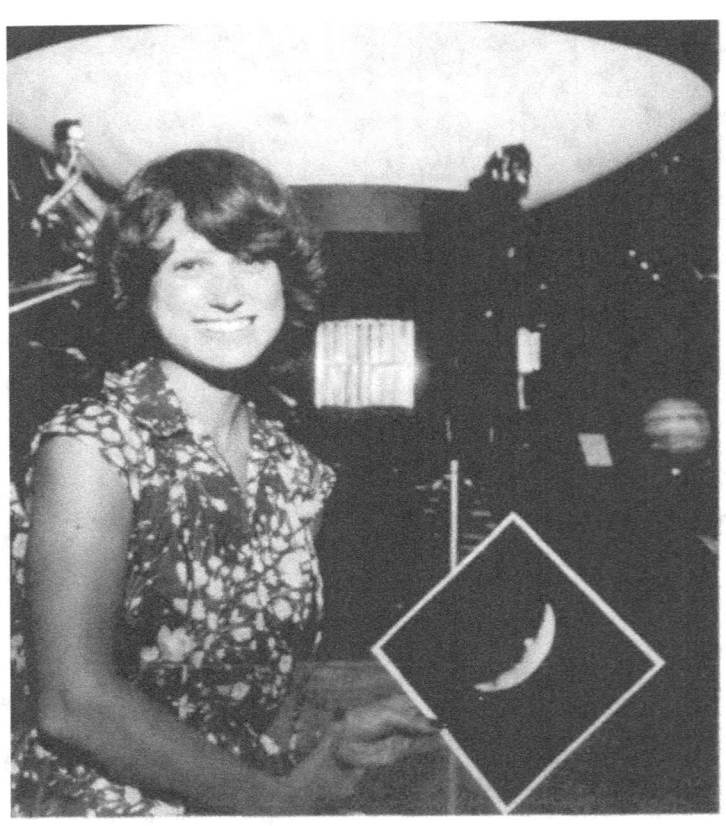

other Imaging Team scientists saw it. As soon as the JPL computers were operating, Joseph Veverka and Robert Strom began working with the two interactive TV terminals to look for evidence in other pictures of ongoing eruptions. Faint clouds or plumes would not show up in normally processed pictures, but could be brought out easily with the computer-controlled displays. By midmorning, several additional volcanic plumes had been found.

Meanwhile, on March 11, John Pearl of the IRIS team had independently drawn the conclusion that volcanic activity must be taking place on Io. He and Rudy Hanel found evidence of strongly enhanced thermal emission from parts of the satellite. The most prominent was a source nearly 200° C hotter than its surroundings. On March 12, Pearl brought his new results to the Imaging Team, and sure enough, the hot spot was located near one of the volcanic plumes! A month later, continuing

Linda Morabito shows the discovery photo of the volcanic eruptions on Io. [P-21718]

The dark side of Jupiter revealed many surprising phenomena to Voyager 1. A wide-angle view, taken on March 5, led to the discovery of a double auroral arc at north-polar latitudes and numerous flashes of lightning illuminating the clouds during this 3-minute, 12-second exposure, taken at a range of half a million kilometers. [260–460]

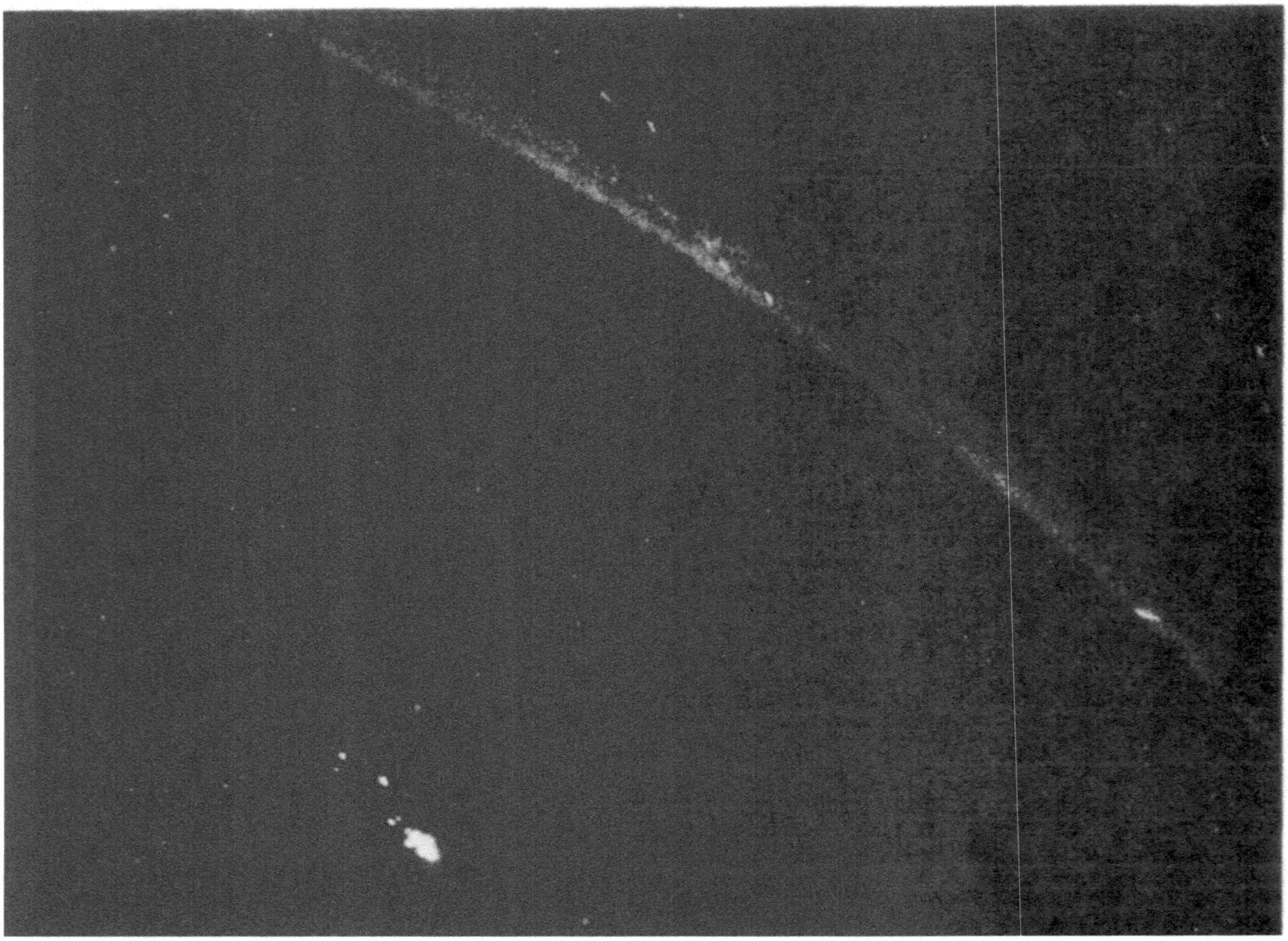

analysis of IRIS spectra yielded identification of sulfur dioxide (SO_2) gas over this same erupting volcano. At last a source had been located for the enigmatic sulfur and oxygen ions in the magnetosphere.

The volcanoes provided a thread with which to weave together the disparate data on Io. A few months earlier there had been a report of a sudden brightening of Io in the infrared; now it seemed plausible that thermal emission from an eruption was the source. The Voyager ultraviolet experimenters had been worrying over the source of the intense sulfur emissions they had seen and had been disturbed by the changes in the gas clouds around Io since the Pioneer 10 and 11 flybys; now a variable source for these gas clouds was identified. In addition, the craterfree surface and bizarre features seen in the Voyager images could be recognized as the product of violent explosive eruptions on Io. It appeared that Peale, Cassen, and Reynolds had found, in their theoretical calculations, the key to the most geologically active body ever encountered in the solar system.

News of the discovery was released to the press on Monday, March 12. During the next few days, a total of eight gigantic eruptions were located in the Voyager pictures of Io. Within a few weeks, scientists all over the world were thinking with renewed energy about this incredible satellite.

With four new planet-sized satellites now photographed, there was a sudden requirement for maps and for names to be assigned to the newly discovered features. The maps were produced from Voyager images at the Astrogeology Branch of the U.S. Geological Survey at Flagstaff, Arizona. The names, proposed by a group of scientists headed by Voyager Imaging Team members Tobias Owen and Hal Masursky, were given official approval by the International Astronomical Union in August. For a time, a dual nomenclature persisted for the erupting volcanoes on Io. The eruption plumes were given numbers, P_1, P_2, etc., while the volcanic features were given names taken from the mythology of fire and volcano legends. Thus the "hoof print" of Io was called Pele, for the Hawaiian volcano goddess, and the 280-kilometer-high plume associated with it was called P_1. By the time of the Voyager 2 encounter, scientists had prepared maps on which to plot their new discoveries.

While the Voyager scientists fanned out across the world to share their findings with colleagues, attention at JPL turned to Voyager 2. In response to the discoveries of the first encounter, changes were required in the sequencing of scientific observations for July. Voyager 2, still troubled by a faulty receiver, might require more coddling from the spacecraft team than had its sister spacecraft, now safely on the way to Saturn.

CHAPTER 7

THE SECOND ENCOUNTER: MORE SURPRISES FROM THE "LAND" OF THE GIANT

Approaching Jupiter

At the beginning of July, the dry summer heat had returned to Pasadena, and so had the press. The scientists had come days or weeks earlier to look at data being transmitted from the second Voyager as it approached Jupiter and its satellites. The mood at JPL seemed quieter than it had been in March for Voyager 1, although the press room would once again be deluged with observers on the day of encounter. This would be our second good, close look at the Jovian system, but it was to be no summer rerun. Voyager 2 would have a different view of each world, and, in addition, both Io and Jupiter had undergone changes, as though to ensure that no one would become bored and fall asleep in front of a TV monitor. In a sense, this encounter was to be another first look at Jupiter and its satellites, with a view of each object that was quite different from what had been seen before.

Changes in Jupiter's cloud formations became noticeable long before July. After a gap of six weeks following the first flyby, Voyager 2's observatory phase began on April 24, 1979, seventy-six days before its July 9 encounter with Jupiter. During this time, the spacecraft's ultraviolet and fields and particles instruments studied the Jovian system and its interaction with the solar wind. Between April 24 and May 27, Voyager 2's imaging system concentrated on the motions in Jupiter's atmosphere, creating another approach time-lapse "movie." From May 27 to 29 photographs were taken in a more rapid sequence, showing the planet during five 10-hour rotations. From these studies it

was apparent that Garry Hunt's prediction had been right—the weather *had* changed by July. A month before the encounter JPL's Voyager Bulletin—Mission Status Report announced that "Jupiter is sporting quite a different face than it did just four months ago. The bright 'tongue' extending upward from the Red Spot is interacting with a thin, bright cloud above it that has traveled twice around Jupiter in four months." The turbulent region west of the Great Red Spot had begun to break up and separate from the Red Spot. The white ovals south of the Red Spot had drifted to the east (about 0.35 degrees a day), while the Red Spot itself had drifted west (about 0.26 degrees a day). The white zone seen just south of the Red Spot by Voyager 1 had become very narrow—like a thin white line just barely outlining the bottom of the spot. The Red Spot had also changed: It had become a more uniform orange-red, perhaps reverting to the color seen by Pioneers 10 and 11. The brown spots that had been seen in the north temperate region at the same longitude as the Red Spot were now on the other side of the planet. A dark brown spot not present during the Voyager 1 flyby had developed along the northern edge of the brown equatorial region on the Red Spot side of the planet. Some of the white markings that seemed to have protruded into the equatorial region at the time of the first flyby were missing in the Voyager 2 photographs.

As Voyager 2 entered the far encounter period on May 29, all instruments on the spacecraft (except for the photopolarimeter) seemed to be in good shape for encounter. As was the case with Voyager 1, the polarization wheel on

The particles of the rings of Jupiter are stronger reflectors of red light than of blue, as can be seen in this view, assembled from two images taken in orange and violet light. Since the images were registered on the rings, not the planet, the bright bands of colors along the edge of Jupiter are an artifact of misalignment. [P-21779]

In early June, as Voyager 2 carried out its observatory phase, additional changes in Jupiter's face began to be apparent. These two images, taken from a distance of 24 million kilometers, have a resolution of about 500 kilometers. (Top) The Great Red Spot and the white oval south of it are seen to be followed on the west by regions of chaotic and turbulent clouds. This is not the same white oval that was near the Red Spot in March; the differential rotation of the planet carried a different oval close to the Red Spot during the intervening three months. (Right) Io is visible to the right of the planet, and the shadow of Ganymede falls on the colored clouds of Jupiter's equatorial belt. [P-21713C and P-21714C]

94

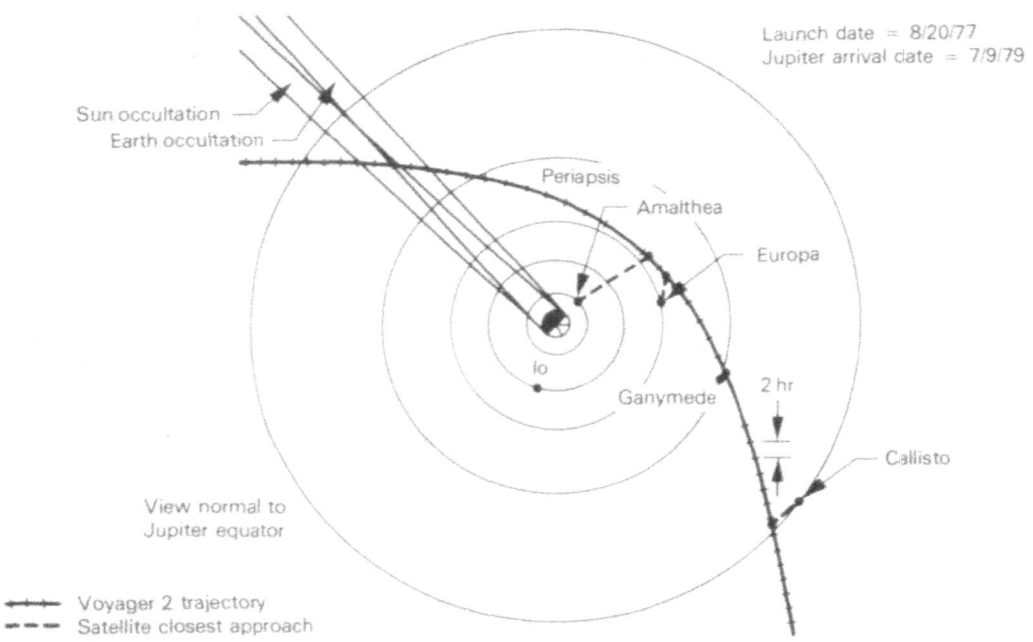

Launch date = 8/20/77
Jupiter arrival date = 7/9/79

Sun occultation

Earth occultation

Periapsis

Amalthea

Europa

Io

Ganymede

2 hr

Callisto

View normal to
Jupiter equator

Voyager 2 trajectory
Satellite closest approach

The Voyager 2 trajectory was complementary to that of Voyager 1. This time, the satellites were encountered before Jupiter, revealing their other hemispheres. As shown in this drawing, the spacecraft flew by first Callisto, then Ganymede, then Europa. The ten-hour Io volcano watch took place immediately after closest approach to Jupiter. [260-533A]

These two faces of Jupiter were photographed by Voyager 2 on May 9 at a distance of 46 million kilometers from the planet. Voyager scientists began to detect significant changes in the cloud patterns since the Voyager 1 encounter two months earlier. [260-507]

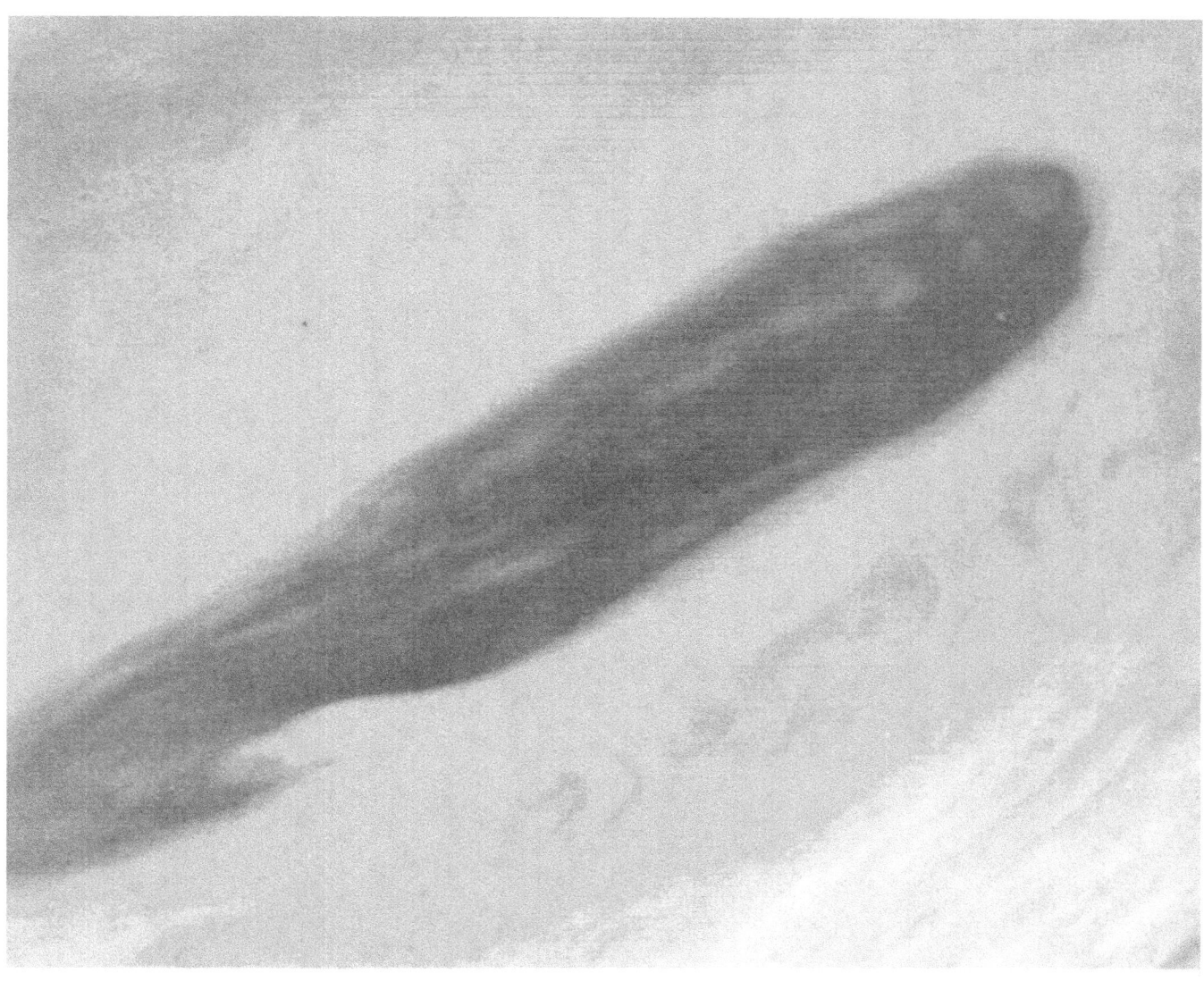

The weather is changing over one of the northern hemisphere brown ovals in this picture taken July 6. The brown ovals are regions in which breaks in the upper layer of ammonia clouds reveal darker clouds below. A high, white cloud is seen moving over the darker cloud, providing an indication of the structure of the cloud layers. Thin white clouds are also seen within the dark cloud. At right, blue areas, free of high clouds, are seen. [P-21753C]

Voyager 2's photopolarimeter was stuck, so the instrument was able to obtain only color photometry measurements.

Although Voyager 2's radio receiver still could not track a Doppler-shifted radio signal from Earth (the problem is that it "hears a monotone," explained Deputy Project Manager Esker K. Davis), the Deep Space Network engineers had learned to work with the spacecraft, determining what frequency the spacecraft would listen to at any particular time. They had discovered that some of the "housekeeping" telemetry signals from the receiver were sensitive to the match between the incoming frequency and the receiver frequency. By monitoring these signals, they could detect a frequency drift in time to correct the transmission, thus keeping the system in tune in spite of slow changes in the receiver. The system was slow and demanding but effective; all the necessary command sequences were successfully loaded into the computer, and communications during the encounter were entirely successful.

The timing offset experienced by Voyager 1 as a result of Jupiter's intense radiation environment was not expected to be a problem on Voyager 2 for two reasons: Even at closest approach, Voyager 2 would still be more than twice as far from Jupiter as Voyager 1 had been, and the Voyager 2 computer was programmed to resynchronize the spacecraft's timing systems automatically every hour. In this

Complex activity in the southern hemisphere of Jupiter continued during the Voyager 2 encounter, although changes had occurred in the region of the Great Red Spot. A white oval, different from the one observed in a similar position at the time of the Voyager 1 encounter, was situated south of the Red Spot. The region of white clouds extended from east of the Red Spot and around its northern boundary, preventing small cloud vortices from circling the feature. The disturbed region west of the Red Spot had also changed since the equivalent Voyager 1 image. The picture was taken on July 3 from a distance of 6 million kilometers. [P-21742C]

way, even if the radiation environment proved to be much higher than anticipated, the image smear that might occur from a timing offset would be prevented.

As a result of the discoveries made by Voyager 1, the project scientists decided to modify some of Voyager 2's preplanned sequences. As early as April 1, the painstaking job of constructing new computer commands began. A ten-hour Io Volcano Watch was added to the spacecraft's program, taking advantage of the fact that shortly after closest approach to Jupiter, the spacecraft would re-

main within about 1 million kilometers of Io for a long period, keeping nearly the same face in view. Provisions were also made to take extensive ultraviolet measurements of the emission from the glowing torus surrounding Jupiter near the orbit of Io. Further studies would be made of the dark side of Jupiter to search for lightning and auroral activity, and there was also the hope that the plasma wave instrument would be able to detect lightning whistlers (radio signals created by lightning bolts) as the Voyager 1 instrument had done. A high priority was given to observations of the

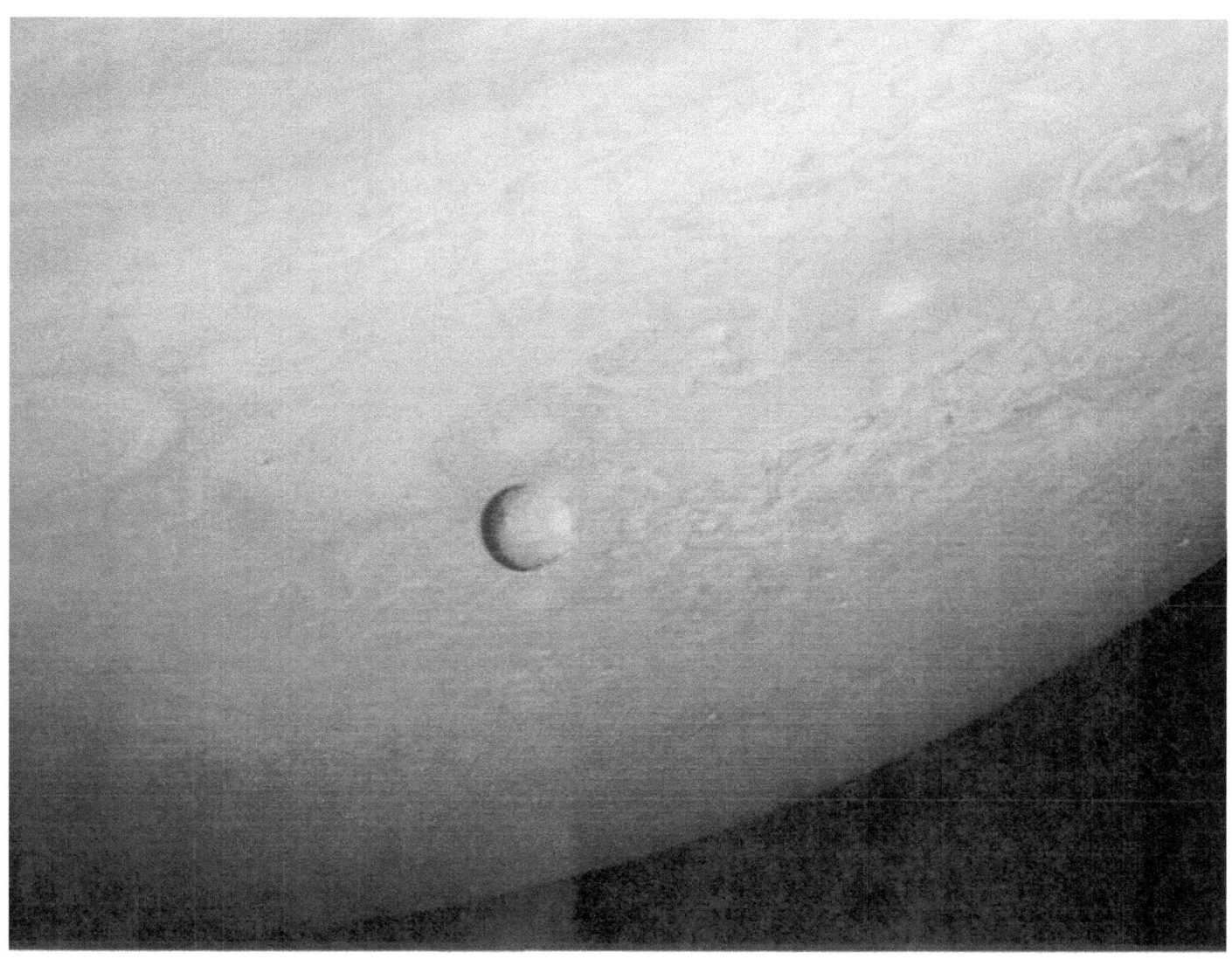

Io appeared in front of Jupiter as seen by Voyager on June 25, at a range of 12 million kilometers. At a resolution of about 200 kilometers, the bright and dark spots on the satellite are just beginning to be resolved, but it was not possible to determine if any eruptions were still in progress. [P-21719C]

newly discovered ring, which had not been in the original Voyager 2 sequence at all. The spacecraft would cross the ring plane twice, photographing the ring during both the inbound and the outbound passages.

As was originally planned, Voyager 2 would make its closest approaches to Callisto, Ganymede, Europa, and Amalthea before encounter with Jupiter. Because of the difference in the trajectories of the two spacecraft, Voyager 2 would see the faces of Callisto and Ganymede not seen by Voyager 1. The most important difference, however, was that the second Voyager would fly much closer to Europa than Voyager 1 did, giving scientists their first good look at the mysterious streaks scratched on the surface of that bright golden world. Voyager 2 would also have a closer flyby of Ganymede, giving us a second chance to examine its strange "snowmobile tracks." The major loss, of course, was Io, which would be seen from Voyager 2 only at distances of a million kilometers or more.

The Encounter

Wednesday, July 4 *(Range to Jupiter, 5.3 million kilometers; range to Earth, 921 million kilometers).* While most of the nation celebrated Independence Day with picnics, sports events, and fireworks, the scientists and engineers at JPL were working around the clock.

99

ENCOUNTER DISTANCES
FOR VOYAGER 2

Object	Range to Center at Closest Approach (kilometers)	Best Image Resolution (km per line pair)
Jupiter	722 000	15
Amalthea	558 000	10
Io	1 130 000	20
Europa	206 000	4
Ganymede	62 000	1
Callisto	215 000	4

Voyager 2 had already entered Jupiter's territory, crossing the bow shock for the first time on July 2 at a distance of 99 R_J from Jupiter, indicating that the magnetosphere had expanded in the interval between the two encounters. At about noon on July 3, the spacecraft encountered the magnetopause, but on July 4, the data from the particles and fields instruments were ambiguous. Apparently the magnetosphere was pulsating in response to changing pressures, and the spacecraft was playing tag with the rapidly shifting boundaries of the bow shock and the magnetopause.

As the low energy charged particle instrument began to measure particles coming from inside the Jovian magnetosphere, it became apparent that some important changes had taken place since Voyager 1's encounter. From the composition of the particles, it appeared that they were largely of solar origin, unlike the heavy concentrations of ions of sulfur and oxygen seen by Voyager 1. Scientists began to speculate that the Io volcanoes, which presumably eject sulfur and oxygen into the magnetosphere, might have declined in activity. In the evening, the first images of Io at a resolution high enough to allow the volcanic plumes to be seen would be beamed back to Earth.

Thursday, July 5 *(Range to Jupiter, 4.4 million kilometers)*. The press room at Von Karman Auditorium opened and the members of the press, most of them veterans of the first encounter, arrived at JPL. Meanwhile, the spacecraft continued to measure fluctuations in the magnetospheric boundary. By noon, JPL had reported at least eleven crossings of the bow shock as the solar wind flirted with Jupiter's magnetosphere. Apparently the solar wind was much more variable in July than it had been in March. At times the bow shock seemed to be

Voyager scientists anxiously awaited the first views of Io that would show whether the volcanic eruptions seen in March were still active. This picture was taken on July 4, at a range of 4.7 million kilometers, about the same as that of the volcano discovery picture on March 8. One large plume is clearly visible, rising nearly 200 kilometers above the surface. At the time of release of this picture on July 6, the scientists wrote, "The volcano apparently has been erupting since it was observed by Voyager 1 in March. This suggests that the volcanoes on Io probably are in continuous eruption." [P-21738B/W]

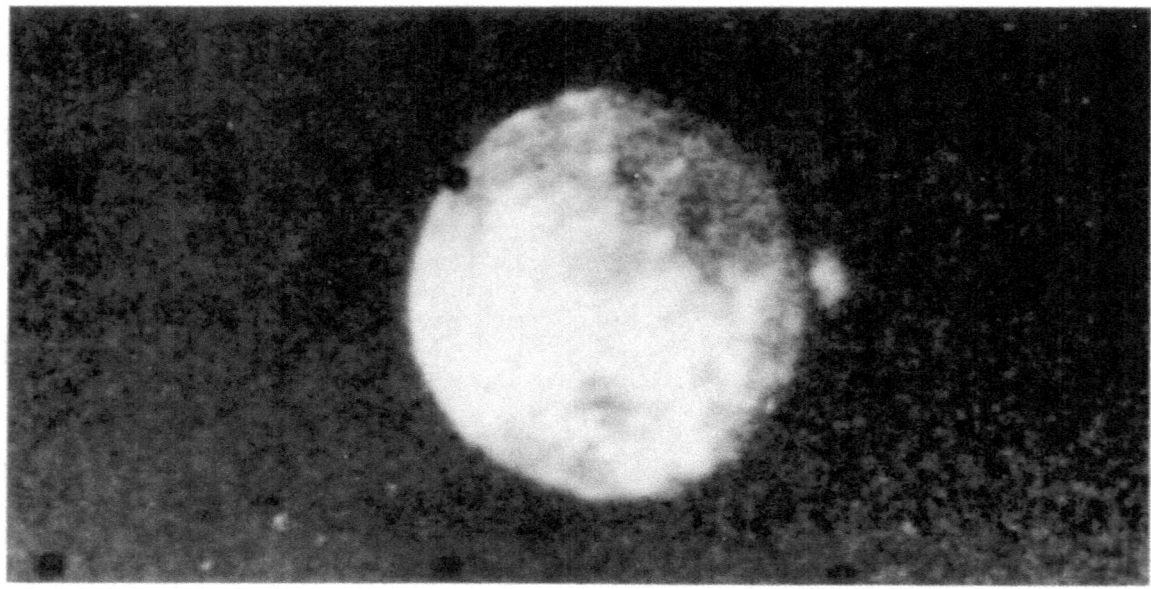

thicker than that experienced by Voyager 1; one Voyager 2 crossing took ten minutes, whereas the longest Voyager 1 crossing was only one minute long. Even though the processes affecting the magnetosphere seemed more complex, the magnetosphere was less compressed; when Voyager 2 actually entered the magnetosphere at a distance of 62 R_J, it was much farther from Jupiter than Voyager 1 had been at its final crossing (47 R_J).

Photos obtained the day before from over 4 million kilometers showed that at least one of Io's volcanoes was still active. A total of eight ongoing eruptions had been seen by Voyager 1, and scientists were anxious to see how many of these were still erupting four months later.

While attention at JPL focused on the unfolding drama of the Jupiter encounter, many members of the world's press seemed more interested in the fate of Skylab, which was nearing its death plunge into the Earth's atmosphere. Launched in 1973, Skylab had been one of NASA's more successful projects. Three crews of astronauts had visited it, carrying out intensive studies of the Sun and breaking one record after another for the duration of manned space flight. Since the departure of the final group of three astronauts in 1974, Skylab

VOYAGER 2 BOW SHOCK (S) AND MAGNETOPAUSE (M) CROSSINGS

Boundary	Day	Distance (R_J)
Inbound		
S	7/02	99 (multiple)
S	7/02	97
S	7/03	87
M	7/04	72 (multiple)
M	7/05	71
S	7/05	69
S	7/05	67
M	7/05	62
Outbound		
M	7/23	169
M	7/23	173
M	7/24	174
M	7/24	175
M	7/24	176
M	7/24	177
M	7/25	184
M	7/25	185
M	7/27	213
M	7/31	253
M	8/01	258
M	8/01	262 (multiple)
M	8/03	279 (multiple)
S	8/03	283 (multiple)

Although Voyager 2 did not come as close to Io as had Voyager 1, some changes in the surface during the four months between encounters were so large that they could still be easily seen. These two pictures of the region of the volcano Pele were taken in early March and early July, respectively. The most dramatic change was the filling in of the indentation in the ejecta ring, turning the hoofprint into a symmetric oval. The oval is about 1000 by 700 kilometers in outermost dimension, and the area that changed amounts to more than 10 000 square kilometers. [260-687AC]

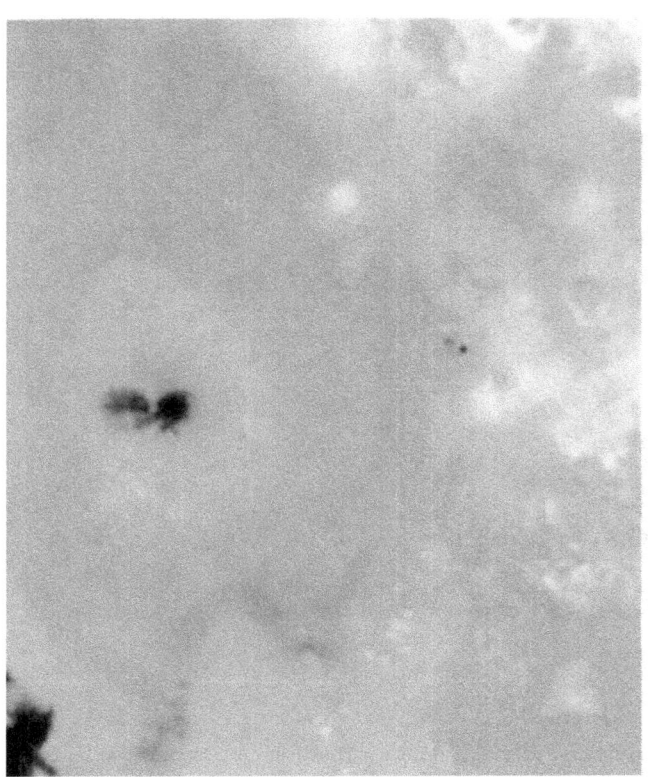
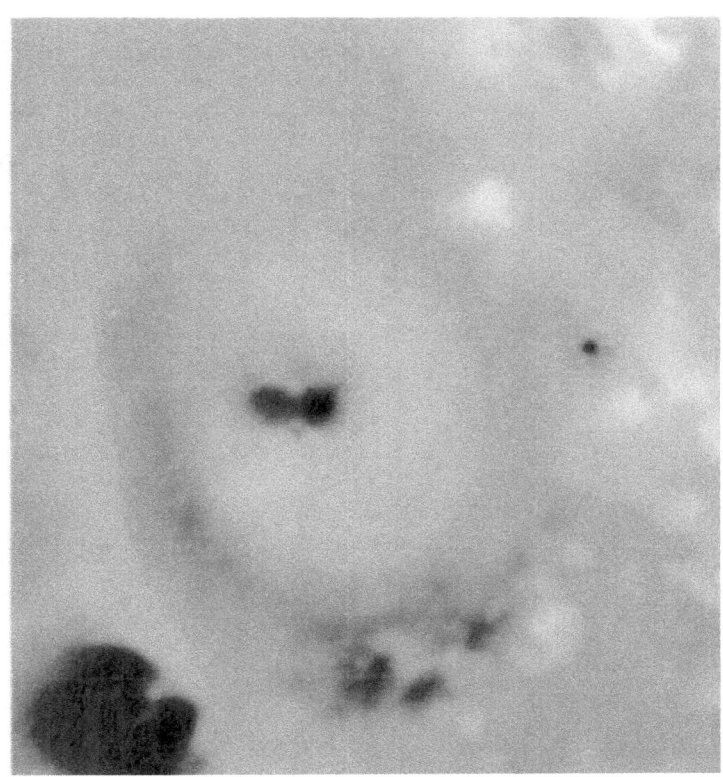

had been sinking gradually lower as a result of friction with the extreme upper atmosphere of Earth. During the past year, higher temperatures in the atmosphere had increased this drag, and now the end was near. With a strange fascination, the world watched the end of this old spacecraft, almost seeming to forget the spectacular new results being transmitted from Jupiter. To the frustration of the Voyager team and the press "camped out" in Von Karman Auditorium for the second encounter, the exaggerated stories of a possible Skylab disaster took precedence over Voyager news. Ultimately, Skylab fell over the Indian Ocean and Australia on Wednesday, July 11, just as the major findings of Voyager 2 were being released.

Friday, July 6 *(Range to Jupiter, 3.5 million kilometers).* With the first satellite encounter still two days away, Voyager 2 continued to make a variety of measurements of Jupiter and all the Galilean satellites. As the distance to Io decreased, it was possible to see detailed surface features as well as to look for the volcanic plumes at the edge of the disk, silhouetted against black space. By the end of the day, the Great Red Spot loomed so large that six imaging frames (a 2 × 3 mosaic) were required to encompass it and its immediate surroundings.

At the first formal press conference of the Voyager 2 encounter, Project Scientist Ed Stone reviewed the progress of the mission. Because the ailing spacecraft receiver was working so well, Ray Heacock, Voyager Project Manager, announced that the major trajectory correction maneuver at Jupiter had been rescheduled to take place only two hours after closest approach. Since the geometry was especially favorable at this time, the 76-minute rocket burn could put the spacecraft on its planned route to Saturn with a minimum expenditure of fuel, thereby preserving the option of sending the spacecraft on to Uranus.

The Io torus was under observation, both directly by the ultraviolet spectrometer, and indirectly by the charged particle instruments. The LECP instrument had begun to pick up sulfur ions, but at lower energies and lower concentrations than those recorded during the first encounter. In the ultraviolet, glows could be seen both from the torus and from aurorae in the polar regions of Jupiter.

Photographs of Io showed that the heart-shaped feature surrounding Pele (P_1), Io's largest volcano, had changed shape. The inden-

tation of the heart had disappeared, making the heart into an oval. Apparently a new deposit of volcanic ejecta had blanketed the surface, altering its color. Perhaps an earlier obstruction in the volcanic vent, or the shape of the vent itself, had caused the area surrounding Pele to look heart-shaped. In any case, whatever had caused the indentation was now gone. At the same time, new photos failed to show a plume above Pele, and there was speculation that changes in this eruption might be related to the altered population of charged particles in the magnetosphere.

Saturday, July 7 *(Range to Jupiter, 2.6 million kilometers).* As the spacecraft rapidly closed on Callisto, better and better photographs were taken of the previously unseen hemisphere. As with the Voyager 1 observations, however, the main impression was one of heavy cratering, unrelieved by other geologic structures. Meanwhile, the coverage of Io had improved as the satellite rotated to the point at which a census

The Voyager 2 pictures of Callisto looked remarkably similar to those obtained of the other side of the satellite by Voyager 1. Seen from a distance of 2.3 million kilometers, the large craters (100 kilometers or more across) appear as light spots. No new major impact features such as Valhalla, discovered by Voyager 1, are visible on the hemisphere seen by Voyager 2. [P-21740C]

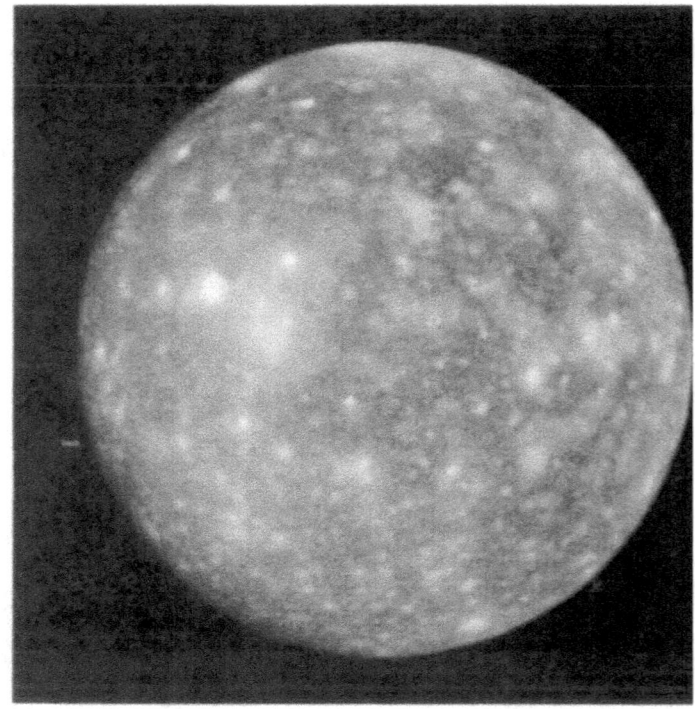

of the volcanic eruptions seen in the first encounter began to emerge.

At the 11:00 a.m. press conference, Larry Soderblom announced that four of the volcanoes discovered by Voyager 1 had been looked at again by Voyager 2, and three of them—Prometheus (P_2), Loki (P_3), and Marduk (P_7)—were still active. However, there was no trace of volcanic activity coming from Pele, the source of the largest plume seen by Voyager 1. P_1 was either greatly subdued or had turned off completely.

Dr. Soderblom also announced that Voyager 2 images had detected another giant ring structure on Callisto, bringing the total to three, and there were probably more. This particular ring feature was estimated to be about 1500 kilometers across. It was also noted that although Callisto generally seemed to be saturated with shoulder-to-shoulder craters, the crater density near the ring structures seemed to be lower.

Saturday was a fairly quiet time for the scientists but not for the spacecraft or the spacecraft team. "We blocked out about 7½ hours,"

explained Michael Devirian, Ground Data Systems Development, Integration, and Test Director, "in which we could send it a set of commands and re-send it if necessary to make sure all close-encounter commands were received by Voyager 2 until all the commands got through. The whole thing went perfectly the first time." So everything was "go" for close encounter. The near encounter phase began at 6:36 p.m. PDT.

Sunday, July 8 *(Range to Jupiter, 1.5 million kilometers)*. At 2:30 a.m. the first long-exposure sequence of ring pictures was taken, and at 3:00 a.m. the intensive period of the Callisto encounter began. Eighty high-resolution images were obtained of the satellite, centered around closest approach (215 000 kilometers) at 6:13 a.m. Incoming photos showed some features that looked like double-walled craters, but no more giant ring structures were seen. It appeared that there was an asymmetry in the distribution of large impact features over Callisto's surface. "Callisto may turn out to be the

A new face of Ganymede was revealed to Voyager 2. This image was taken July 7 from a distance of 1.2 million kilometers and clearly shows the large dark area Regio Galileo, as well as much of the lighter grooved terrain discovered by Voyager 1. The bright spots are impact craters. This image also shows what appear to be polar caps, extending down to about latitude 45° in both the northern and southern hemispheres. [260-670]

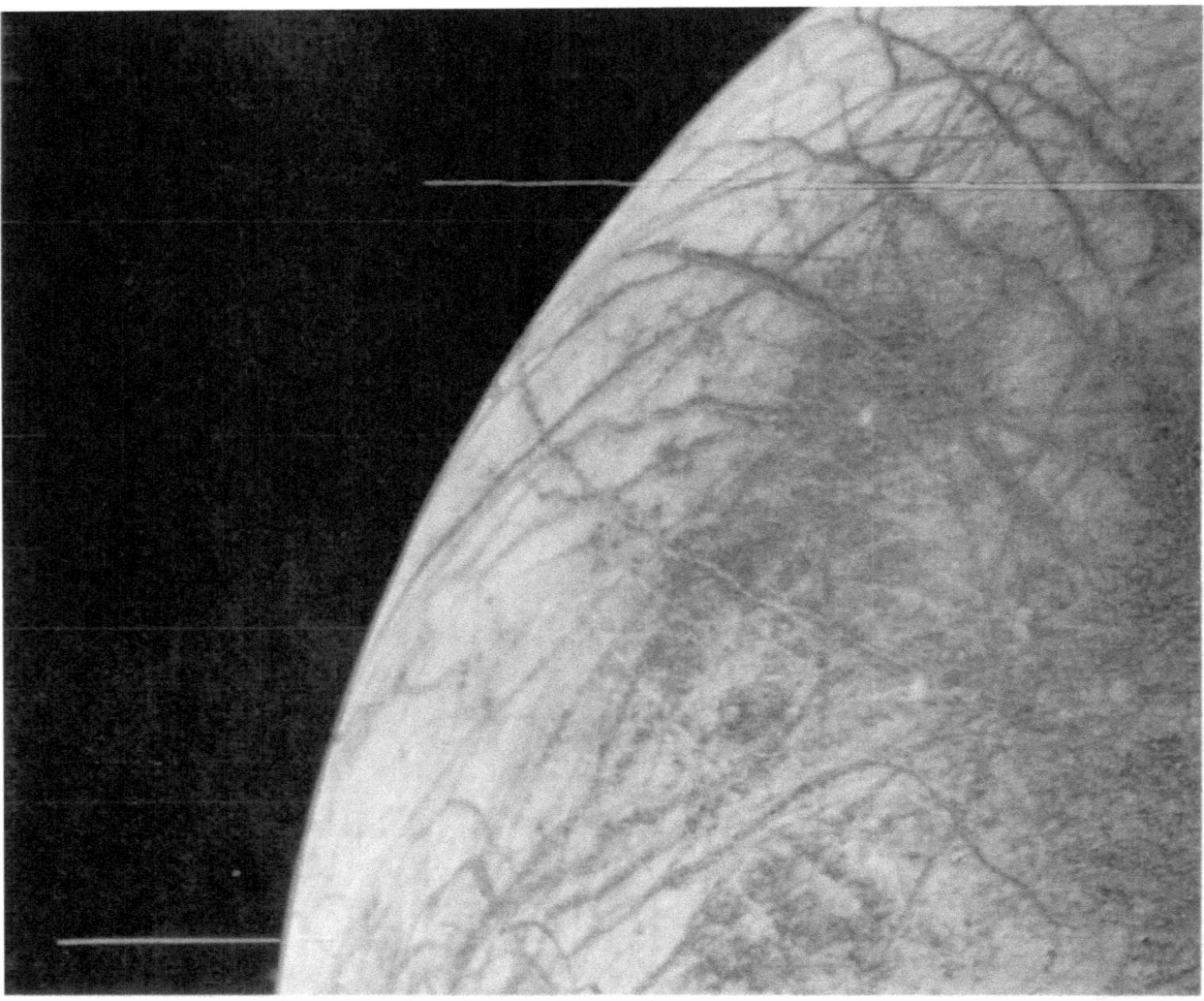

most heavily cratered body in the solar system," Torrence Johnson remarked. Garry Hunt was to add later on, "There's just not room for another crater on that body—it's totally full."

At the press conference, Brad Smith confirmed the earlier finding that Io's volcano Pele was quite dead—at least for now. Although P_4 had not yet been looked at, all other volcanoes discovered by Voyager 1 were still active, but no new plumes had been found. However, new ultraviolet images of P_2 (Loki) suggested that the eruption had increased in size. (In a later report, the imaging team announced that P_2 had increased in height to 175 kilometers and had changed to a two-column plume.) The new photographs of Jupiter's ring showed it to be quite narrow and ribbonlike, Dr. Smith announced. The artist's drawing (released during the Voyager 1 encounter), intended to show the outer edge of the ring, turned out to be a good representation of the actual ring, Dr. Smith said.

There seemed to be less high-speed sulfur and oxygen inside Jupiter's magnetosphere than there had been during the Voyager 1 encounter, George Gloeckler announced. Voyager 2's low energy charged particle instrument was finding substantial amounts of carbon, silicon, magnesium, and other elements of solar origin, but the Io-associated elements were almost depleted. The ultraviolet instrument had found as much glowing sulfur in Io's torus as before, but less of it seemed to be raised to energies high enough to leave the torus and be detected elsewhere in the magnetosphere.

There were other indications of Jupiter's changing weather. In a Voyager report Sunday evening, Garry Hunt remarked, "One very exciting observation came the other day which caused major excitement down in the imaging area. We actually saw a white cloud starting to intrude across a dark barge [large brownish oval-shaped feature in Jupiter's northern

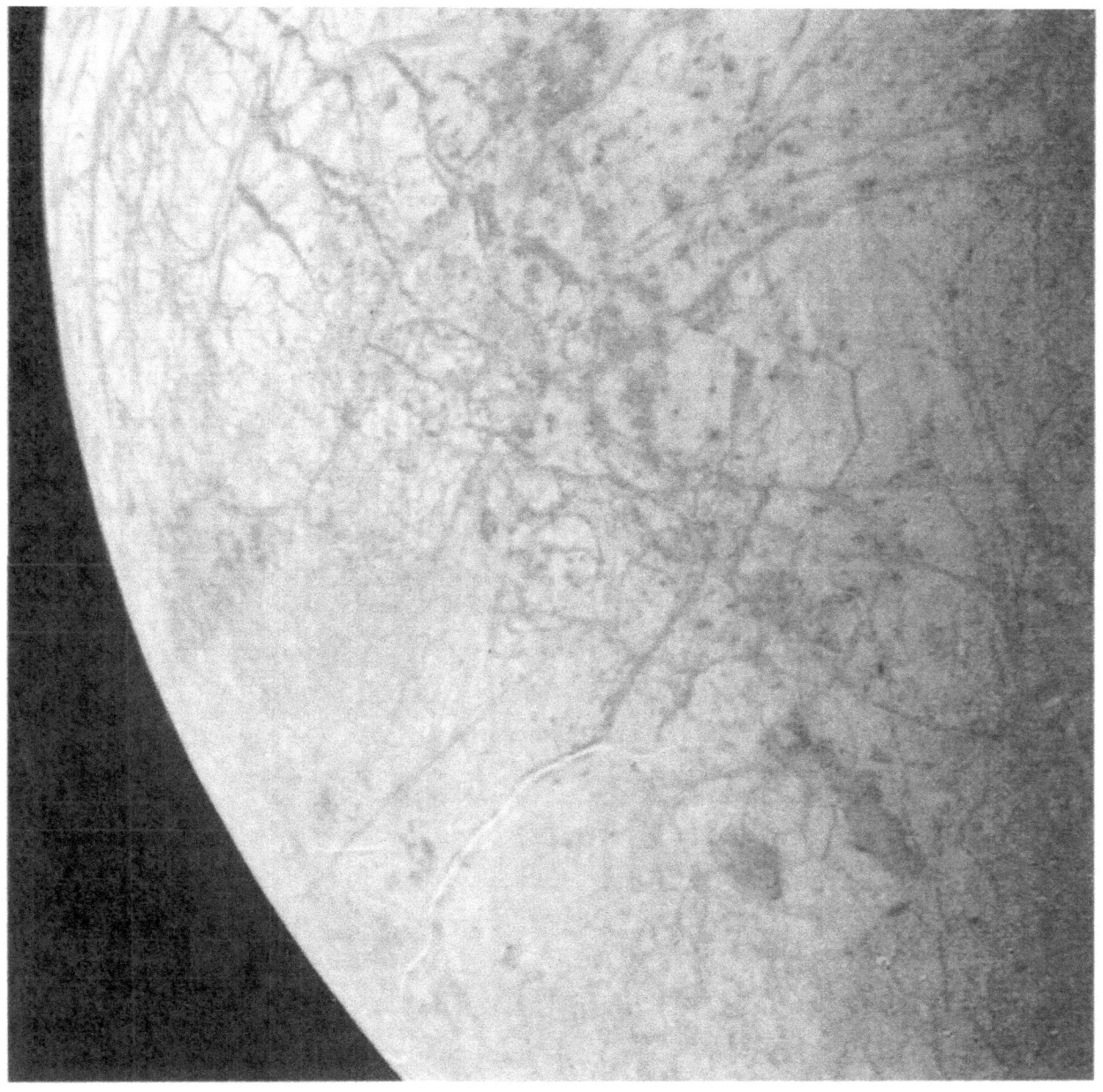

The first close-up views of Europa were both exciting and perplexing to Voyager scientists. The best Voyager 1 resolution had been only about 30 kilometers, but the Voyager 2 trajectory permitted a much closer flyby. These pictures, taken on July 9 at a range of 240 000 kilometers, have a resolution of about 5 kilometers. The bright icy crust of Europa is covered with a spectacular series of dark streaks, giving the satellite a cracked appearance. In a few cases, narrower light lines run down the centers of the dark streaks, which are typically a few tens of kilometers in width. Very few, if any, impact craters are visible on Europa. [P-21760C and P-21764C]

hemisphere]. Atmospheric scientists get very excited by that because this is showing us how the colors layer themselves up—that white cloud is clearly above the dark brown. We're desperately trying to understand the relationship of colors on Jupiter.''

Monday, July 9 *(Range to Jupiter at encounter, 722 000 kilometers).* Encounter day! And not just one encounter, but a whole sequence: Ganymede, Europa, Amalthea, Jupiter, and Io. By early Sunday evening, a wealth of new data on Ganymede was pouring in. Not only was this a side of the satellite not seen before, but Voyager 2 would pass closer to Ganymede than had Voyager 1. Encounter took place at 1:06 a.m., at a range of 62 000 kilometers. Between 9:00 p.m. Sunday night and 1:30 a.m. Monday morning a total of 217 photos, plus infrared and ultraviolet spectra, were scheduled.

Sixty-nine photos were sent back in real time; others were recorded for playback later.

As the Ganymede pictures appeared on the TV screens, they revealed a world of tremendous variety. Some regions were heavily cratered: "Ganymede looks like Mercury or the highlands of the moon," one Voyager scientist remarked. Other parts of the surface, however, showed very different features: long, parallel mountain ridges that looked like grooves made with a giant's rake; narrow, segmented lines; white ejecta blankets from impacts that looked like a dazzling, snow-covered landscape. Some of the pictures suggested cracking and slipping of Ganymede's crust, while others showed what appeared to be remnants of ancient terrain unaffected by subsequent intense geologic activity. Many of the highest resolution frames were not seen at this time; they were recorded on the spacecraft for playback later.

Starting at about 8 a.m., Earth began receiving the first closeup views of Europa. Europa "could be the most exciting satellite in the whole Jovian system," said Larry Soderblom, "because it's sort of the transition body between the solid silicate body, Io, and the ice balls, Ganymede and Callisto." The icy crust looked as though it "had been ruptured all over—as though it was in pieces—just as though it had been broken in place and left there." At 11:43 a.m., closest approach took place at a range of 206 000 kilometers. By this time the scientists were dazzled by what they had seen; some were calling Europa the most bizzare of all the Galilean satellites. In the Imaging Team viewing area, David Morrison compared Europa's surface to "a cracked egg," and Gene Shoemaker said, "It looks like sea ice to me." When someone commented that the canal-like streaks were reminiscent of Mars, Torrence Johnson replied, "It looks like some pictures of Mars I've seen, but only on the walls of Lowell Observatory." Another quipped, "Where is Percival Lowell, now that we need him!"

There were to be two press conferences: one to present spacecraft and scientific results and one to celebrate the second successful flyby and to talk of new goals—Saturn, and perhaps Uranus.

At the first conference, Ed Stone began by discussing the radiation Voyager was experiencing. One of Jupiter's surprises was that the radiation environment was greater than had been anticipated, and this caused problems with the radio receiver. The receiver frequency was shifting "more rapidly than we had anticipated," said Ray Heacock, "and we have not been able to keep an uplink continuously with the spacecraft." The solution was to keep sending up commands at different frequencies until a frequency the spacecraft would accept was found. Just how bad were the radiation levels? Ed Stone commented, "The penetrating radiation at a given distance is more intense now at this distance than it was when Voyager 1 flew by." From a preliminary analysis it seemed that, overall, Voyager 2 would still be subjected to lower radiation levels than Voyager 1 had been, but to higher levels than had been expected. In addition, Voyager 2's radio receiver was much more sensitive. The higher than expected radiation intensity also led Voyager scientists to have the ultraviolet spectrometer shut off, since that instrument was also quite sensitive to the radiation.

The fourth member of the Galilean satellites had finally been seen, and Larry Soderblom happily introduced Europa. "Well, some few months ago, before the Voyager 1 encounter, we thought we had some idea of what planets were like—at least the planets in the inner solar system: Mars, Mercury, the Moon, the Earth. And we've discovered many times over in the last couple of months how narrow our vision really was. Included in the Jovian collection of satellites are the oldest (Callisto), the youngest (Io), the darkest (Amalthea), the brightest (Europa), the reddest (Amalthea and Io), the whitest (Europa), the most active (Io), and the least active (Callisto). Today we found the flattest (Europa)."

In spite of the appearance of a cracked or broken surface, Europa showed no topography at all. Toward the sunset line, where the low angle of illumination should reveal even low relief, "the surface disappears—as if it were the surface of a billiard ball." It seemed clear that Europa has much less relief than the other two icy satellites, Ganymede and Callisto. But why can't Europa's surface support relief? Perhaps Europa has a thick ice mantle—on the order of 100 kilometers. If Europa is affected by tidal heating, then such an ice mantle might be "sort of soft and slushy" rather than rigid as are the crusts of Ganymede and Callisto. "The fact that the surface of Europa cannot support relief of any substantial amount suggests that the surface must be soft." But, Dr. Soderblom added, there does not appear to be much lateral motion

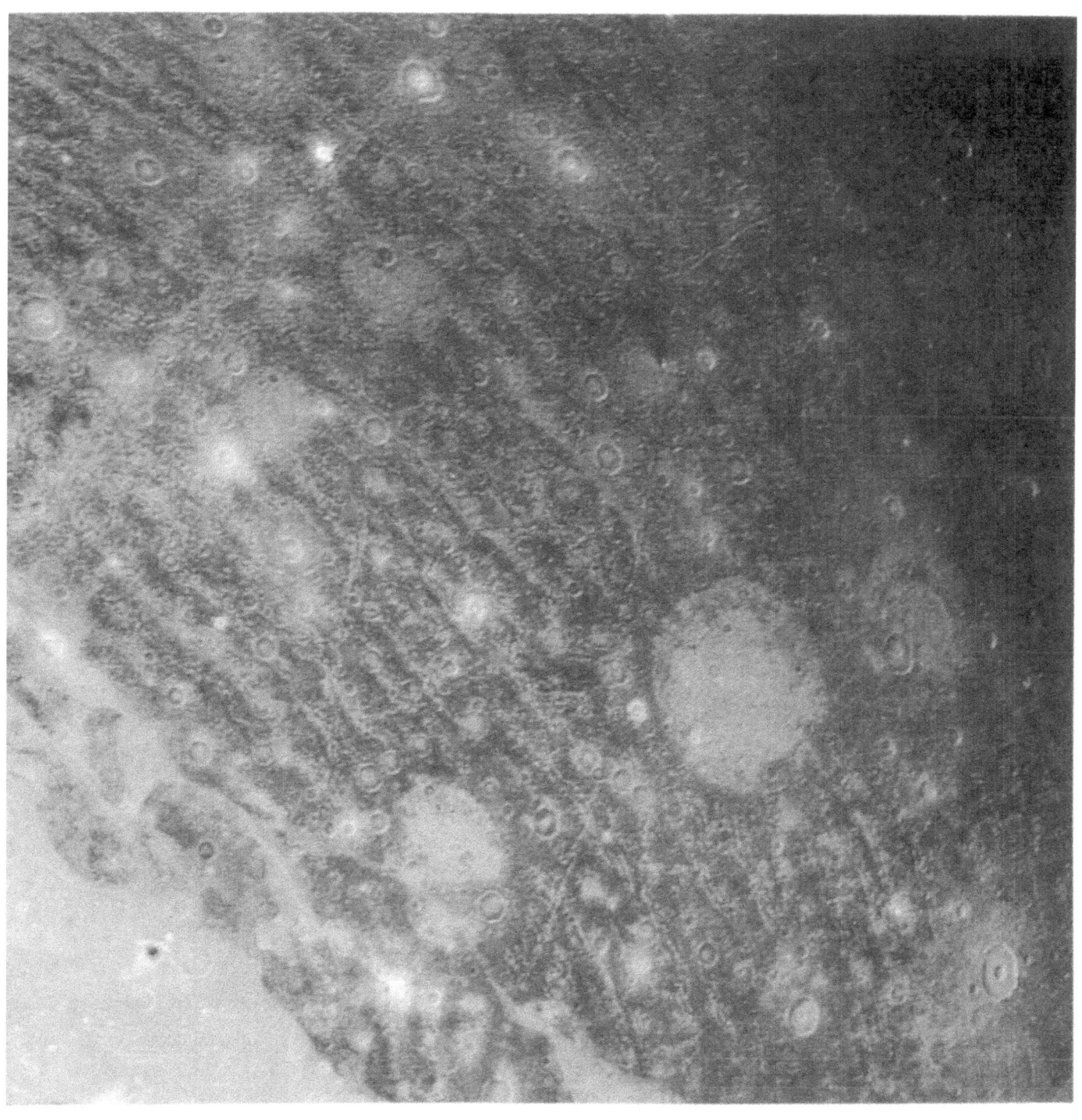

Regio Galileo is the largest remnant of the ancient, heavily cratered crust of Ganymede. This Voyager 2 color reconstruction was made from pictures taken at a range of 310 000 kilometers; the scene is about 1300 kilometers across. Numerous craters, many with central peaks, are visible. The large bright circular features have little relief and are probably the remnants of old large craters that have been annealed by the flow of icy near-surface material. The closely spaced, arcuate linear features are analogous to features on Callisto, such as the "ripple" marks surrounding the ancient impact feature Valhalla. [P-21761C]

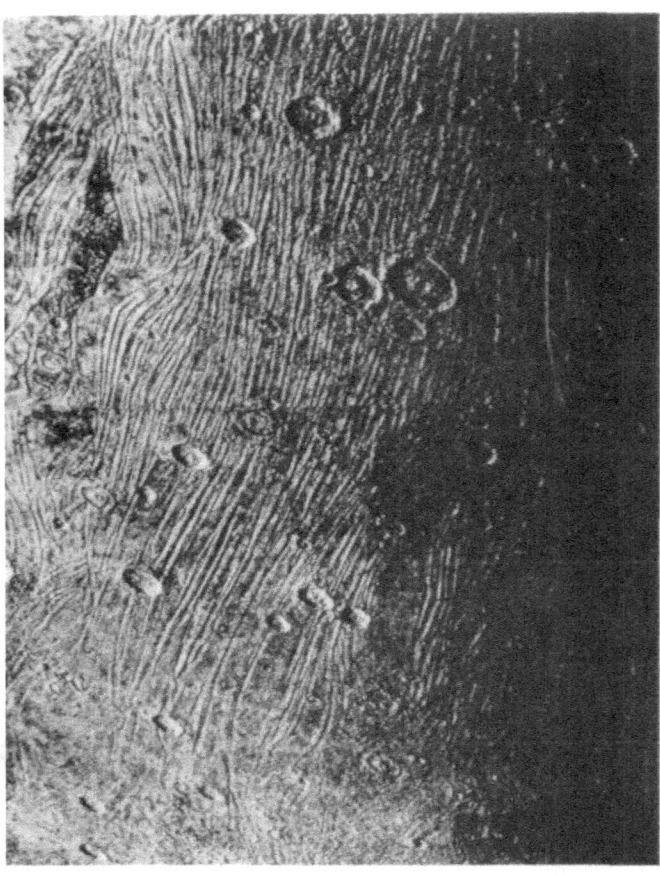

At high resolution, the grooved terrain on Ganymede shows a wonderful complexity. Surface features as small as 1 kilometer across can be seen in this mosaic of Voyager 2 images taken July 9. The grooves are basically long, parallel mountain ridges, 10 to 15 kilometers from crest to crest—about the same scale as the Appalachian mountains in the Eastern United States. The numerous impact craters superposed on the mountain ridges indicate that they are old—probably formed several billion years ago. [260-637]

or rotation causing the surface markings—they don't seem to be offset—rather, "it's as if Europa had been cracked, broken, by some process which crushed it like an eggshell and just left the pieces sitting there. Expansion and contraction of ice and water are a good way to crunch up the surface."

The other side of Ganymede presented quite a different face from the one Voyager 1 had seen. Here were the dark ancient cratered terrains, the shoulder-to-shoulder craters reminiscent of Callisto, and there was a huge circular feature on Ganymede looking like the remnant of a Callisto-style ringed basin, preserved in the ancient, dark terrain. The very

large dark feature revealed by Voyager in the northern hemisphere which bears these impact scars was later named "Regio Galileo," for the discoverer of the Galilean satellites. It was seen in the low-resolution Pioneer 10 picture of Ganymede taken in 1973, but its nature was not understood. It is so large it has even been glimpsed on occasions of exceptionally stable "seeing" with ground-based telescopes.

3:29 p.m. PDT—Jupiter Encounter! In the press room half a dozen cameras clicked in unison as the universal clock declared the Voyager 2 had made its closest approach to Jupiter—650 000 kilometers from the cloud tops, zipping by at about 73 000 kilometers per hour—neither as close nor as fast as Voyager 1. By the time of the special press conference at 4:30 p.m., everyone at JPL was in a party mood. Thomas A. Mutch, who had replaced Noel Hinners as NASA Associate Administrator for Space Science, Robert Parks, and Rodney Mills were the speakers.

The Jovian system is a place of "incredible beauty and mystery. Jupiter has been a nice place to go by, but we wouldn't want to stop there—we're going on to Saturn," Rod Mills explained, and Bob Parks agreed.

Tim Mutch had a different perspective. "Although we have just heard Jupiter somewhat downgraded in favor of Saturn, nonetheless what we have been witnessing, first in March and now, in July, is a truly revolutionary journey of exploration. We have gone beyond the familiar part of the solar system to objects that are so exotic that their very existence, at least as far as I'm concerned, was something I'd accepted intellectually, but didn't really accept in an immediate sense. We're starting out in our own space program on a new stage of space exploration—on our own long journeys beyond the solar system to distant lands. We never like to think, or rather, it's statistically unlikely, that we're at a turning point in history. But if you look back at history books, such events are clearly read into the record. And I submit to you that when the history books are written a hundred years from now, two hundred years from now, the historians are going to cite this particular period of exploration as a turning point in our cultural, our scientific, our intellectual development."

Although everyone was already celebrating another successful mission, the encounter was far from over. Data continued to come in; there was still the ten-hour Io Volcano Watch, which had begun at 4:31 p.m.; there were more obser-

108

vations of Jupiter, including scheduled ring observations and dark side searches for aurorae and lightning bolts. There was a lot of work and excitement yet to come. Jupiter had another surprise in store for Voyager 2.

Tuesday, July 10 *(Range to Jupiter, 1.4 million kilometers).* The Io watch continued through the night. As time passed, the satellite rotated in the same direction as the motion of the spacecraft, keeping nearly the same side in view. Because of this, a few volcanoes could be closely watched, but most would be missed entirely. During the sequence, the illuminated crescent steadily shrank, until at the end, volcanic plumes could be seen on both edges,

During the 10-hour Io volcano watch on July 9, the spacecraft kept nearly the same face of Io in view. Most of the surface was turned away from the Sun, however, and only a thin crescent could be seen, shrinking as the observations continued. These four frames were all photographed with identical exposures from a range of about 1 million kilometers. These images show Amirani (P_5) and Maui (P_6) on the west edge, brightening as the Sun illuminates them more nearly from behind. Masubi (P_8) is faintly visible in the crescent (top right and left); (bottom right) Loki (P_2) rises 250 kilometers above the surface, catching the morning sunlight on the east edge of Io. [260-677]

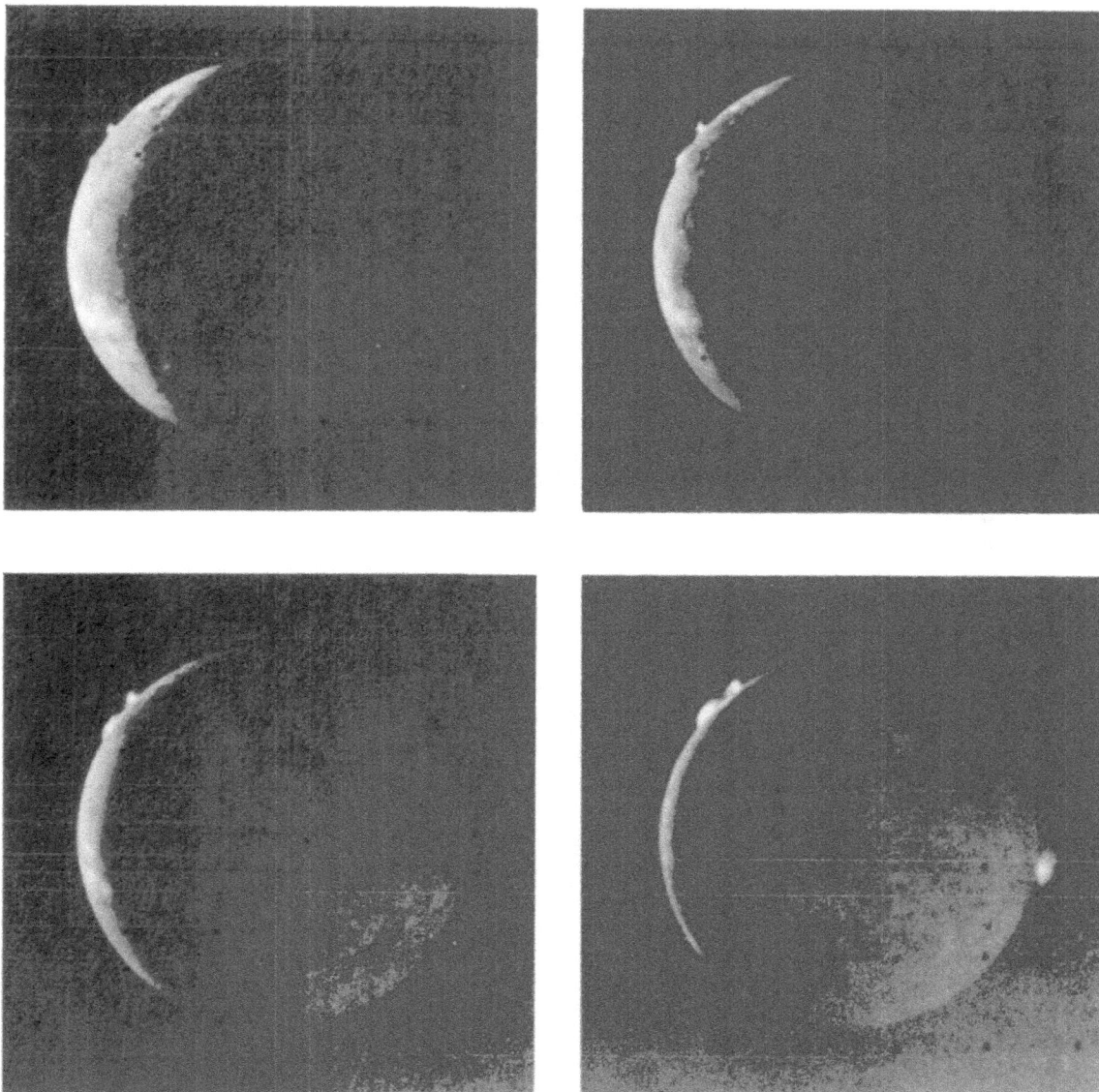

one illuminated by the setting Sun, the other shining in the dawn light.

At the 11 a.m. press conference, Esker Davis announced that engineers had lost contact with the spacecraft radio receiver Monday evening (probably due to Jupiter's radiation) and had to "chase it around most of the night," sending commands at various frequencies until they locked on to the frequency the spacecraft would accept. The major trajectory correction maneuver, begun at about the same time contact with the receiver was lost, was successful. The 76-minute thruster firing, done at periapsis instead of two weeks after encounter, enabled the spacecraft to get a bigger "boost" from Jupiter than was originally planned, amounting to a fuel saving of about 10 kilograms of hydrazine, enough to preserve the option of going on from Saturn to Uranus.

Andrew Ingersoll discussed some results of the analysis of the Jovian atmosphere. "At first, Voyager seemed to do nothing but emphasize the chaos, not the order." But, with the help of ground-based observations, Reta Beebe found that there is a "regular alternation of eastward and westward jets" underlying the seemingly chaotic visible features. "The turbulence we see in the visible clouds seems to be a minor side show, or a process without much energy or mass compared to the very great energy and mass that might be moving around in the deep atmosphere."

"We're continuing to operate in our panic mode to try to get pictures to the press," Brad Smith said as he introduced new photographs of the satellites. In earlier photos, Ganymede had seemed to have two different kinds of terrain—an ancient, cratered, Callisto-like surface, and the stranger, grooved terrain—terrains that might be representative of two very different types of major episodes in Ganymede's history. The most recent images showed a much more confused picture, with several additional types of surface geology.

At the daily project science briefing, another interpretation was being discussed. Lyle Broadfoot reported that new measurements of the position of the ultraviolet aurora demonstrated that it was caused by charged particles from the Io torus, not from the outer parts of the magnetosphere. Apparently these plasma particles arise in the volcanic eruptions,

are trapped for a time in the torus, and then fall into the polar regions of Jupiter, where they excite auroral emissions. A terrestrial aurora, in contrast, is caused by particles that originate in the solar wind. Jim Sullivan of the plasma investigation estimated that about two tons of material each second are fed from Io into the plasma torus. This plasma, driven by the rotation of the Jovian magnetic field, appears to be able to supply the million-million watts of power radiated in the ultraviolet.

By 5:00 p.m. the excitement had died down; many of the scientists had parties to attend that evening, and some members of the press were planning parties of their own. The schedule of spacecraft activities also seemed to have slowed. There were dark-side observations planned to search for lightning and aurorae. There would be a few more ring pictures—not too much to see on the monitors that night . . . or so many people thought. But a few people were waiting around, perhaps to catch a glimpse of lightning or auroral activity, or to wait for another look at Jupiter's faint ring.

Between 5:52 and 6:16 p.m., six long-exposure, wide-angle photographs of the dark side of Jupiter had been scheduled to search for aurorae and lightning. The spacecraft was 1 450 000 kilometers from Jupiter and about two degrees below the equatorial plane.

Shortly after 6 p.m., the first of these ring photographs appeared on the TV monitors with unexpected brilliance. Taken in orange and violet light, the images showed the outline of Jupiter and, protruding from it, two narrow lines—one reaching all the way to Jupiter's limb, the other broken off, apparently hidden by the shadow of the giant planet. Seen from the new perspective of the shadow of Jupiter, the tenuous rings were remarkably clear. A sudden renewal of excitement surged through the devotees remaining in the press room. About 6:15, Brad Smith came down to join the press to watch the remainder of this series of pictures come in. "Hey Brad, are you going to burn out the camera with the ring?" someone joked. "Well, the rings do forward scatter nicely, don't they?" Dr. Smith replied. As the wide-angle pictures were followed by narrow-angle views, more and more detail became apparent. For the first time, a definite width for the ring could be seen, and there was even a hint of additional material inside the main ring. All in all,

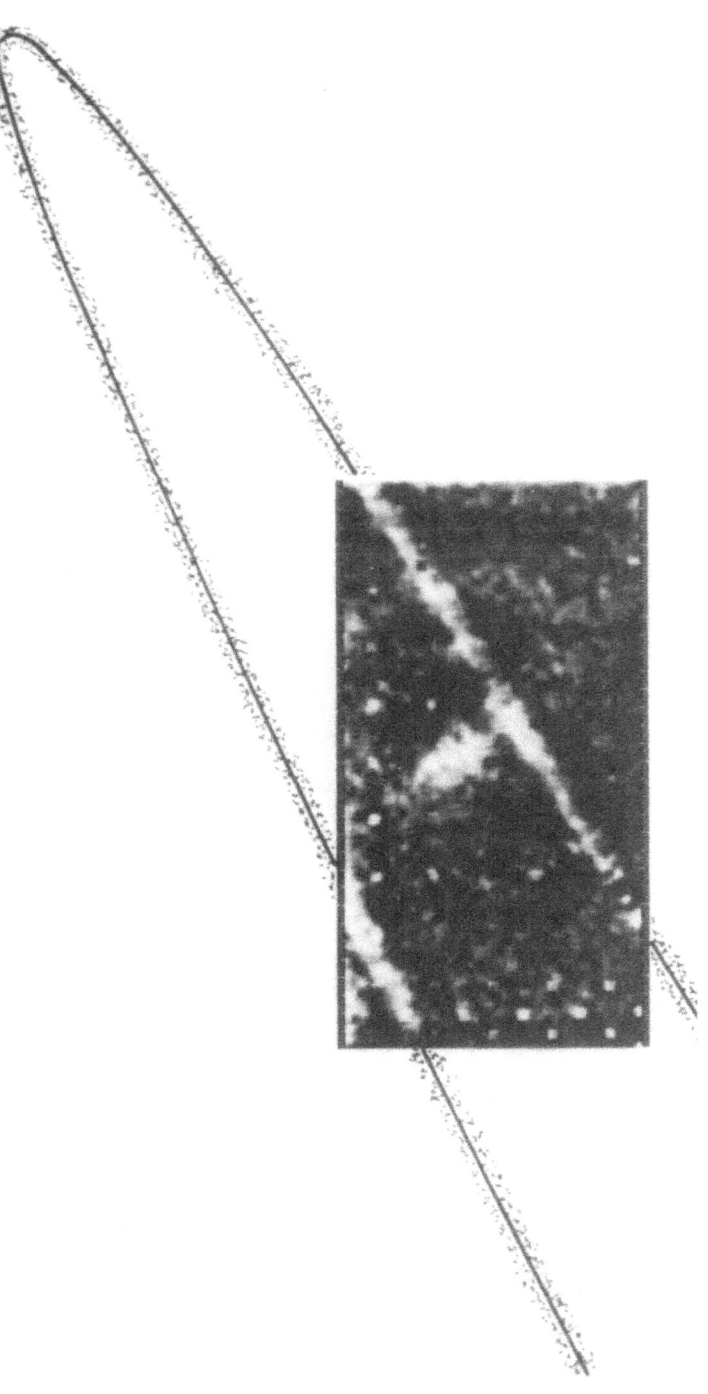

From a vantage point 2.5 degrees above the ring plane, Voyager 2 was able for the first time to determine the width of Jupiter's ring. This picture shows that the ring is ribbon-like and only a few thousand kilometers wide, quite unlike the broad rings of Saturn. [P-21757B/W]

Voyager had provided one more splendid series of pictures before it took off for Saturn.

Wednesday, July 11. JPL Public Information Officer Frank Bristow opened the 10 a.m. press conference with an announcement: "We'll have the report from the Imaging Team including the tremendous pictures that we received here last night of the Jupiter ring that excited the entire team."

Brad Smith showed the ring pictures. "As many of you who were here last night know, we got some rather nice pictures of the ring of Jupiter. It's as though Voyager 2 was fearful that we might be becoming just a little bit apathetic after this series of marvelous discoveries and felt that it had to dazzle us one more time before it left for Saturn. The rings appear very much brighter than we had expected them to be." The outer ring is about 6500 kilometers wide. There is material inside the ring. There is a rather sharp outer boundary and a somewhat diffuse inner region. "And it is now our belief that the material in the ring goes all the way down to the surface of Jupiter." There is a very narrow relatively bright outer ring and an extremely faint inner ring that goes all the way down to Jupiter's cloud tops.

Larry Soderblom summarized the satellite data: With respect to the Galilean satellites, "We're in a relatively high state of ignorance."

The Io Volcano Watch images seemed to indicate that plume P_2 was now the highest volcano on Io, since P_1 seemed to have become quiet. Io may be the easiest Galilean satellite to try to understand, because we can actually see the geological processes that are shaping the planet. Io's "twin," Europa, seems to be where "our highest state of ignorance" lies. "The faint bright streaks which show some relief are evidently different from the diffuse dark bands which don't seem to show topography, but the similarity of these forms [that both the light and dark markings are of planetary scale] suggests that they must be related."

Ed Stone speculated about the other two Galilean satellites. Ganymede and Callisto are essentially identical in size, mass, and probably composition. By examining them, we can perhaps learn what happens when bodies with very similar chemistry have different "life histories" and different surface properties (there are indications that Ganymede's crust may not have been as rigid as Callisto's). Going

111

One of the most spectacular of the Voyager 2 images was obtained from inside the shadow of Jupiter. Looking back toward the planet and the rings with its wide-angle camera, the spacecraft took these photos on July 10 from a distance of 1.5 million kilometers. The ribbon-like nature of the rings is clearly shown. The planet is outlined by sunlight scattered from a haze layer high in the atmosphere. On each side, the arms of the ring curving back toward the spacecraft are cut off by the planet's shadow as they approach the brightly outlined disk. [P-21774B/W]

The rings of Jupiter proved to be unexpectedly bright when seen with the Sun nearly behind them. Strong forward scattering of sunlight is characteristic of small particles. These two views were obtained by Voyager 2 on July 10 from a perspective inside the shadow of Jupiter. The distance of the spacecraft from the rings was about 1.5 million kilometers. Although the resolution has been degraded by camera motion during the time exposures, these images reveal that the rings have some radial structure. [260-610B/W and 260-674]

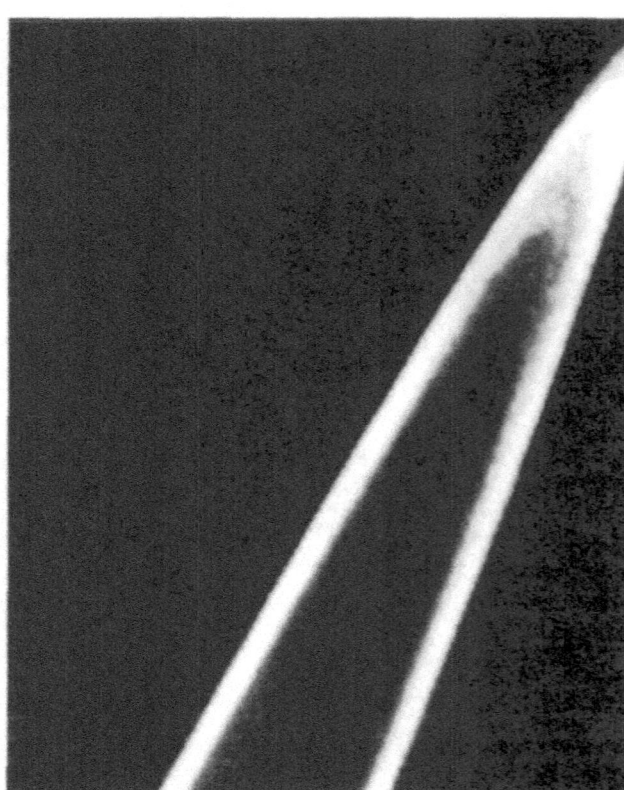

HIGHLIGHTS OF THE VOYAGER 2 SCIENTIFIC FINDINGS

Atmosphere

The main atmospheric jet streams were present during both Voyager encounters, with some changes in velocity.

The Great Red Spot, the white ovals, and the smaller white spots at 41°S, appear to be meteorologically similar.

The formation of a structure east of the Great Red Spot created a barrier to the flow of small spots which earlier were circulating about the Great Red Spot.

The ethane to acetylene abundance ratio in the upper atmosphere appears to be larger in the polar regions than at lower latitudes and appears to be 1.7 times higher on Voyager 2 than on Voyager 1.

An ultraviolet map of Jupiter shows the distribution of absorbing haze. The polar regions are surprisingly dark, suggesting that the absorbing material must be at high altitudes.

Equatorial ultraviolet emissions indicate planet-wide precipitation of charged particles into the atmosphere from the magnetosphere.

The high-latitude ultraviolet auroral activity is due to charged particles that originate in the Io torus.

Satellites and Ring System

The ring consists of a bright, narrow segment surrounded by a broader, dimmer segment, with a total width of about 5800 kilometers.

The interior of the ring is filled with much fainter material that may extend down to the top of the atmosphere.

Images of Amalthea in silhouette against Jupiter indicate that the satellite may be faceted or diamond shaped.

Volcanic activity on Io changed somewhat, with six of the plumes observed by Voyager 1 still erupting.

The largest Voyager 1 plume (Pele) had ceased, while the dimensions of another plume (Loki) had increased by 50 percent.

Several large-scale changes in Io's appearance had occurred, consistent with surface deposition rates calculated for the large eruptions.

Europa is remarkably smooth with very few craters. The surface consists primarily of uniformly bright terrain crossed by linear markings and very low ridges.

There are four basic terrain types on Ganymede, including younger, smooth terrain and a rugged impact basin first observed by Voyager 2.

Callisto's entire surface is densely cratered and is likely to be several billion years old.

Equatorial surface temperatures on the Galilean satellites range from 80 K (night) to 155 K (the subsolar point on Callisto).

Magnetosphere

The outer region of the magnetosphere contains a hot plasma consisting primarily of hydrogen, oxygen, and sulfur ions.

The hot plasma generally flows in the corotation direction out to the boundary of the magnetosphere.

Beyond about 160 R_J, the hot plasma streams nearly antisunward.

Outbound the spacecraft experienced multiple magnetopause crossings between 204 R_J and 215 R_J.

The abundance of oxygen and sulfur relative to helium at high energy increases with decreasing distance from Jupiter.

Measurements of high energy oxygen suggest that these nuclei are diffusing inward toward Jupiter.

The ultraviolet emission from the Io plasma torus was twice as bright as four months earlier and the temperature had decreased by 30 percent to 60 000 K.

The low-frequency (kilometric) radio emissions from Jupiter have a strong latitude dependence and often contain narrowband emissions that drift to lower or higher frequencies with time.

A complex magnetospheric interaction with Ganymede was observed in the magnetic field, plasma, and energetic particles up to about 200 000 kilometers from the satellite.

---

*Adapted from a summary prepared by E. C. Stone and A. L. Lane for the Voyager 2 Thirty-Day Report.

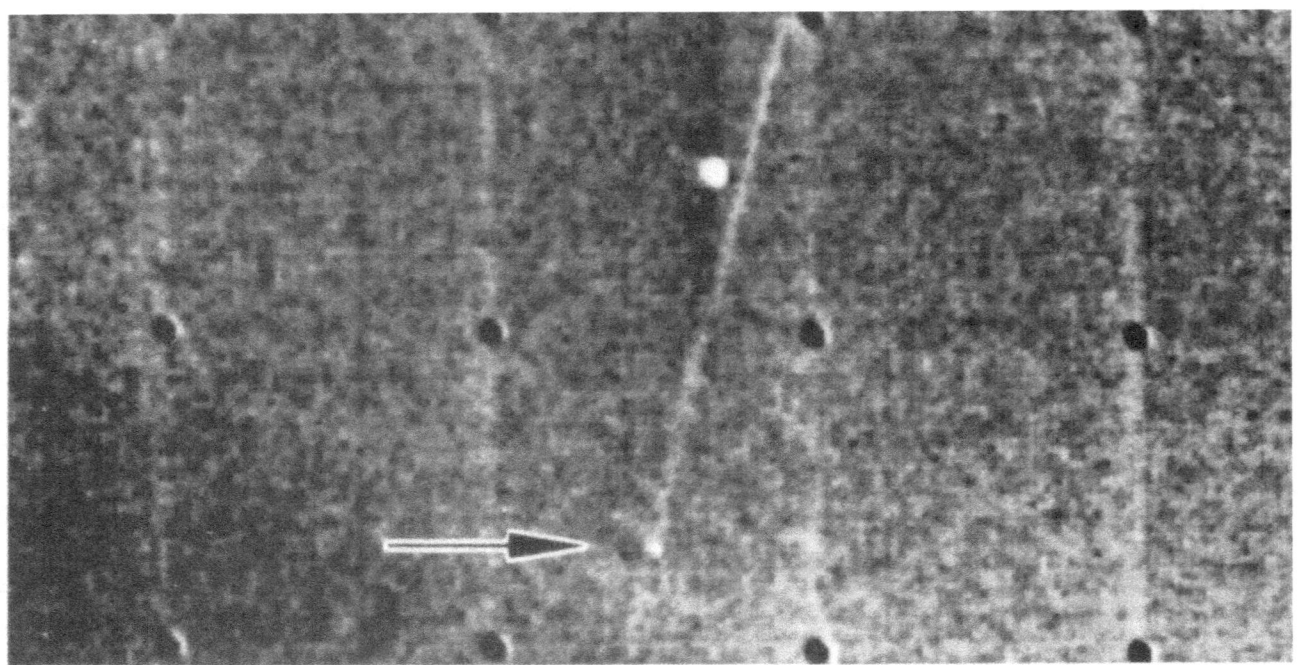

A new inner satellite of Jupiter, provisionally designated 1979J1, was discovered by David Jewitt and Ed Danielson of Caltech in these Voyager 2 ring photographs. (Top) In a 15-second exposure with the wide-angle camera, the edge-on ring shows as a faint line, and the satellite is the dot indicated by the arrow. (Bottom) In a narrow angle 96-second exposure, the motion of the satellite can be seen. Again, the faint band is the ring, blurred by camera motion, and the arrow indicates the streak due to the satellite. A star streak is located above and to the left of the satellite; note that the length and angle of the two trails are different, owing to satellite motion. [260-807 and P-22172]

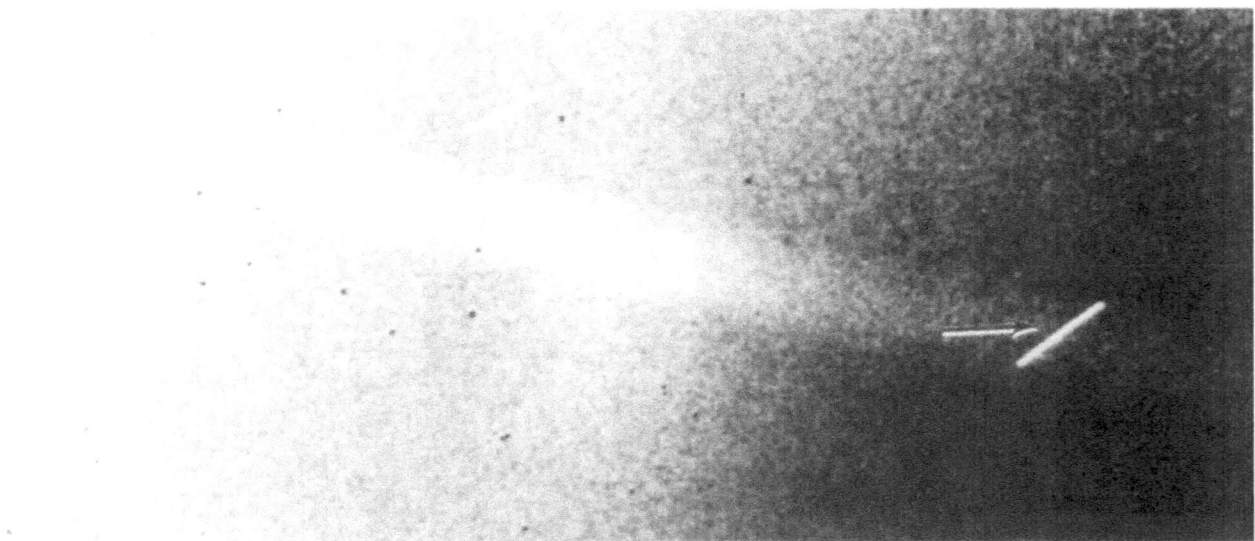

further, he added that Callisto and Mercury, the least dense and the most dense, respectively, of the terrestrial-style planets, although totally different in composition and density, seem to have similar surfaces and similar histories. What would have happened to Mercury if it had been made of ice, water, and rock as Callisto is? Would it have evolved as Callisto did?

A New Satellite

One of the most fascinating discoveries of Voyager 2 was not recognized at first. Graduate student David Jewitt of the California Institute of Technology, working with Imaging Team member Ed Danielson, began a detailed analysis of all the ring photos in late summer. In early October he determined that a short

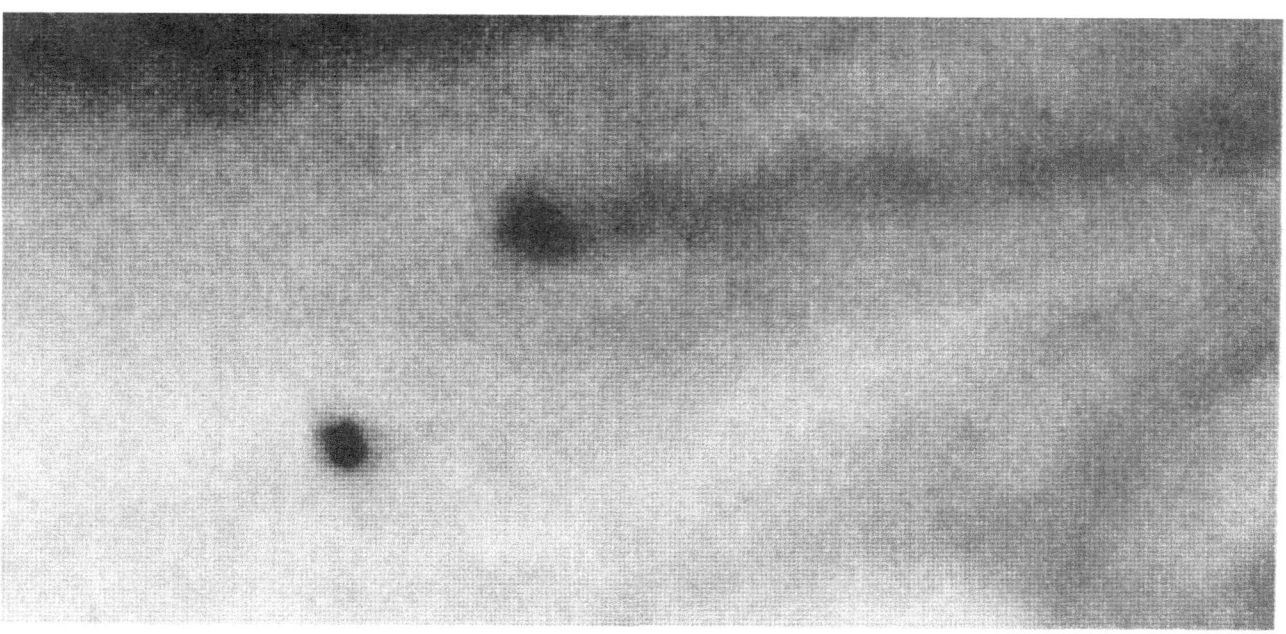

A fifteenth satellite of Jupiter was discovered in the spring of 1980 by Steven Synnott of JPL. It was first seen on this Voyager 1 image taken March 5, 1979, in which the 75-kilometer-diameter satellite shows as a dark oval against the planet. Also visible is the shadow of the satellite, designated 1979J2. This satellite orbits between Io and Amalthea with a period of 16 hours and 11 minutes. [P-22580B/W]

streak on a photo taken July 8, previously presumed to be an image of a star trailed by the time exposure, did not correspond to any known star position. Perhaps this was a new satellite! Additional sleuthing turned up a second image of the same part of the ring that also showed the anomalous object, together with trails due to known stars. The differing angles and lengths of the trails of the object and the stars confirmed that this was indeed a 14th satellite of Jupiter. Following the guidelines of the International Astronomical Union, it was designated 1979J1, pending later assignment of a mythological name. The proposed name is Adrastea, a nymph who nursed the infant Zeus in Greek legend.

The newly discovered satellite orbits Jupiter at a distance of 58 000 kilometers above the equatorial cloud tops, placing it just at the outer edge of the ring and much closer to the planet than is Amalthea, previously thought to be the innermost satellite. It travels at 30 kilometers per second (nearly 70 000 miles per hour), circling Jupiter in just seven hours and eight minutes. From its brightness, scientists guessed that it might be 30–40 kilometers in diameter.

The proximity of Adrastea to the ring suggests a relationship between the two. When the discovery was announced to the press in mid-October, it was speculated that the ring material might originate on the satellite,

perhaps eroded away by the energetic charged particles in the inner Jovian magnetosphere. Once again, Voyager had added to our perspective on planetary processes, suggesting that undiscovered but similar small satellites might also be associated with the rings of Saturn and Uranus.

Voyager 2 had certainly added a few years' of data of its own to Voyager 1's "ten years' worth of data." It had given a different view of the Jovian system, helping to solve some of the mystery surrounding Jupiter and its satellites, and creating new mysteries. As Voyager 2 sped out away from Jupiter, riding along the giant planet's huge magnetotail, attention turned to Saturn: What would Pioneer 11, the Pathfinder, discover in September 1979? What would the Voyagers learn in November 1980 and August 1981? Would all go well? Would Voyager 2 fly on to Uranus?

There was also a yearning to examine more closely, with the Galileo Project, what had been unknown for so long, yet had become so familiar in only a few months' time—the little dark, red "potato" Amalthea, the volcano-covered world Io, the mysterious "cracked billiard ball" Europa, cratered and groovy Ganymede, ancient Callisto, and the king of the planets itself, a colorful, banded world of stable climate and ever-changing weather patterns.

CHAPTER 8

JUPITER – KING OF THE PLANETS

A Star That Failed

More massive than all the other planets combined, Jupiter dominates the planetary system. The giant revealed by Voyager is a gas planet of great complexity; its atmosphere is in constant motion, driven by heat escaping from a glowing interior as well as by sunlight absorbed from above. Energetic atomic particles stream around it, caught in a magnetic field that reaches out nearly 10 million kilometers into the surrounding space, embracing the seven inner satellites. From its deep interior through its seething clouds out to its pulsating magnetosphere, Jupiter is a place where forces of incredible energy contend.

At its birth, Jupiter shone like a star. The energy released by infalling material from the solar nebula heated its interior, and the larger it grew the hotter it became. Theorists calculate that when the nebular material was finally exhausted, Jupiter had a diameter more than ten times its present one, a central temperature of about 50 000 K, and a luminosity about one percent as great as that of the Sun today.

At this early stage, Jupiter rivaled the Sun. Had it been perhaps 70 times more massive than it was, it would have continued to contract and increase in temperature, until self-sustaining nuclear reactions could ignite in its interior. If this had happened, the Sun would have been a double star, and the Earth and the other planets might not have formed. However, Jupiter did not make it as a star; after a brief flash of glory, it began to cool.

At first Jupiter continued to collapse. Within the first ten million years of its life, the planet was reduced to nearly its present size, with only a few percent additional shrinkage during the past 4.5 billion years. The luminosity also dropped as internal heat was carried to the surface by convection and radiated away to space. After a million years Jupiter emitted only one-hundred thousandth as much radiation as the Sun, and today its luminosity is only one-ten billionth of the Sun's.

Jupiter's internal energy, although small by stellar standards, has important effects on the planet. About 10^{17} watts of power, comparable to that received by Jupiter from the Sun, reach the surface from the still-luminous interior. The central temperature is still thought to be about 30 000 K, sufficient to maintain the interior in a molten state. Scientists generally agree that Jupiter is an entirely fluid planet, with no solid core whatever.

Composition and Atmospheric Structure

Because of its great mass, Jupiter has been undiscriminating in its composition. All gases and solids available in the early solar nebula were attracted and held by its powerful gravity. Thus it is expected that Jupiter has the same basic composition as the Sun, with both bodies preserving a sample of the original cosmic material from which the solar system formed.

The primary constituents of Jupiter have long been suspected to be hydrogen and helium, the two simplest and lightest atoms. However, it has proved impossible to derive accurate measurements of the abundance of these two elements from astronomical observations. On

The Jupiter seen by the Voyager cameras is a cloud-belted world of rapid jet streams and complex cloud forms. Prominent in this Voyager 1 image, taken February 5 at a range of 28.4 million kilometers, is the alternating structure of light zones and dark belts, and the Great Red Spot and numerous smaller spots. Also easily visible are the two inner Galilean satellites, Io and Europa. The resolution in this picture is 500 kilometers, about five times better than can be obtained from Earth-based telescopes. Callisto can be faintly seen at the lower left. [P-21083C]

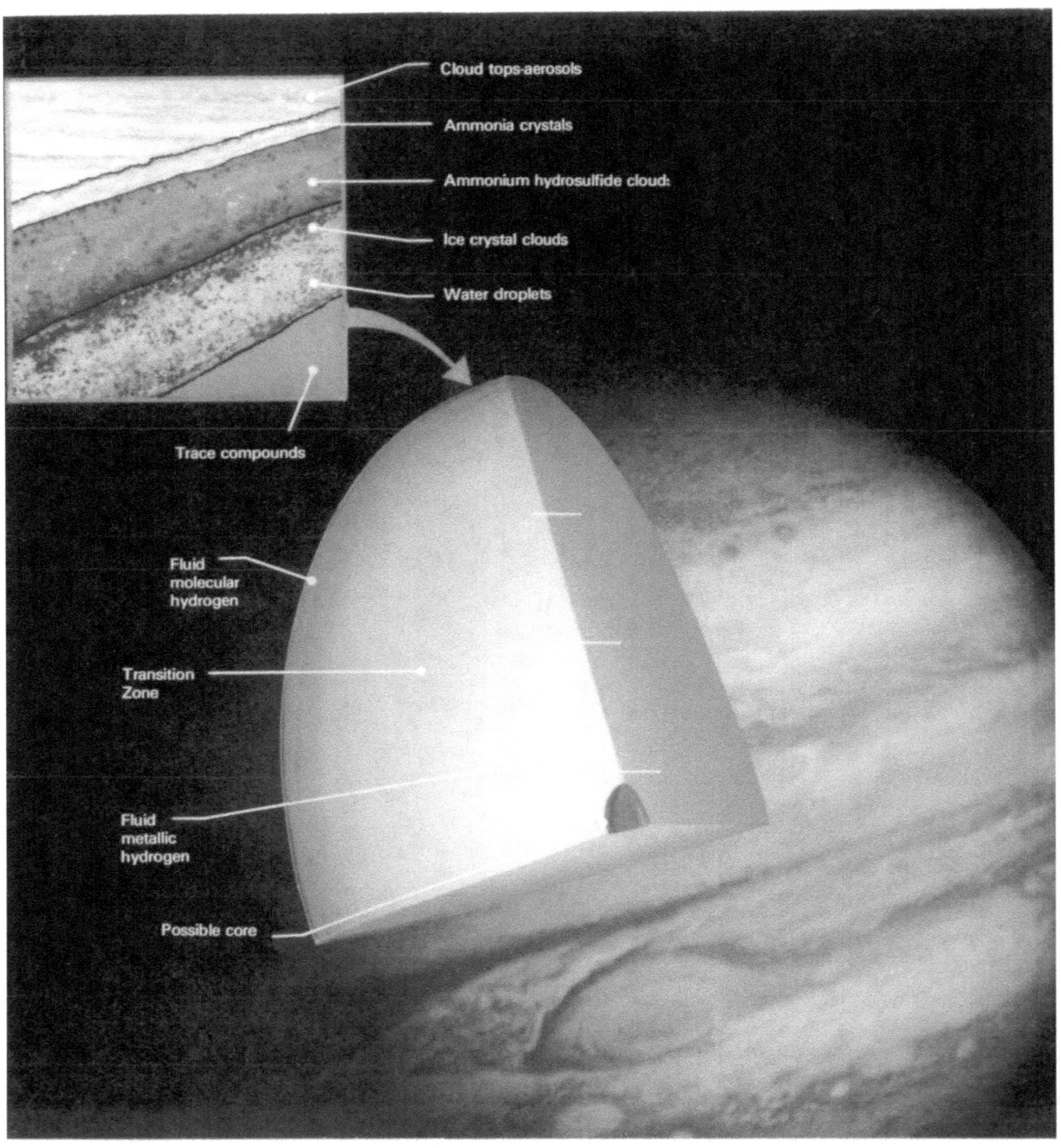

Cloud tops-aerosols

Ammonia crystals

Ammonium hydrosulfide cloud:

Ice crystal clouds

Water droplets

Trace compounds

Fluid molecular hydrogen

Transition Zone

Fluid metallic hydrogen

Possible core

Jupiter is a gas giant, composed of the same elements as the Sun and stars—primarily hydrogen and helium. Its internal structure is dominated by the properties of hydrogen, its most abundant constituent and by the high temperatures in the deep interior that remain from its luminous youth. Most of the interior is liquid: metallic hydrogen at great depths and high pressures, and normal hydrogen nearer the surface. In the upper few thousand kilometers, the hydrogen is a gas. The primary known or suspected cloud layers are, from the top down, thin hydrocarbon "smog"; ammonia; ammonium hydrosulfide; water-ice, and liquid water. [260-828]

118

the basis of a rather simple infrared measurement, Pioneer investigators found $He/H_2 = 0.14 \pm 0.08$. On Voyager, IRIS was able to obtain much improved infrared spectra, yielding an initial value of $He/H_2 = 0.11 \pm 0.3$. Voyager scientists expect that further analysis will reduce the uncertainty to about ± 0.01. The ratio of 0.11 is in excellent agreement with the solar value of about 0.12, supporting the idea that Jupiter and the Sun have similar elemental compositions.

Astronomers have known for a long time that, in addition to hydrogen and helium, the compounds methane (CH_4) and ammonia (NH_3) are present in the visible atmosphere of Jupiter. In the 1970s, additional spectra in the infrared resulted in the discovery of water (H_2O), ethane (C_2H_6), germane (GeH_4), acetylene (C_2H_2), phosphine (PH_3), carbon monoxide (CO), hydrogen cyanide (HCN), and carbon dioxide (CO_2). All these are trace constituents, with two of them, ethane and acetylene, apparently formed at high altitudes by the action of sunlight on methane.

A total of approximately 100 000 infrared spectra, many of small regions on the disk, were obtained by IRIS. These spectra generally show hydrogen, helium, methane, ammonia, phosphine, ethane, and acetylene. In addition, excellent spectra were obtained in "hot spots," regions in which breaks in the upper clouds permit radiation from deeper layers to escape. (The hot spots generally correspond to dark brown regions on photographs of the planet.) IRIS measured temperatures in the hot spots up to $-13°$ C but no higher; apparently this temperature corresponds to the top of a deeper cloud deck. Spectral features indicative of the presence of water vapor and germane were clearly seen in the hot spots.

Further analysis of the IRIS spectra will be required to derive the abundances of the gases detected. However, even the preliminary data showed how variable Jupiter can be, especially in its upper atmosphere. The two hydrocarbons, ethane and acetylene, vary in relative abundance with latitude; there is less acetylene near the poles. In addition to this planetwide trend, smaller variations were seen from place to place and between the observations in March and July. All the variations will eventually provide information on the processes of formation, transportation, and destruction of hydrocarbons in the upper atmosphere.

Voyager did not make any direct measurements of the chemical composition of the

ELEMENTS DETECTED IN THE JOVIAN MAGNETOSPHERE

Element	Atomic Number	Instruments
Hydrogen (H)	1	UVS, Plasma, LECP, CRS
Helium (He)	2	Plasma, LECP
Carbon (C)	6	LECP
Nitrogen (N)	7	CRS
Oxygen (O)	8	UVS, Plasma, LECP, CRS
Neon (Ne)	10	CRS
Sodium (Na)	11	LECP, CRS
Magnesium (Mg)	12	CRS
Silicon (Si)	14	CRS
Sulfur (S)	16	UVS, Plasma, LECP, CRS
Iron (Fe)	26	CRS

clouds, but theorists generally agree that the uppermost clouds are ammonia cirrus, and that layers of ammonium hydrosulfide (NH_4SH) and water exist at deeper levels. All these clouds are formed in the troposphere, the layer of the atmosphere in which convection takes place. The top of the ammonia cloud deck is thought to have a pressure of about 1 atmosphere and a temperature of about $-113°$ C.

Ammonia cirrus is white, yet Jupiter's clouds display a spectacular range of colors. Voyager did not determine the nature of the coloring agents; they may be minor constituents—trace impurities in a sea of white clouds. Perhaps organic polymers, formed from atmospheric chemicals such as methane and ammonia that have reacted with lightning, are responsible for the oranges and yellows. The color of the Red Spot could be caused by red phosphorus (P_4). According to this theory, phosphine (PH_3) from deep in Jupiter's atmosphere is brought to high altitudes by the upwelling of the Great Red Spot. Ultraviolet light, penetrating the upper reaches of the Red Spot, splits the phosphine molecules, and, through a series of chemical reactions, converts the phosphine into pure phosphorus. However, this theory fails to explain the existence of the smaller red spots on Jupiter; these spots are not at such high altitudes as the Great Red Spot (which is the highest and coldest of Jupiter's visible clouds), so it is unlikely that ultraviolet light could react with any phosphine in these areas to produce red phosphorus.

Various forms of elemental sulfur might be responsible for the riot of color we see on

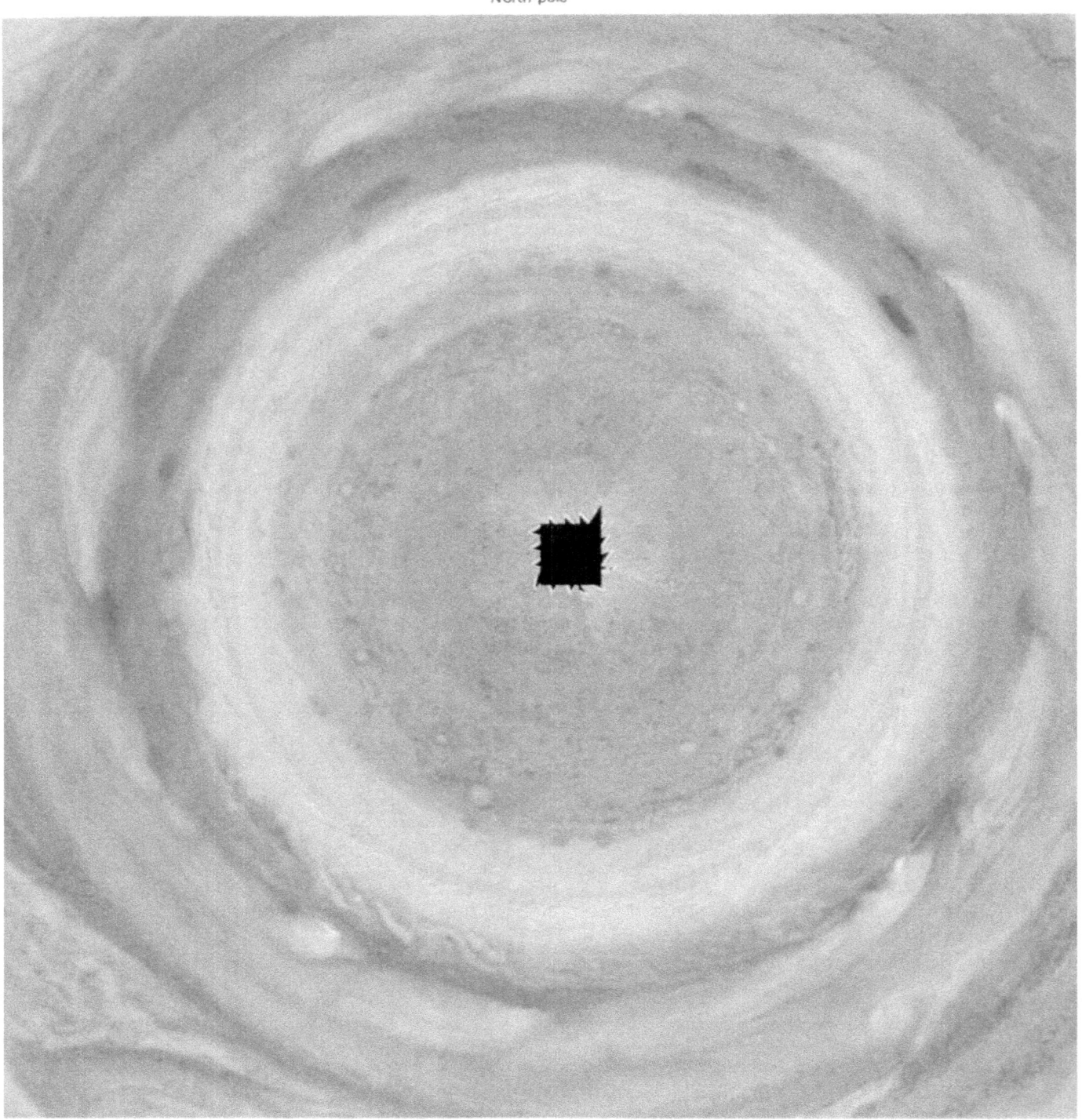

Although the Voyager spacecraft never flew over the poles of Jupiter, it is possible to reconstruct from several images the view that would be seen from directly above or below the planet. The north pole is shown above and the south pole at right. Note the absence of a strong banded structure near both poles. The regular spacing of cloud features is obvious. In the southern hemisphere, the three white ovals are 90 degrees apart in longitude, but a fourth oval at the other quadrant is missing. The irregular black areas at each pole are places for which no Voyager data exist. The resolution of the original pictures from which these polar projections were made was about 600 kilometers. [P-21638C and P-21639C]

120

Jupiter. Sulfur forms polymers (S_3, S_4, S_5, S_8,) that are yellow, red, and brown, but no sulfur in any form has been detected on Jupiter. "We never promised you we were going to identify the colors on Jupiter with this mission," one of the atmospheric scientists remarked, "but we will have a probe that is going into the atmosphere in the mid-1980s—Galileo." Perhaps the mystery of the Jovian clouds will have to wait till then.

Temperature maps of Jupiter were obtained by IRIS in radiation arising at different levels above the clouds. Maps show temperatures at pressures of 0.8 atmosphere near the clouds, and 0.2 atmosphere near the top of the troposphere. In addition to the low temperatures over the bright zones and the higher temperatures over dark belts, there is a great deal of smaller scale structure. It is interesting that a cold area corresponding to the Great Red Spot is clearly visible even near the top of the troposphere, indicating that this feature disturbs the atmosphere to very high altitudes.

The structure of the atmosphere of Jupiter above the troposphere was investigated through the radio occultation experiment as well as by IRIS. The level in which the minimum temperature of about −173° C occurs has a pressure of 0.1 atmosphere. Above this point lies the stratosphere, in which temperatures increase with altitude as a result of sunlight absorbed by the gas or by aerosol particles resembling smog. At 70 kilometers above the ammonia clouds, the temperature is about −113° C. Above this level, the temperature stays approximately constant, although at extreme altitudes the temperature again rises in the ionosphere.

Weather on Jupiter

The Voyager pictures reveal a planet of complex atmospheric motions. Spots chase after each other, meet, whirl around, mingle, and then split up again; filamentary structures curl into spirals that open outward; feathery cloud systems reach out toward neighboring regions; cumulus clouds that look like ostrich plumes may brighten suddenly as they float toward the east; spots stream around the Red Spot or get caught up in its vortical motion—all in an incredible interplay of color, texture, and eastward and westward flows. Such changes can be noticed in the space of only a few Jovian days.

If one could "unwrap" Jupiter like a map, views such as these would be obtained. The planet as it appeared about March 1 is shown at the top and as it was in early July at the bottom. The comparison between the pictures shows the relative motions of features in Jupiter's atmosphere. It can be seen, for example, that the Great Red Spot moved westward and the white ovals eastward during the time between the acquisition of these pictures. Regular plume patterns are equidistant around the northern edge of the equator, while a train of small spots moved eastward at approximately latitude 80° S. In addition to these relative motions, significant changes are evident in the recirculating flow east of the Great Red Spot, in the disturbed region west of the Great Red Spot, and as seen in the brightening of material spreading into the equatorial region from the more southerly latitudes. [P-21771C]

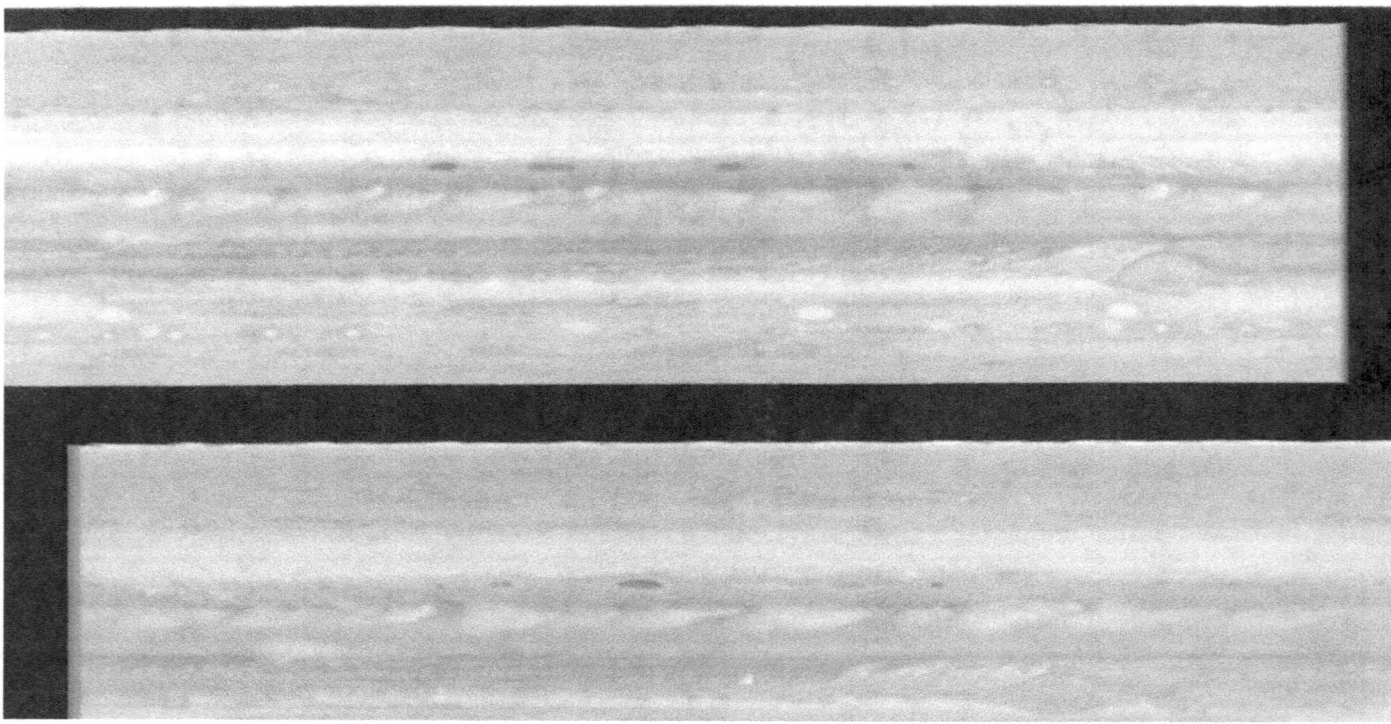

On a broader time scale, greater changes on the face of Jupiter can be seen. Features drift around the planet; even the large white ovals and the Great Red Spot slide along in their respective latitudes. Belts or zones intrude upon each other, resulting in one of the banded structures splitting up or seeming to squeeze together and eventually disappear. Small structures form, then die. The largest spots may slowly shrink in size, and the Red Spot itself changes its size and color.

The Jupiter of Pioneers 10 and 11 was quite unlike the planet seen by Voyager 1. At the time of the Pioneer exploration, the Great Red Spot, embedded in a huge white zone, was more uniformly colored, and pale brown bands circled the northern hemisphere. In the intervening years, the south temperate latitudes have changed completely, developing the complex turbulent clouds seen around the Red Spot by Voyager 1. Yet, even between the two Voyagers, Jupiter appeared to be undergoing a dynamic "facelift." At a quick glance, Voyager 2 photographs showed the visage that had been familiar since early in 1979, but a closer look showed that it is not quite the same. The white

Differing characteristics of Jupiter's meteorology are apparent in high-resolution images, such as this one taken by Voyager 1 on March 2 at a range of 4 million kilometers. The well-defined pale orange line running from southwest to northeast (north is at the top) marks the high-speed north temperate current with wind speeds of about 120 meters per second. Toward the top of the picture, a weaker jet of approximately 30 meters per second is characterized by wave patterns and cloud features which have been observed to rotate in a clockwise manner at these latitudes of about 35°N. These clouds have been observed to have lifetimes of one to two years. [P-21193C]

 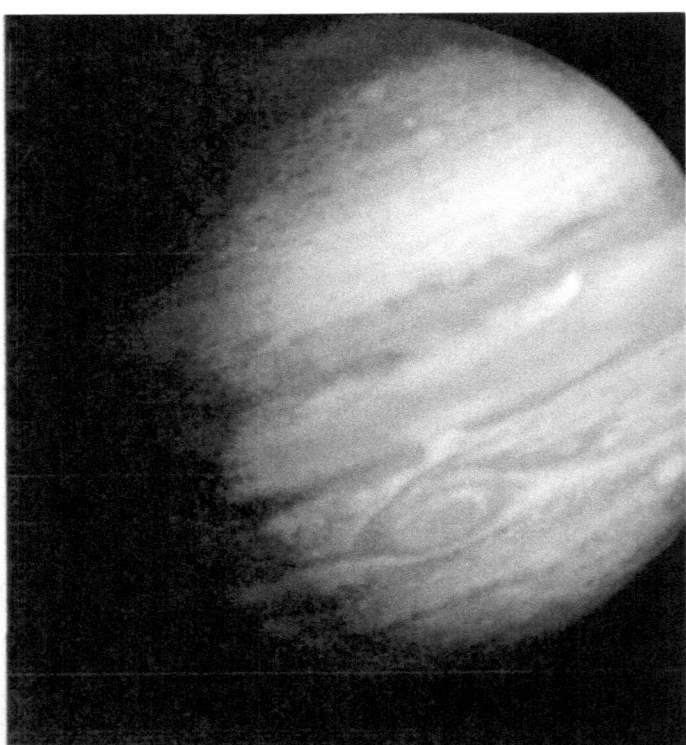

Jupiter's cloud patterns changed significantly in the few months between the two Voyager flybys. Most of the changes are the result of differential rotation, in which the prevailing winds at different latitudes shift long-lived features with respect to those north or south. Thus, for example, the three large white ovals shifted nearly 90 degrees in longitude, relative to the Great Red Spot, between March and July. [P-21599]

band below the Great Red Spot, fairly broad during the first flyby, had become a thin white ribbon where it rims the southern edge of the Spot. The turbulence to the west of the Red Spot had stretched out and become "blander" than it was before. Small rotating clouds seemed to be forming out of the waves in this region. The cloud structure that had been east of the Red Spot during the Voyager 1 flyby spread out, covering the northern boundary and preventing small clouds from circling the huge red oval. The Red Spot itself also changed. Its northern boundary seemed—at least visually—to be more set off from the clouds that surround it, and the feature appeared to be more uniform in color, perhaps reverting back to the personality it had in Pioneer days.

The most obvious features in the atmosphere of Jupiter, after the banded belts and zones, are the Great Red Spot and the three white ovals. These have often been described as "storms" in Jupiter's atmosphere. The ovals are about the size of the Moon, and the Red

Spot is larger than the Earth. Voyager has revealed that in many respects the white ovals, which formed in 1939, resemble their ancient red relative. All four spots are southern hemispheric anticyclonic features that exhibit counterclockwise motion; hence they are meteorologically similar. Other smaller bright elliptical and circular spots also exhibit anticyclonic motion, rotating clockwise in the northern hemisphere and counterclockwise in the southern hemisphere. In general, these features are circled by filamentary rings that are darker than the spots they surround. Hints of interior spiral structure can be seen in some of these spots. All the elliptical features in the southern hemisphere lie to the south of the strong westward-blowing jet streams. The spots tend to become rounder the closer they are to the poles.

Along the northern edge of the equator are a number of cloud plumes, which appear to be regularly spaced all around the planet. Some of the plumes have been observed to brighten rapidly, which may be an indication of convective activity; indeed, some of the plume struc-

The Great Red Spot of Jupiter is a magnificent sight, whether viewed in normal or exaggerated color. These pictures were taken by Voyager 1 at a range of about 1 million kilometers; the area shown is about 25 000 kilometers, with features visible on the originals that are as small as 30 kilometers across. The Red Spot is partly obscured on the north by a thin layer of overlying ammonia cirrus cloud. South of the Red Spot is one of the three white ovals, which are also anticyclonic vortices in the atmosphere. The frame at the top right is in natural color, while the red and blue have been greatly exaggerated in the frame at the right to bring out fine detail in the cloud structure. [P-21430C and P-21431C]

tures seem to resemble the convective storms that form in the Earth's tropics. The plumes travel eastward at speeds ranging from about 100 to 150 meters per second, but they do not move as a unit.

The most visible cloud interactions take place in the region of the Great Red Spot. Material within the Red Spot rotates about once every six days. Infrared measurements show that the Red Spot is a region of atmospheric upwelling, which extends to very high altitudes; however, the divergent flow suggested by this upwelling seems to be very small—one bright feature was observed to circle the Red Spot for sixty days without appreciably changing its distance from the spot's center. During the Voyager 1 flyby, spots were seen to move toward the Red Spot from the east, flow along its northern border, then either flow on to the west past the Red Spot or into the outer regions of its vortex. A spot caught on the outer edge of the Red Spot flow might break in two as it reached the eastern edge of the spot, with one piece remaining in the vortex and the other moving off to the east. Alternatively, a spot floating toward the Red Spot from the east might be pushed northward to join the eastward current flowing north of the giant red oval.

By the time of the second Voyager encounter, a ribbon of white clouds curled around the northern border of the Red Spot, blocking the motion of small spots that might otherwise have been caught up in the vortex. Spots approaching the Red Spot from the east just turned around and headed back in the direction from which they had come.

In the northern hemisphere, small brown anticyclonic features speed around the planet, often colliding with one another. On collision,

Voyager 2 captured the Red Spot region four months after Voyager 1, when some changes had taken place in the cloud circulation pattern around it. This is a mosaic of Voyager 2 frames, taken on July 6. The white oval to the south is not the same one that was present in a similar location during the Voyager 1 flyby, because of differential rotation at the two latitudes. [260-606]

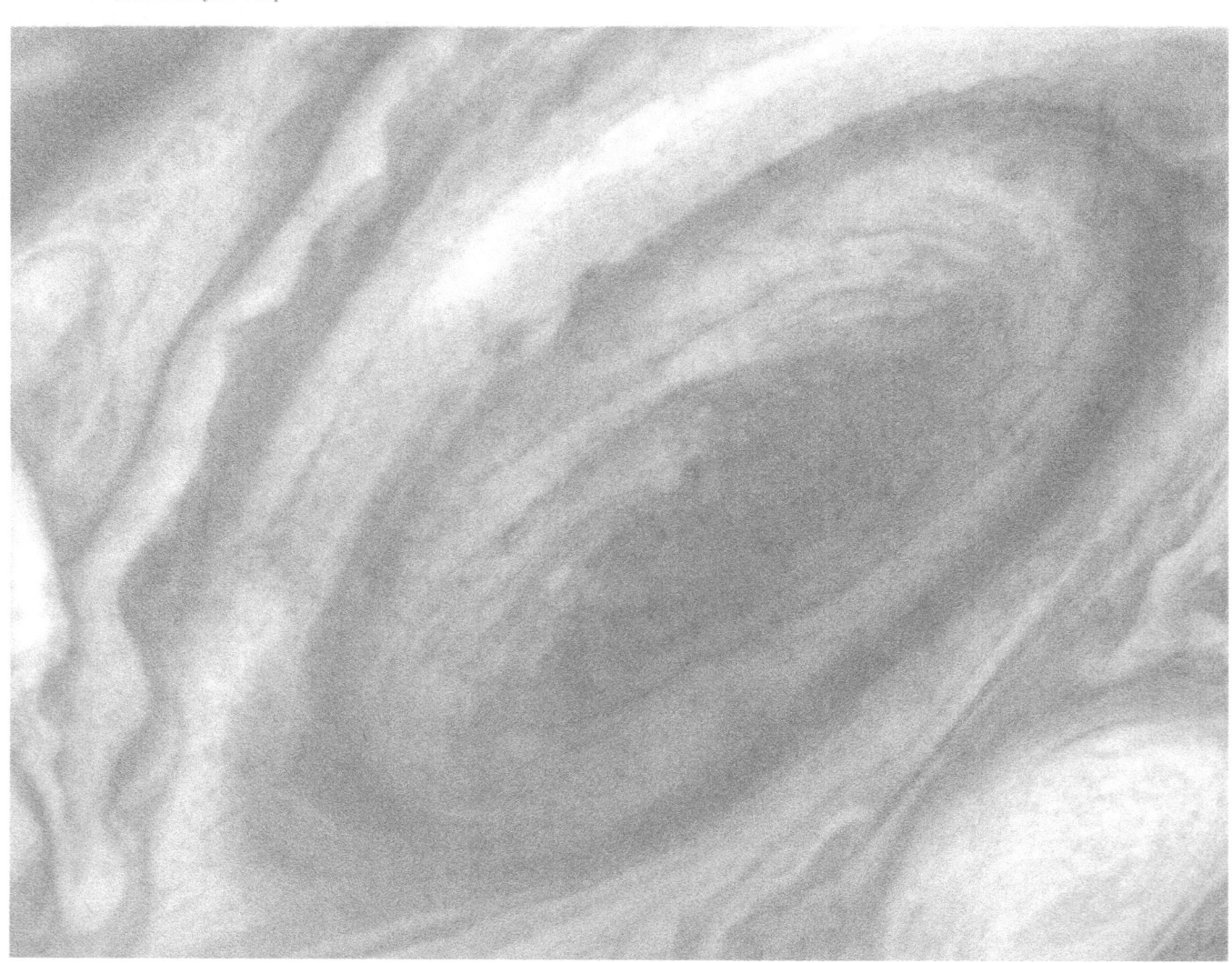

the spots may combine and roll around together for a while. Ultimately, part of the mass of the combined spots is ejected as a streamer, and the remaining material continues on its eastward path.

Order out of Chaos

Despite all the turbulence in Jupiter's atmosphere—this ever-changing chaotic mixture of cyclonic and anticyclonic flows, of ovals and filaments, of reds, browns, and whites—a pattern may be emerging: There is an underlying order to the seemingly random mixing of patterns we see in the Jovian atmosphere.

First, the changing weather patterns are in some sense cyclic. The fact that Jupiter may be reverting to the appearance it presented at the time of Pioneers 10 and 11 is not surprising. From Earth-based studies, astronomers have found that the face of Jupiter often goes through a major change every few years. The transition is very rapid, but the planet maintains its new "look" for some time—until the next major transition. At the time of the Pioneer flybys, "The Red Spot was prominent, but it was surrounded by intense cloud. There was no visible structure at all in the south tropical zone—it was totally bland. You couldn't see the turbulent area to the west," explained Garry Hunt. "And, I believe that the buildup of cloud that we're seeing to the east of the Red Spot is the beginning of the transition that will produce the Pioneer look."

There is even more order underlying Jupiter's changing atmosphere. This order is revealed in part in the alternating belts and zones. It is believed that the cloud-covered zones are regions of rising air, and the belts are regions of descending air. The internal energy of Jupiter provides the power to maintain this pattern of slow vertical circulation. In addition, there are horizontal or zonal flows that are much more regular than the changing cloud patterns.

The three large white ovals are the longest-lived features in Jupiter's atmosphere, after the Great Red Spot. Like the Red Spot, they are anticyclonic, or high-pressure, regions. During Voyager 2 encounter, one of the ovals was just south of the Red Spot. This picture shows the other two ovals as they looked in early July. The clouds show very similar internal structures. To the east of each of them, recirculating currents are clearly seen. In the lower frame, a similar structure is seen to the west of the cloud. [P-21754C]

127

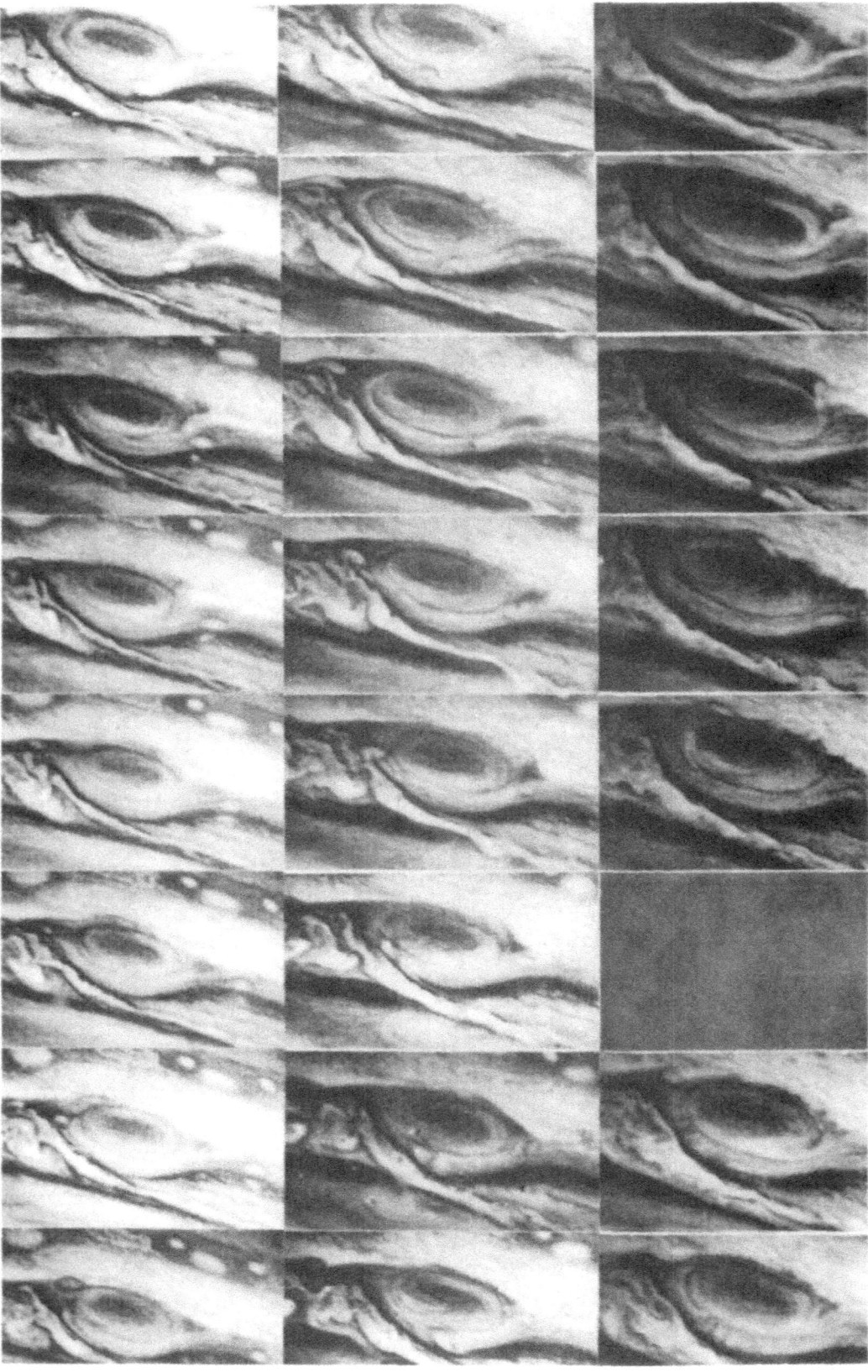

A sequence of pictures of Jupiter, taken once per rotation (about ten hours), can be used to construct a time-lapse movie of the circulation of the Jovian atmosphere. These frames are from the Voyager 1 "Blue Movie" of the Great Red Spot region. Every odd Jovian rotation is shown, so the 24 frames correspond to 48 Jupiter days, or about 20 Earth days. White spots can be seen entering the Red Spot from the upper right and being carried around by its six-day rotation until they are ejected toward the lower right. Above the Red Spot, the flow is toward the right; below it, toward the left. The rotation of the Spot is counterclockwise, or anticyclonic. [260-449]

"At first, Voyager seemed to do nothing but emphasize the chaos, not the order in Jupiter's atmosphere," Andy Ingersoll stated. "There are turbulent regions in which individual little spots seem to change every Jovian rotation. And the whole texture in certain turbulent regions is unrecognizable in one earth day." With the Voyager spacecraft, more detail could be seen than ever before; in addition, changes could be observed on a small timescale, as they happened. "It became much more of a mystery how large-scale order could exist in the face of all this small-scale chaos. But I think we are beginning to see the order underneath. What we are looking at when we observe Jupiter are minute cloud particles representing only a small fraction of the mass of the atmosphere." The large-scale order the scientists had found was a regular alternation of eastward and westward jets. "If we take all the measurements from Earth-based observations over the last 75 years, we find that every current that has ever been seen from the Earth over 75 years is visible in one ten-hour rotation. They're all there—they were just invisible." The ever-changing appearance was dancing above a regular, almost stationary pattern of alternating flows, which may come from deep within Jupiter's atmosphere. Why this alternating pattern persists remains a mystery. Even if the underlying pattern can be thought of as a sort of Jovian climate, it still does not explain the mechanics of "the minor sideshow"—the changing weather patterns. Analysis of the Voyager pictures is sure to keep planetary meteorologists busy for many years to come.

Lights in the Night Sky

Toward the end of the first encounter period, Voyager 1 flew behind Jupiter, and the spacecraft's wide-angle camera scanned the northern hemisphere on the nightside of the planet, searching for aurorae and lightning bolts. The most impressive darkside feature found was a tremendous aurora in the north polar region. But this was not the first time Jovian aurorae had been detected. Very-high-energy auroral emissions resulting from ultraviolet glows of atomic and molecular hydrogen had been detected prior to encounter on the bright side of Jupiter by the ultraviolet spectrometer. The ultraviolet observations indicate that atmospheric temperatures in the auroral regions are at least 1000 K. In both the visible and the ultraviolet spectra, the aurorae are confined to the polar regions and result from charged magnetospheric particles striking the upper atmosphere. The ultraviolet aurorae are created when high-energy particles from the Io plasma torus spiral in toward Jupiter on magnetic field lines.

Several meteor trails were also evident in the darkside pictures of Jupiter's atmosphere. Traveling at roughly 60 kilometers per second as they entered, these fireballs brightened

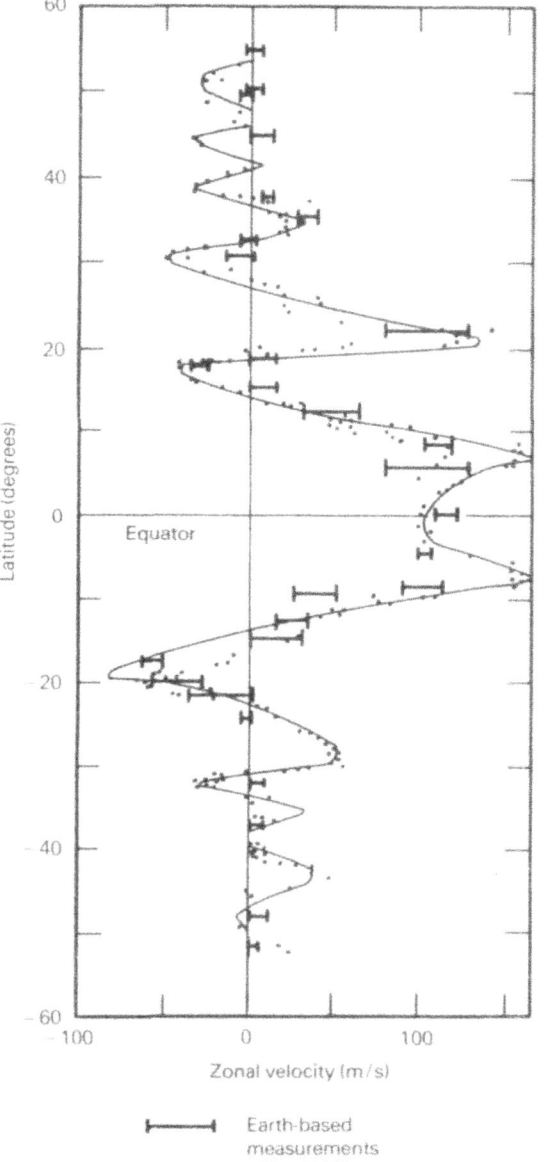

Earth-based measurements (1974-1979)

At different latitudes, the strong prevailing zonal winds produce different apparent rotation rates. Plotted here are the horizontal velocities measured from a pair of Voyager images taken one rotation (about ten hours) apart. Also shown for comparison are older ground-based measurements obtained from careful timings of the apparent rotation rate at different latitudes. The excellent agreement of the two plots indicates the stability of the zonal winds. Also, the wind pattern shows much greater symmetry between northern and southern hemispheres than do the more superficial cloud patterns.

The night side of Jupiter is not dull. A large aurora (northern light) arcs across the northern horizon, while farther south about twenty large bolts of lightning illuminate electrical storms in the clouds. Similar pictures also revealed fireballs, or large meteors, burning up in the atmosphere of Jupiter. [P-21283B/W]

quickly and seemed to survive for about 1000 kilometers before they died.

Clusters of lightning bolts—indicative of electrical storms—were also discovered on Jupiter's nightside. This particular phenomenon does not seem to depend on latitude. The Voyager 1 photograph that captured the huge Jovian aurora also caught the electrical discharges of 19 superbolts of lightning, and Voyager 2 photographs located eight additional flashes. Radio emission (whistlers) from lightning discharges were also detected by the Voyager radio astronomy receivers and the plasma wave instrument.

Magnetic Field

Deep in the interior of Jupiter, the pressures are so great that hydrogen becomes an electrical conductor, like a metal. Currents driven by the rapid rotation of the planet are thought to flow in this metallic core. The result is a magnetic field that penetrates the space around Jupiter.

Direct measurements of the Jovian magnetic field were first made by the Pioneers, and Voyager results generally confirm the initial findings. The strength of the Jovian field is about 4000 times greater than that of the Earth. The dipolar axis is not at the center of Jupiter, but offset by about 10 000 kilometers and tipped by 11 degrees from the axis of rotation. Each time the planet spins, the field wobbles up and down, carrying with it the trapped plasma of the radiation belts. The Voyager particles and fields instruments concentrated not on the planetary magnetic field but on the processes taking place in the magnetosphere.

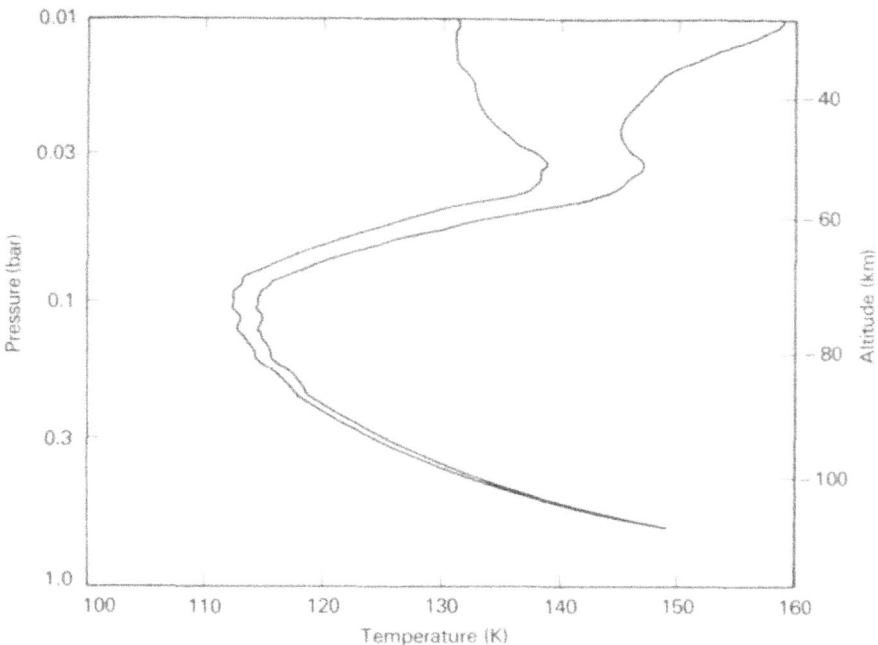

A probe of the Jovian atmosphere is obtained each time a spacecraft passes behind the planet as seen from the Earth. Passage through the ionosphere and atmosphere alters the phase of the radio telemetry signal, and subsequent computer analysis allows members of the Radio Science Team to reconstruct the profile of the atmosphere. Shown here is the atmospheric temperature as a function of pressure as derived from the Voyager 1 X-band occultation data, corresponding to a point at latitude 12°S, longitude 63°. The two curves represent extreme interpretations of the same data; the best fit lies somewhere between. Accuracy is high at greater depths but poor at levels above a pressure of about 0.03 bar. Clearly shown is the temperature minimum near 0.1 bar and the steady increase of temperature with depth as the radio beam probed toward the cloud level near 1.0 bar.

At a wavelength near 5 micrometers, the primary gases in the atmosphere of Jupiter are particularly transparent, and the infrared radiation from the planet comes from relatively great depths. At these depths, it is possible to see evidence of gases such as water vapor that condense at higher altitudes where the temperatures are lower. This IRIS spectrum in the 5-micrometer spectral region shows features identified with water (H_2O), germane (GeH_4), and deuterated methane (CH_3D), as well as the more easily detected ammonia (NH_3).

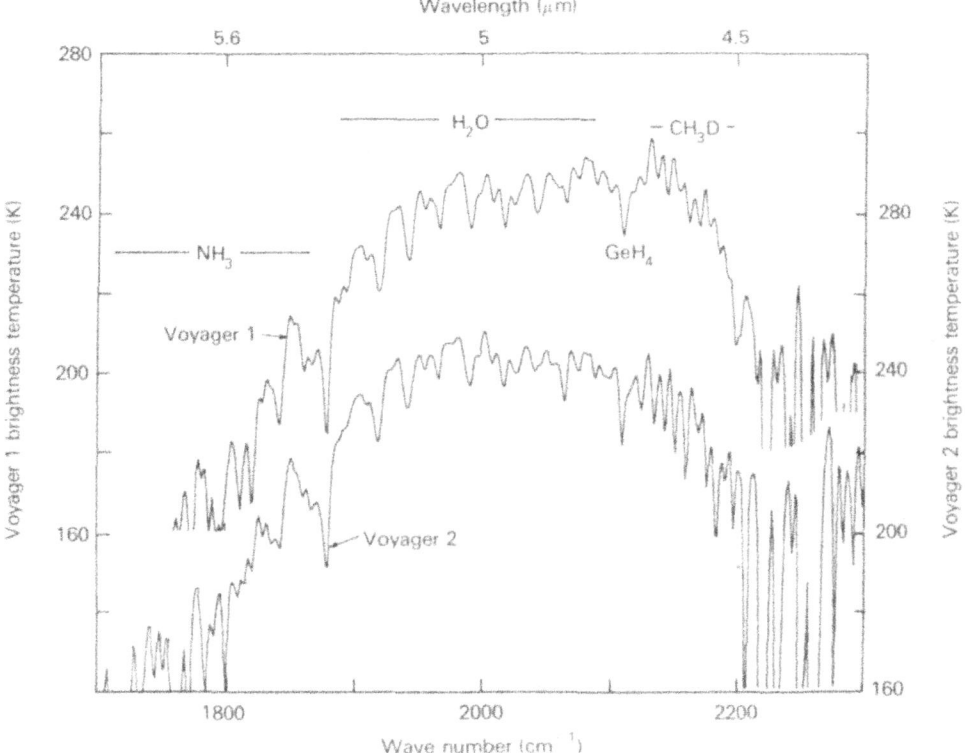

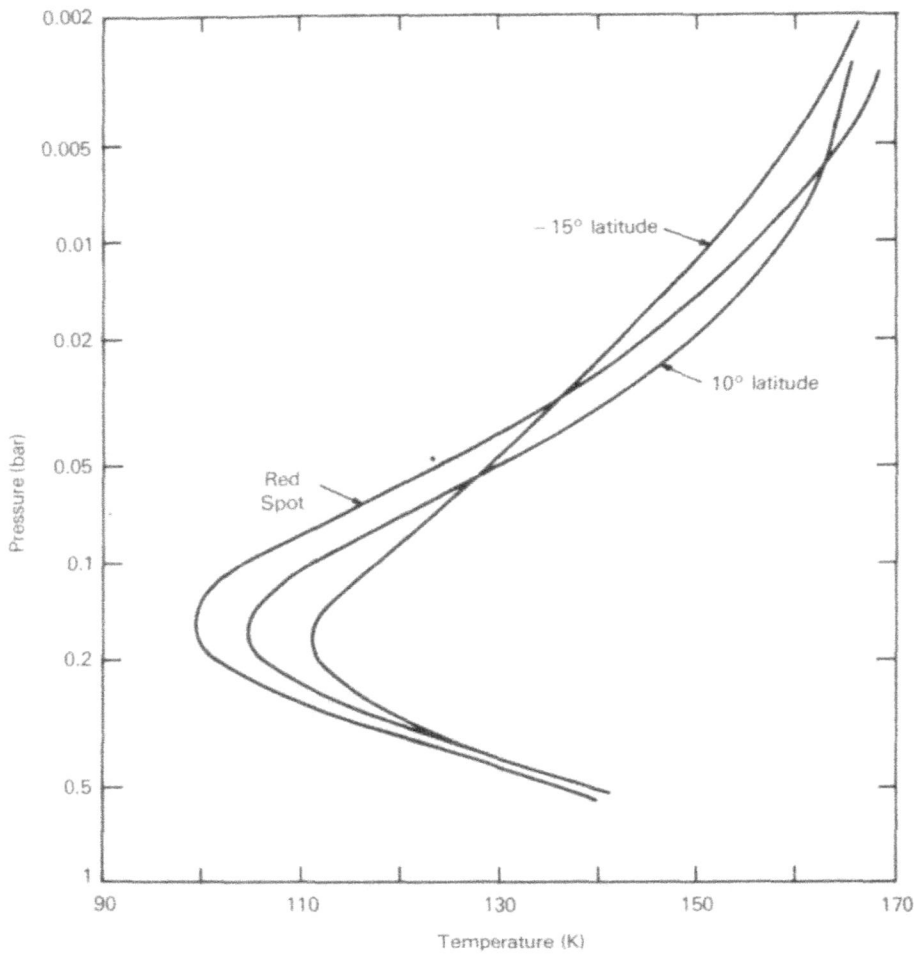

The structure of the Jovian atmosphere can be derived from infrared spectra as well as from the radio occultation data. This profile of temperature as a function of pressure covers the same range of altitudes as does the preceding figure and is in good agreement. Both profiles locate a temperature minimum of about 110 K near a pressure of 0.1 bar (1000 mb = 1 bar = 1 atm). The IRIS data also show variation of structure with position, including a cooler minimum temperature (about 100 K) over the Great Red Spot.

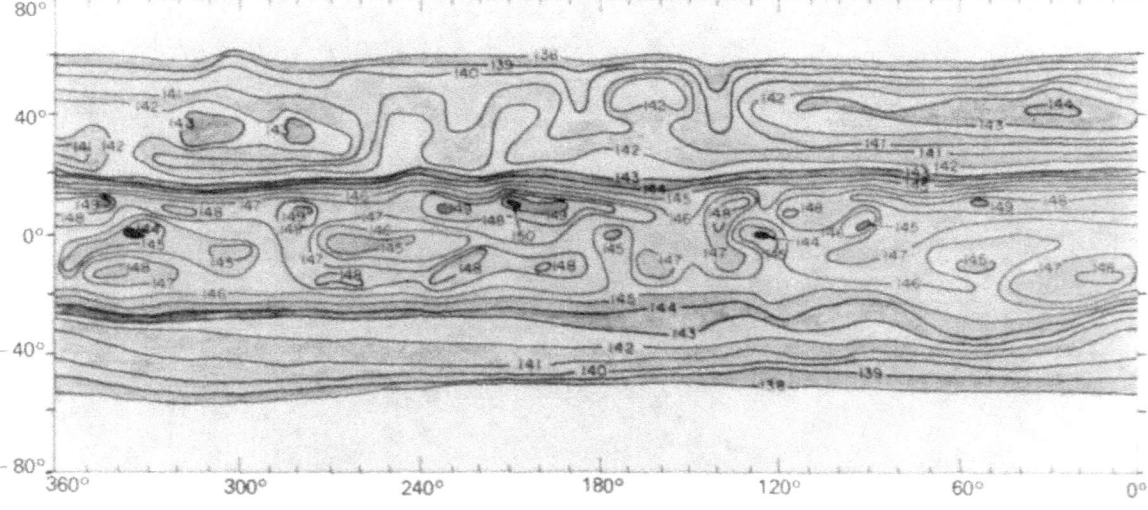

The Magnetosphere

Giant Jupiter has an enormous realm—from the size of its satellite system to its tremendous aurorae and superbolts of lightning, to the huge planet-sized cloud features that surround its atmosphere. The most gargantuan Jovian feature is its magnetosphere, which envelopes the satellites and constantly changes in size, pumping in and out at the whim of the solar wind. The Pioneer and Voyager spacecraft provided four cuts through this dynamic region, showing that its borders in the upwind solar direction lie between 50 R_J and 100 R_J from Jupiter. Downwind, away from the Sun, the magnetosphere extends much farther; some scientists postulate that a magnetotail may reach as far as the orbit of Saturn.

Charged particles in the magnetosphere are subject to powerful forces. Tightly embedded in Jupiter, the magnetic field spins with a ten-hour period as the planet rotates. The particles are caught in the spinning field and accelerated to high speeds. The result is a co-rotating plasma in the magnetic equator of Jupiter, extending outward to at least 20 R_J. Beyond this distance, the flow breaks up and the magnetosphere is more unstable. Within the co-rotation region, the spinning plasma sets up a powerful electric current girdling the planet.

Charged particles can be accelerated in the magnetosphere to high energies, corresponding to speeds tens of thousands of kilometers per second. Some of these particle streams escape from the inner parts of the magnetosphere and can penetrate the magnetopause and be ejected from the Jovian system. On Voyager 1, the low energy charged particle instrument began detecting these streams of "hot" plasma on 22 January, when Voyager 1 was still 600 R_J (almost 50 million kilometers) from the planet. Voyager 2 first detected Jovian particles at an even greater distance, 800 R_J. Hydrogen and helium ions (protons and alpha particles) dominate the magnetosphere at great distances from Jupiter, but increasing amounts of sulfur and oxygen appeared as the spacecraft crossed the magnetopause. The heavier ions presumably originate from Io.

In the inner magnetosphere, the Galilean satellites have a powerful influence on the populations of fast-moving particles. During the Voyager 1 encounter, the primary effect was observed at Io, where the satellite apparently sweeps up energetic electrons. In the million-volt energy range, these particles are depleted near Io, with peaks observed both inside and outside the satellite's orbit. Voyager 2 passed close to Ganymede, and here also major effects were seen, with the satellite apparently

The structure of the atmosphere can be inferred from IRIS spectra at many locations over the disk of Jupiter. Scientists are beginning to assemble this vast amount of information into maps that show the temperatures at a given pressure. Shown here are the observed temperatures (bottom left) at a depth near the cloud tops (0.8 bar) and (below) at an altitude about 30 kilometers higher (0.15 bar). The temperature contours are labeled in degrees Kelvin. The banded structure, with higher temperatures near the dark equatorial belt, is most clearly evident at the lower altitude. Surprisingly, the cool region associated with the Great Red Spot (latitude 23°S) is more apparent at high altitude.

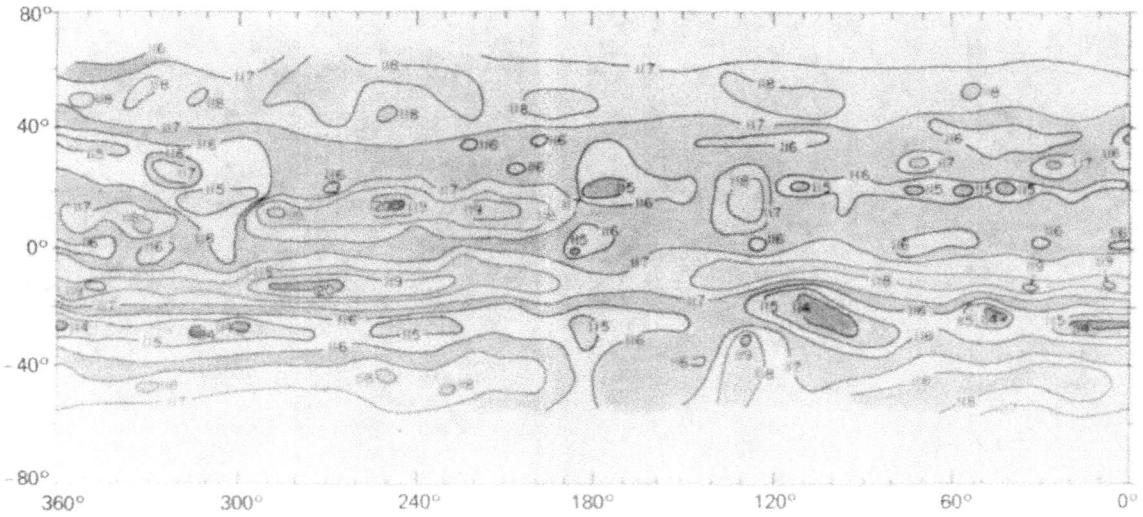

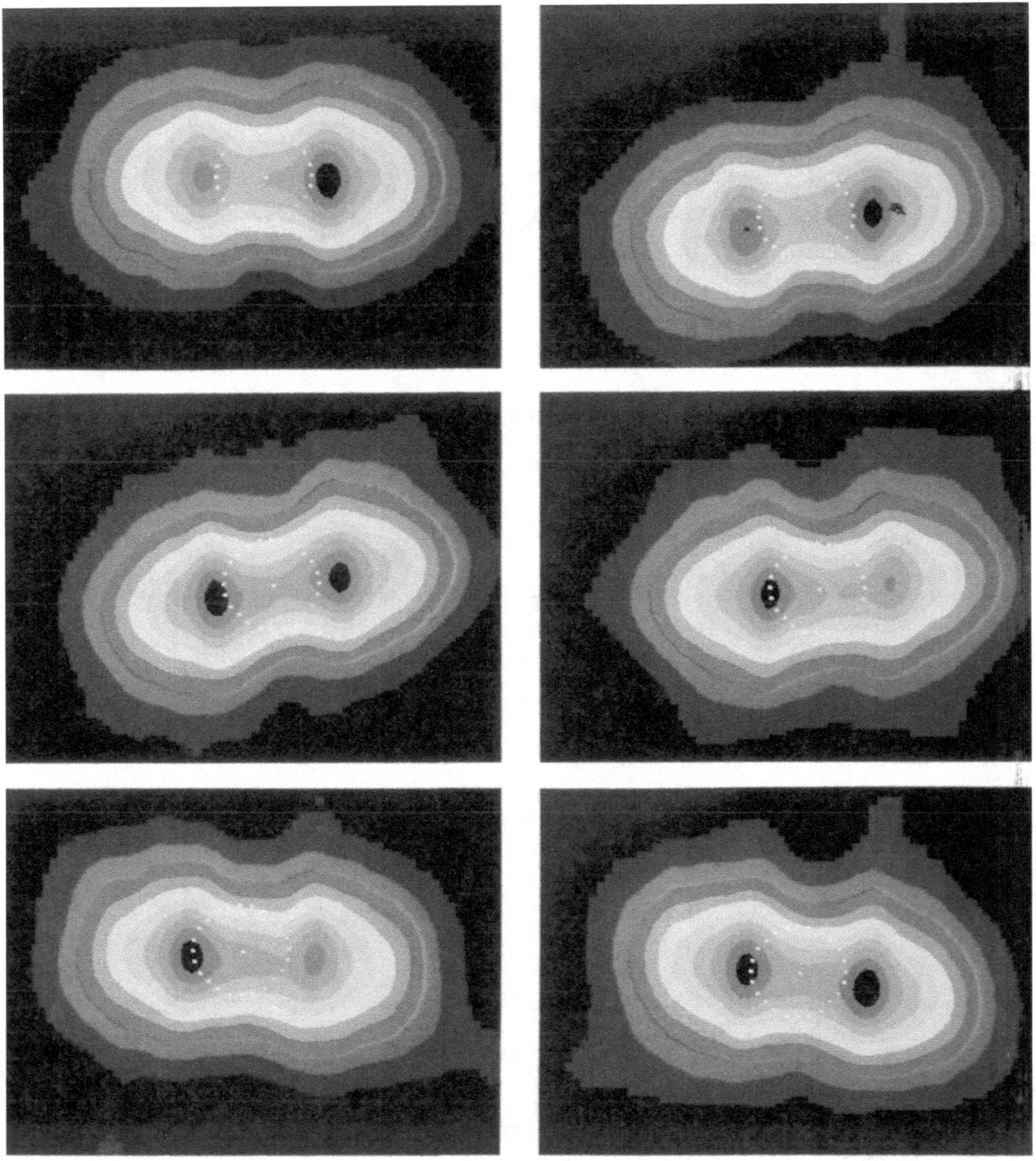

The magnetosphere of Jupiter can be "seen" from Earth by its emissions at radio wavelengths. The recent development of imaging radio telescopes in Great Britain, The Netherlands, and the United States allows frequent mapping of the large-scale features of the innermost magnetosphere, inside the orbit of Io. These six images showing one rotation of Jupiter were obtained near the time of the Voyager 1 flyby by Imke de Pater at the Westerbork Radio Observatory near Leiden. The brightest emission is shown by dark red or black. The size of the planet is indicated by a dotted white circle. The tilt of the magnetosphere relative to the rotation axis of the planet can be seen by the wobble of the magnetosphere as Jupiter rotates.

absorbing electrons. In the wake created by the motion of the co-rotating plasma past Ganymede, the particle populations showed large and complex variations.

Just inside the Jovian magnetosphere is the "hot spot" of the solar system: a 300–400 million degree plasma detected by Voyager 1 while it was still about 5 million kilometers from Jupiter. T. P. Armstrong commented, "Even the interior of the Sun is estimated to be less than 20 million degrees." S. M. Krimigis added that the temperature of this plasma is "the highest yet measured anywhere in the solar system." Fortunately for Voyager, this region of incredibly hot plasma is also one of the solar system's best vacuums. The spacecraft was in little danger because the bow shock protects this region from the solar wind, and most of the particles in Jupiter's magnetosphere are held in much closer to the planet.

The very hot plasma in the outer magnetosphere discovered by Voyager is thought to play an important role in establishing the size of the Jovian magnetosphere. Although the density is low, only about one charged particle per hundred cubic centimeters, this plasma actually carries a great deal of energy because of the high speed of the particles. It is this plasma pressure, rather than the magnetic field pressure, that appears to hold off the pressure of the solar wind. However, the balance between hot plasma inside the magnetopause and the solar wind outside is not very stable. The Voyager experimenters suggest that a small change in solar wind pressure can cause the

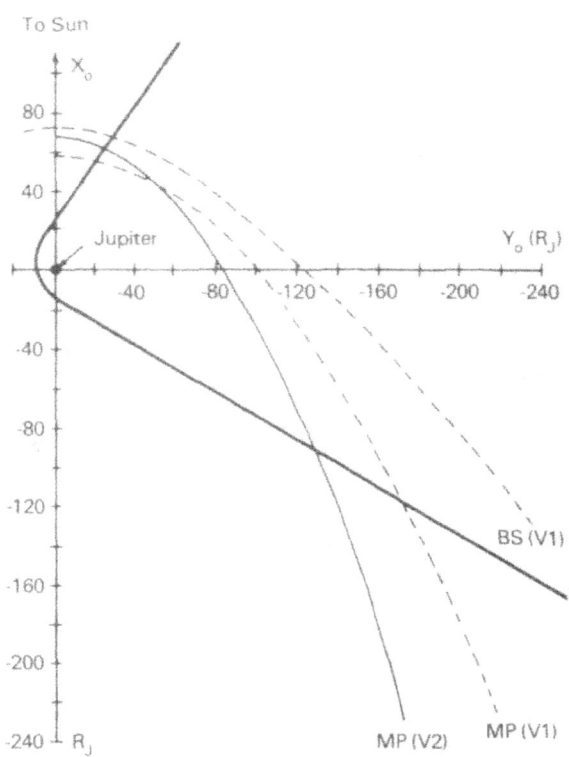

Each Voyager passed through the boundaries of the magnetosphere—the bow shock (BS) and the magnetopause (MP)—on both the inbound and the outbound legs of its passage through the Jovian system. In this diagram, the heavy solid line represents the spacecraft trajectory, as seen looking down from the north. Also shown are the positions of the bow shock in March and of the magnetosphere in both March and July.

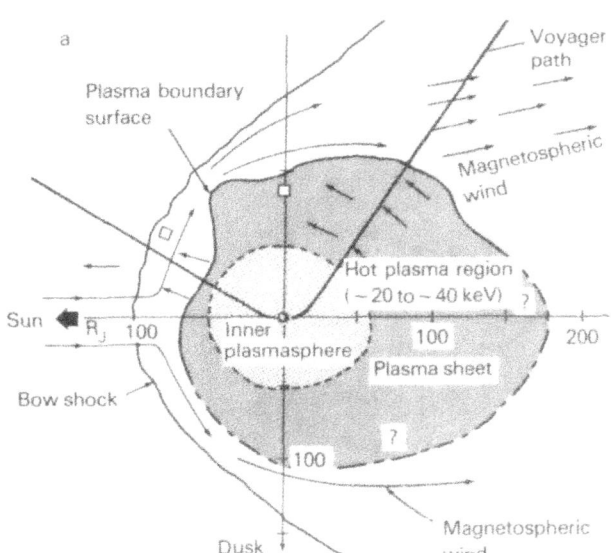

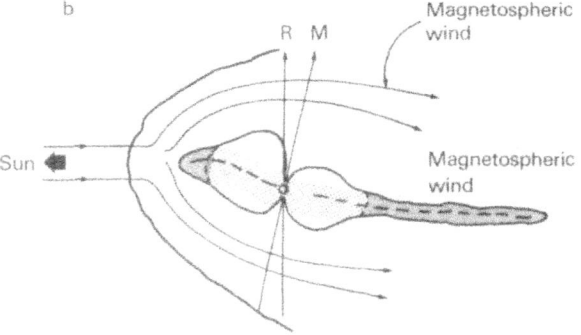

Several regions of plasma (charged particles) make up the Jovian magnetosphere. These sketches, based on Voyager data, show the magnetosphere as viewed from above (left) and as seen from the Jovian equatorial plane (above). Most of the plasma co-rotates with the planet and is confined near the magnetic equator, where it forms a broad plasma sheet about 100 R_J across.

135

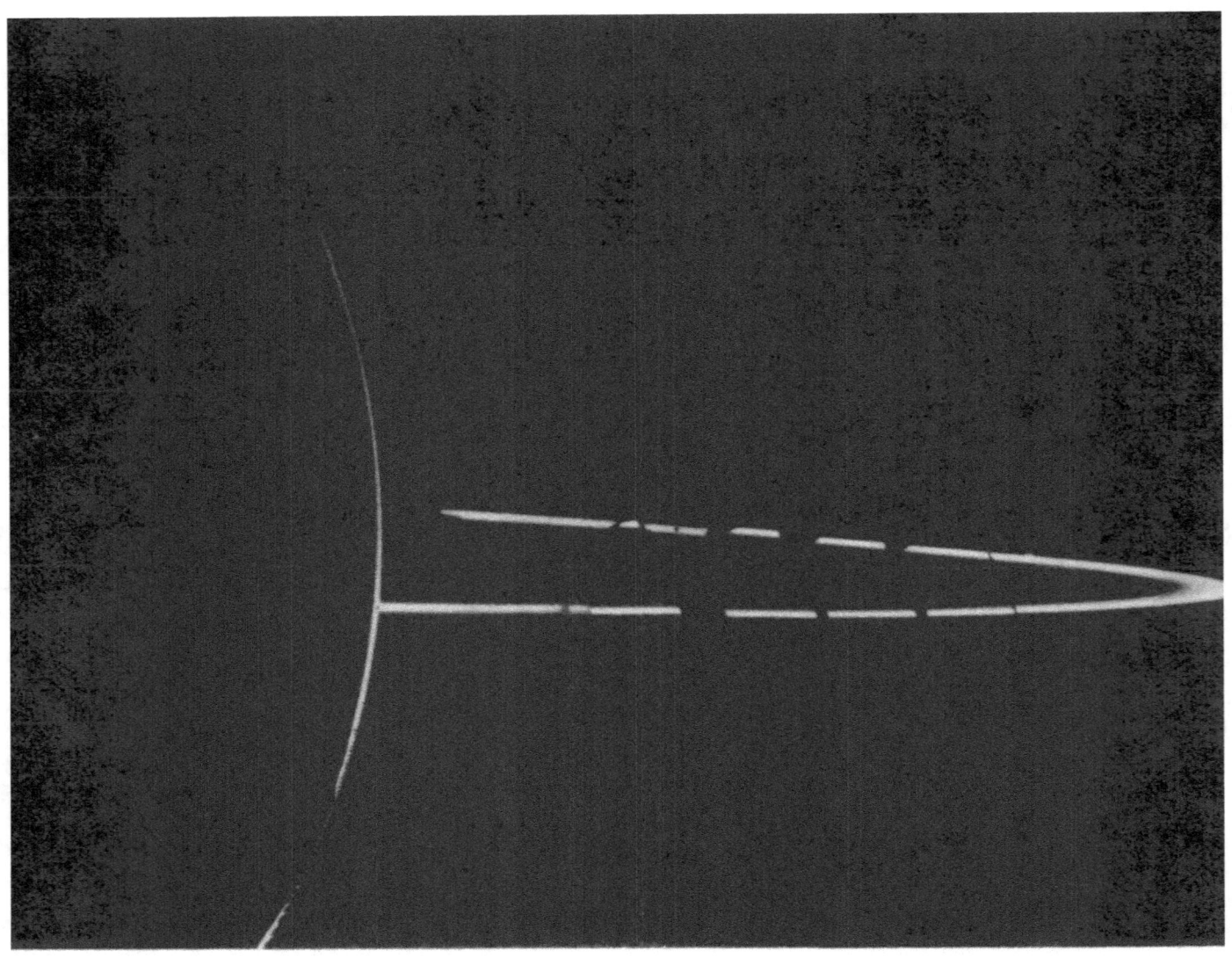

The rings of Jupiter are best seen when looking nearly in the direction of the Sun, since the small particles that comprise them are good forward scatterers of sunlight. This mosaic is of Voyager 2 images (two wide angle and four narrow angle) obtained from a perspective behind the planet and inside the shadow of Jupiter. The spacecraft was 2 degrees below the equator of Jupiter and 1.5 million kilometers from the rings. The shadow of the planet can be seen to obscure the near segment of the ring near the edge of the planet. The brightest region of the ring is about 1.8 R_J from the center of Jupiter. [260-678B]

boundary to become suddenly unstable. A large quantity of the hot plasma can then be lost, producing the bursts seen at large distances and permitting a sudden collapse of the outer magnetosphere. Continued injection of hot plasma from within would then reinflate the magnetosphere, which would expand like a balloon until another instability developed. Processes of this sort may be the cause of the rapidly varying magnetospheric boundaries observed by both Voyager spacecraft.

Rings of Jupiter

One of the spectacular discoveries of Voyager was the existence of a ring system of small particles circling Jupiter. Saturn and Uranus were known to have rings, but none had been seen before at Jupiter.

As revealed by the Voyager cameras, the rings extend outward from the upper atmosphere to a distance of 53 000 kilometers above the cloud tops, 1.8 R_J from the center of the planet. The main rings, however, are much narrower, spanning from 47 000 to 53 000 kilometers above Jupiter. There are two main rings, a 5000-kilometer-wide segment, and a brighter, outer 800-kilometer segment. The thickness of the rings is unknown, except that it is certainly less than 30 kilometers, and probably under 1 kilometer.

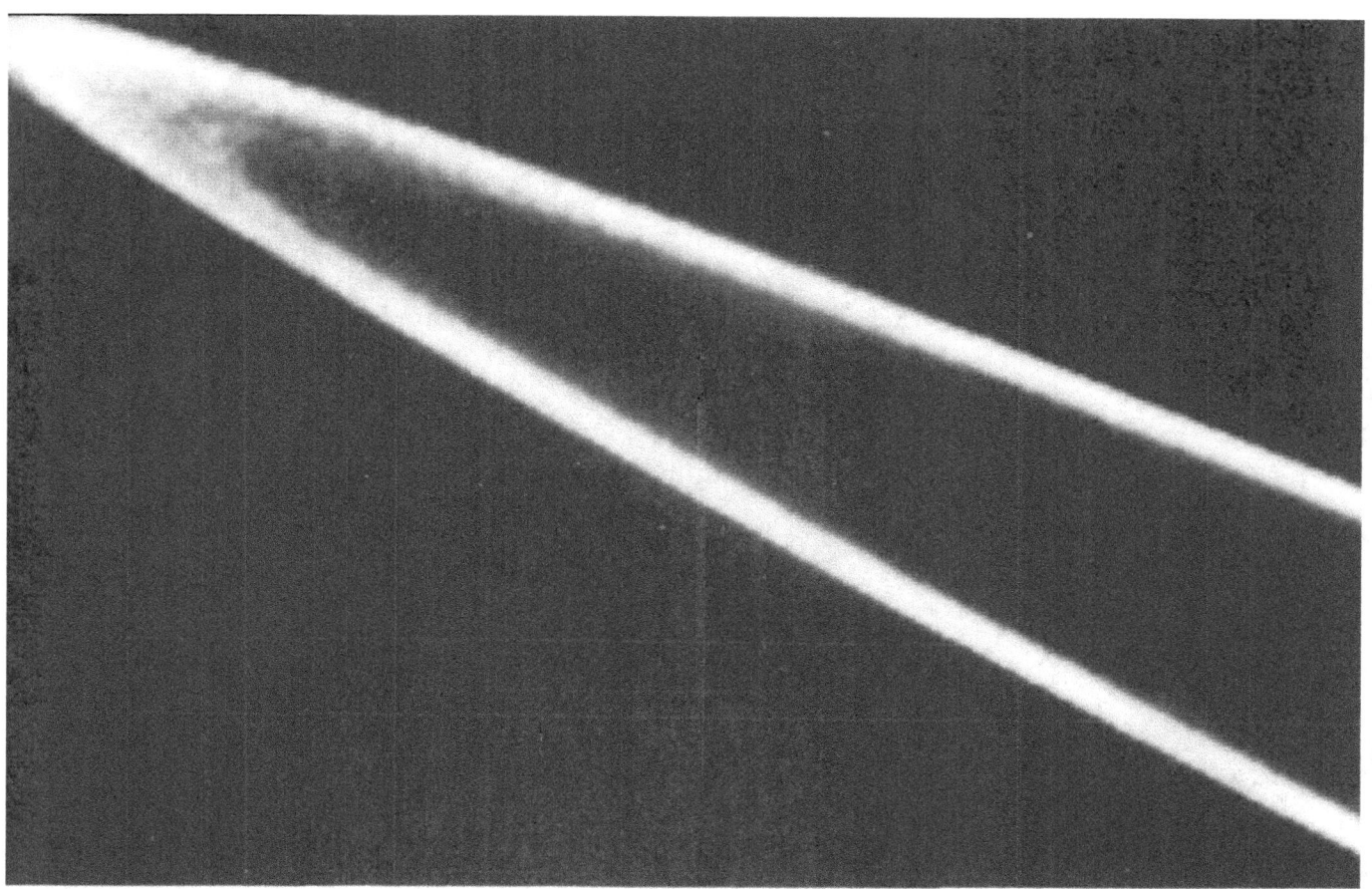

Structure within the ring can be seen in the best Voyager 2 images, taken about 27 hours after closest approach to Jupiter. This enlarged portion of a wide-angle picture taken with a clear filter shows a bright core about 800 kilometers across with a dimmer region a few thousand kilometers across on the inside, and a narrow dim region on the outside. [260-674]

The rings of Jupiter are quite tenuous, which explains why they are invisible from Earth [although they have been detected from Earth since Voyager]. Seen face on, the brightest part of the ring blocks less than one part in ten thousand of the light passing through, making it essentially transparent. In fact, the ring does not even offer much resistance to a spacecraft; Pioneer 11 traversed the ring in 1974 with no obvious ill consequences. Apparently the individual particles that make up the ring are widely dispersed. They can be seen only when the rings are viewed nearly edge-on, or toward the Sun, where they show up well in forward scattered light. It is this extra brilliance when backlit that created the excellent photos taken by Voyager 2 from inside the shadow of Jupiter.

The individual ring particles are probably dark, rocky fragments that are very small— essentially dust grains. They move around Jupiter in individual orbits, circling the planet in 5 – 7 hours. Scientists postulate that such orbits are not stable and that the particles fall slowly in toward Jupiter. Apparently the rings are constantly renewed from some source, which may be the satellite Adrastea (J14), discovered by Voyager 2. There has also been speculation that Adrastea may influence the ring structure by sweeping particles out of the ring. At present the rings of Jupiter remain mysterious. They are clearly very different from the rings of Saturn and Uranus, and reaching an understanding of their origins and dynamics presents many challenges to planetary scientists.

The four Galilean satellites of Jupiter are planet-like worlds, revealed by Voyager to be as diverse and fascinating as the terrestrial planets Mercury, Venus, Earth, and Mars. In this Voyager 1 composite, all four are shown in their correct relative size, as they would appear from a distance of about 1 million kilometers. Relative color and reflectivity are also approximately preserved, although it is not possible to show on a single print the full range of brightness from the dark rocky surface of Callisto to the brilliant white of Europa or orange of Io. Clockwise from the upper left, the satellites are Io (longitude 140°), Europa (longitude 300°), Callisto (longitude 350°), and Ganymede (longitude 320°). [260-499C]

CHAPTER 9

FOUR NEW WORLDS

Jupiter's Satellite System

In a sense, the Voyager Mission revealed a new planetary system. Astronomers had long been fascinated by the large Galilean satellites of Jupiter, but they had only looked from afar, watching the dancing points of light in their telescopes, and, occasionally, as the atmosphere steadied, seeing these points resolve themselves into tiny disks before dissolving again in the turbulence of the terrestrial atmosphere. Much had been learned from telescopic studies, but not until the Voyager flights had we truly seen the Galilean satellites. The historic hours as Voyager 1 cruised past each satellite on March 5 and 6, 1979, fundamentally altered our perspective. Four new worlds were revealed, as diverse and fascinating as the more familiar terrestrial planets. Although not yet household words, the names Io, Europa, Ganymede and Callisto have now been added to Mercury, Venus, Moon, and Mars in the lexicon of important "Earth-sized" bodies in the solar system.

Jupiter has fifteen known satellites, counting the two new satellites discovered by Voyager. These moons vary greatly in size, composition, and orbit. The four outermost satellites, Sinope, Pasiphae, Carme, and Ananke, circle the planet in retrograde orbits of high inclination; their distances from Jupiter vary between 20 and 24 million kilometers (290 R_J to 333 R_J). These small bodies, none more than 50 kilometers in diameter, require nearly two years for each orbit of Jupiter. It is possible that they are captured asteroids, but so little is known about them that astronomers cannot tell if their surface properties resemble those of asteroids, or if these four satellites are even similar to each other.

The next group of Jovian satellites consists of four small difficult-to-observe objects.

These are Lysithea, Elara, Himalia, and Leda, the latter discovered by Charles Kowal of Hale Observatories in 1974. They have similar orbits, varying in distance from Jupiter between 11 and 12 million kilometers (about 160 R_J). Like the outer group, these satellites have orbits of high inclination; unlike the outer group, they move in the proper, prograde direction around Jupiter. The largest, Himalia (170 kilometers in diameter) and Elara (80 kilometers diameter), are known to be very dark, rocky objects, and it seems probable that the others are similar. It is unlikely that the census of the outer groups of irregular satellites is complete, and new satellites less than 10 kilometers in diameter will probably be discovered.

The Jovian system is dominated, of course, by the large Galilean satellites, which vary in size from just smaller than the Moon (Europa) to nearly as large as Mars (Ganymede). These satellites are in regular, nearly circular orbits in the same plane as the equator of Jupiter, and all four lie within the inner magnetosphere of Jupiter, where they interact strongly with energetic charged particles and plasma. Most of this chapter will be devoted to a discussion of these fascinating worlds.

We now know of three additional small satellites inside the orbit of Io, orbiting close to Jupiter. The first, Amalthea, was discovered in 1892; it orbits Jupiter in just twelve hours at a distance of 181 000 kilometers (2.55 R_J). A smaller object, Adrastea (officially 1979J1 for the first new satellite of Jupiter discovered in 1979), is much closer, at 134 000 kilometers (1.76 R_J). As described in Chapter 7, it skirts the outer edge of the ring, circling Jupiter in just over seven hours. The inner satellite moves faster than Jupiter's rotation; seen from the planet, it would rise in the west and set in the east. Both Amalthea and Adrastea are buried

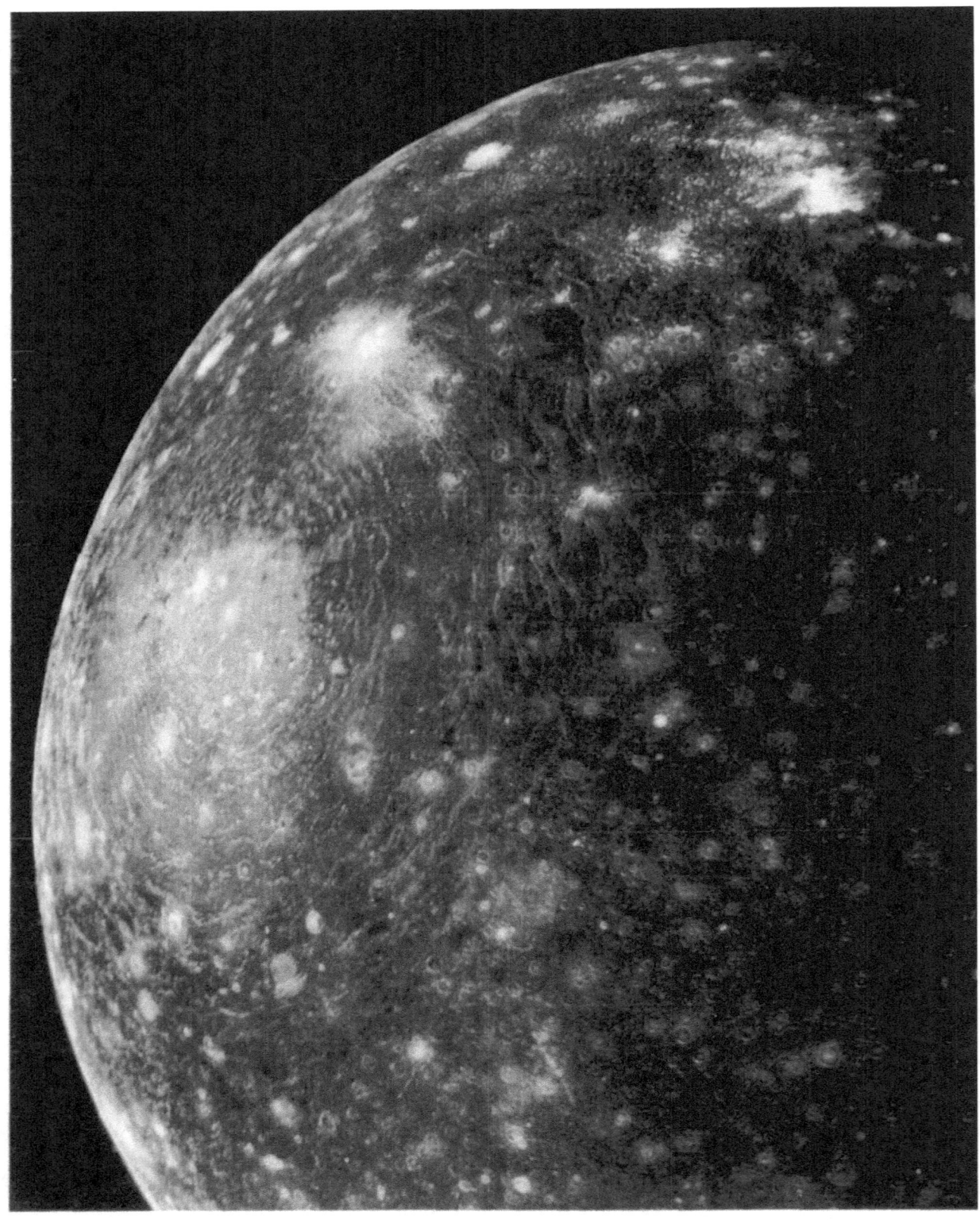

Callisto was revealed by the Voyager cameras to be a heavily cratered and hence geologically inactive world. This mosaic of Voyager 1 images, obtained on March 6 from a distance of about 400 000 kilometers, shows surface detail as small as 10 kilometers across. The prominent old impact feature Valhalla has a central bright spot about 600 kilometers across, probably representing the original impact basin. The concentric bright rings extend outward about 1500 kilometers from the impact center. [260-450]

Name		Distance From Jupiter		Period (days)	Year of Discovery
		10^3 kilometers	Jupiter Radii		
Adrastea	J14	134	1.76	0.30	1979
Amalthea	J5	181	2.55	0.49	1892
1979J2	J15	222	3.11	0.67	1980
Io	J1	422	5.95	1.77	1610
Europa	J2	671	9.47	3.55	1610
Ganymede	J3	1070	15.10	7.15	1610
Callisto	J4	1880	26.60	16.70	1610
Leda	J13	11 110	156	240	1974
Himalia	J6	11 470	161	251	1904
Lysithea	J10	11 710	164	260	1938
Elara	J7	11 740	165	260	1904
Ananke	J12	20 700	291	617	1951
Carme	J11	22 350	314	692	1938
Pasiphae	J8	23 300	327	735	1908
Sinope	J9	23 700	333	758	1914

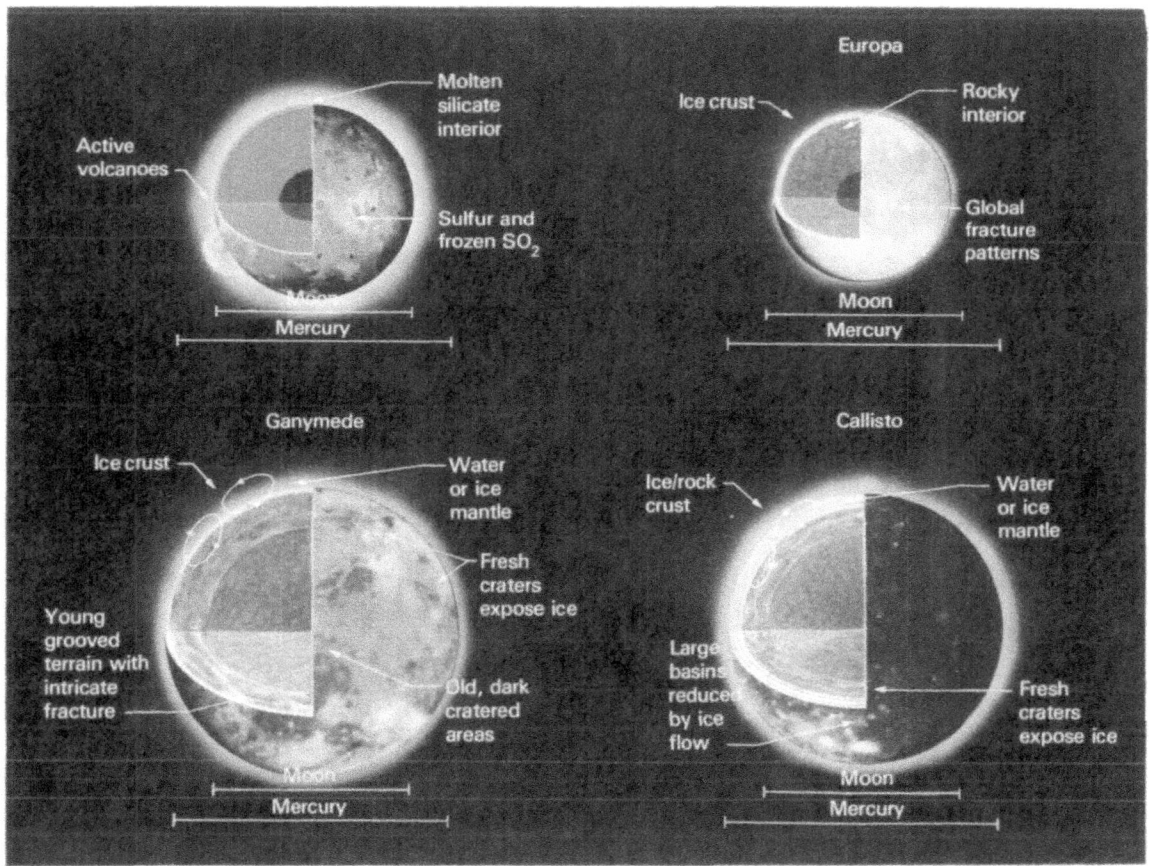

The state of the interiors of the Galilean satellites can be judged from their sizes and densities. These cross-sectional views represent the best guess following the Voyager flybys as to the composition and structure of the objects. Io, with a density equal to that of the Moon and a long history of volcanic activity, is a dry, rocky object. Europa is less dense, and it probably has a global ocean of ice as much as 100 kilometers thick over a rocky interior. Ganymede and Callisto both have densities near 2 grams per cubic centimeter, suggesting a composition about half water and half rock. There is probably a rocky core surrounded by an icy mantle.

deep within the inner magnetosphere where they are continually bombarded by energetic electrons, protons, and ions. Depletion of the Jovian radiation belt particles was observed at the orbits of both satellites by Pioneer 11, which went much closer to Jupiter than the Voyagers, testifying to the intensity of the interaction between these objects and their surroundings.

Long after the flybys of Jupiter, continued analysis of Voyager images revealed another new satellite, Jupiter's fifteenth. Intially designated 1979J2, the unexpected new satellite orbits the planet at $3.17 R_j$, between Io and Amalthea. Stephen Synnott of the JPL Optical Navigation Team discovered the satellite on pictures taken during the Voyager 1 events on March 5, 1979, while searching for additional images of satellite 1979J1. It is about 75 kilometers in diameter, but nothing else is known about its physical properties.

Together, the 15 satellites circling giant Jupiter form a mini-solar system. Perhaps the outer, irregular satellites were captured or resulted from the catastrophic collisions of one or more larger satellites with passing asteroids. The inner seven satellites constitute a coherent system, almost certainly formed together with Jupiter and sharing a common 4.5-billion-year history. They are fascinating as individual worlds, and also as brothers and sisters, and the study of their interrelationships undoubtedly will provide insights into the general problems of planetary formation and evolution.

SIZES AND DENSITIES OF THE GALILEAN SATELLITES

Name	Diameter (kilometers)	Density (grams per cubic centimeter)
Io	3640	3.5
Europa	3130	3.0
Ganymede	5270	1.9
Callisto	4840	1.8

Callisto

Callisto is the least active geologically of the Galilean satellites. Basically a dead world, it bears the scars of innumerable meteoric impacts, with virtually no sign of major internal activity. Callisto is a world of craters, and to understand it we must explore the role that cratering plays in molding planetary surfaces.

The space between the planets is filled with debris, ranging from the larger asteroids, hundreds of kilometers in size, down to microscopic grains of dust. Inevitably, each planet collides with some of these fragments. The smaller particles do little damage; in the case of a planet with an atmosphere, like Earth, they burn up as meteors before reaching the surface, whereas on an airless planet, they erode the surface by sandblasting the exposed rock. The larger impacts are another matter, and the craters they produce can be the dominant features on the surface of a planet.

Voyagers 1 and 2 photographed most of the surface of Callisto at resolutions of a few kilometers or better. Shown here is a preliminary shaded relief map. Additional measurements will improve the accuracy of the coordinate system. [260-672]

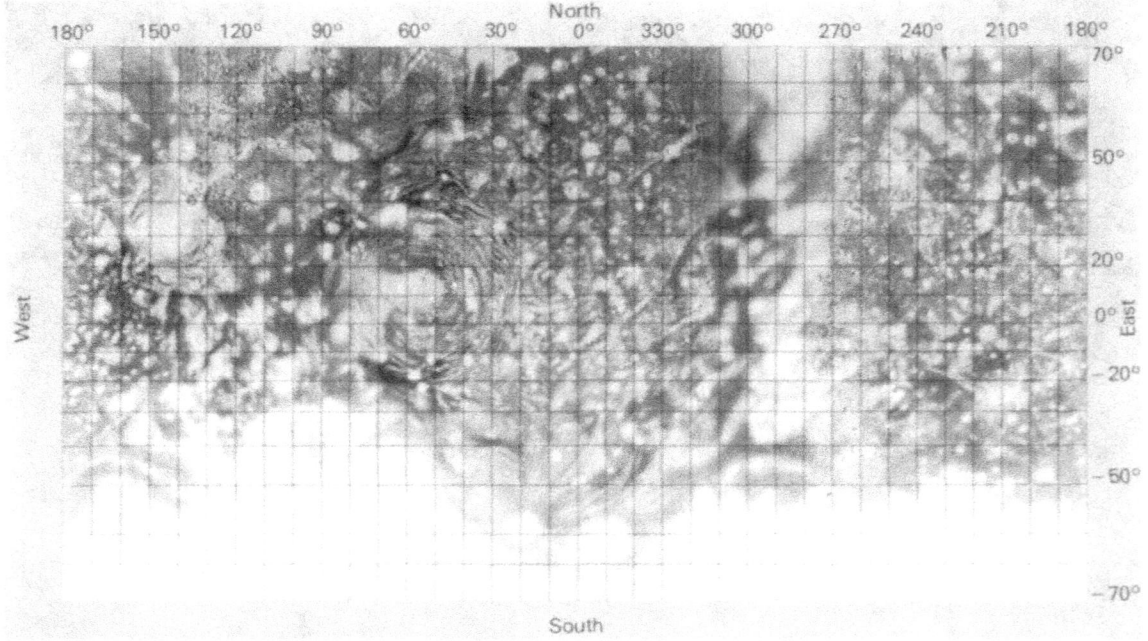

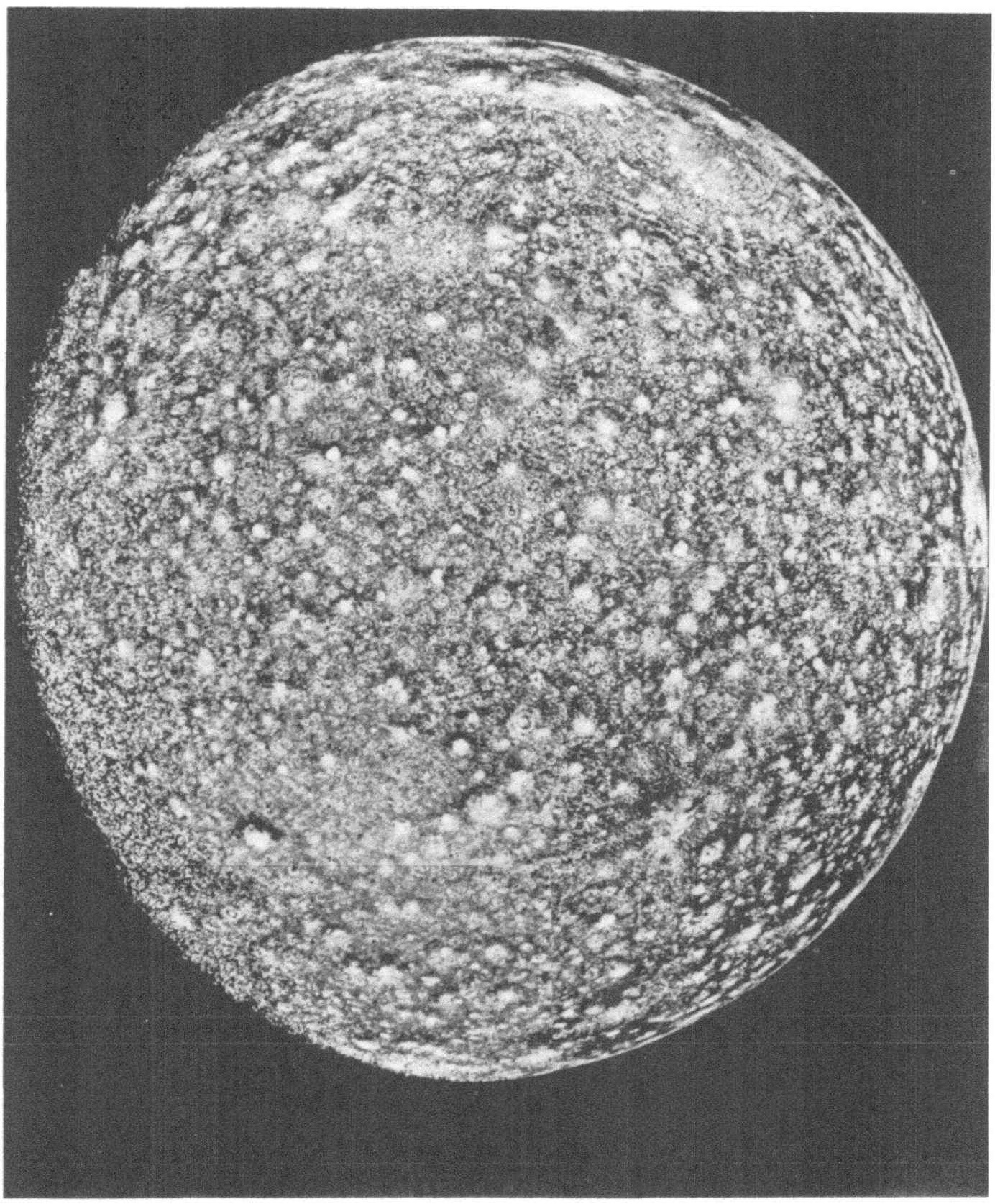

Callisto is a world of craters, as is well shown in this Voyager 2 photomosaic taken from a distance of 400 000 kilometers. Craters about 100 kilometers in diameter cover the surface uniformly. Many have bright rims, perhaps composed of exposed water-ice. There are very few craters larger than 150 kilometers in diameter, however, indicating that the scars of very large impacts do not survive on the surface of Callisto. [P-21746B/W]

We who live on Earth tend not to realize the importance of cratering, for the simple reason that our planet has very few craters, and these are frequently of volcanic rather than meteoric origin. Why are we so favored? Is there an invisible shield to protect us from the cosmic shooting gallery? Clearly not; the Earth has experienced just as many cratering impacts as has the Moon or other planets. The difference is not that craters are formed less often, but that the great geological activity of Earth—erosion, volcanism, mountain building, continental drift, etc.—erases craters as fast as they are formed. On the average, a 10-kilometer-wide crater is formed on Earth about once every million years, but all those older than a few million years have been eroded away, filled in, or crushed beyond recognition by crustal motion.

If a planet lacks great internal geologic forces, large craters can survive almost indefinitely. Such is the case for the Moon. Most of the volcanism and other activity on the Moon ceased 3½ billion years ago, as the dating of lunar samples obtained by the Apollo astronauts showed. Since that time, the lunar surface has been passively accumulating impact scars.

The longer any particular surface area has been exposed, the more densely packed are the craters. Thus crater density is the first thing a planetary geologist looks for in photos of a new world. Craters are the touchstone of this field, revealing the degree of internal activity and allowing the determination of the relative ages of different surface units.

On Callisto the density of craters is very high. In some places they are packed as closely as one can imagine, particularly for craters several tens of kilometers in diameter. Although no one knows the exact rate of formation of impact craters on the Jovian satellites, geologists on the Voyager Imaging Team estimate that it would require several billion years to accumulate the number of craters found on Callisto. They therefore conclude that Callisto has been geologically inactive almost since the time of its formation.

Although superficially similar to the heavily cratered surfaces of the Moon and Mercury, Callisto is far from identical to these rocky worlds. One of the most obvious differences is a lack of craters larger than about 150 kilometers on Callisto, together with a tendency for large craters to have much shallower depths.

The concentric rings surrounding Valhalla are perhaps the most distinctive geological feature on Callisto. This Voyager 1 close-up shows a segment of the ridged terrain. The presence of superposed impact craters shows that the rings formed early in Callisto's history; however, the density of craters is less here than on other parts of the satellite, where the surface is older. [P-22194]

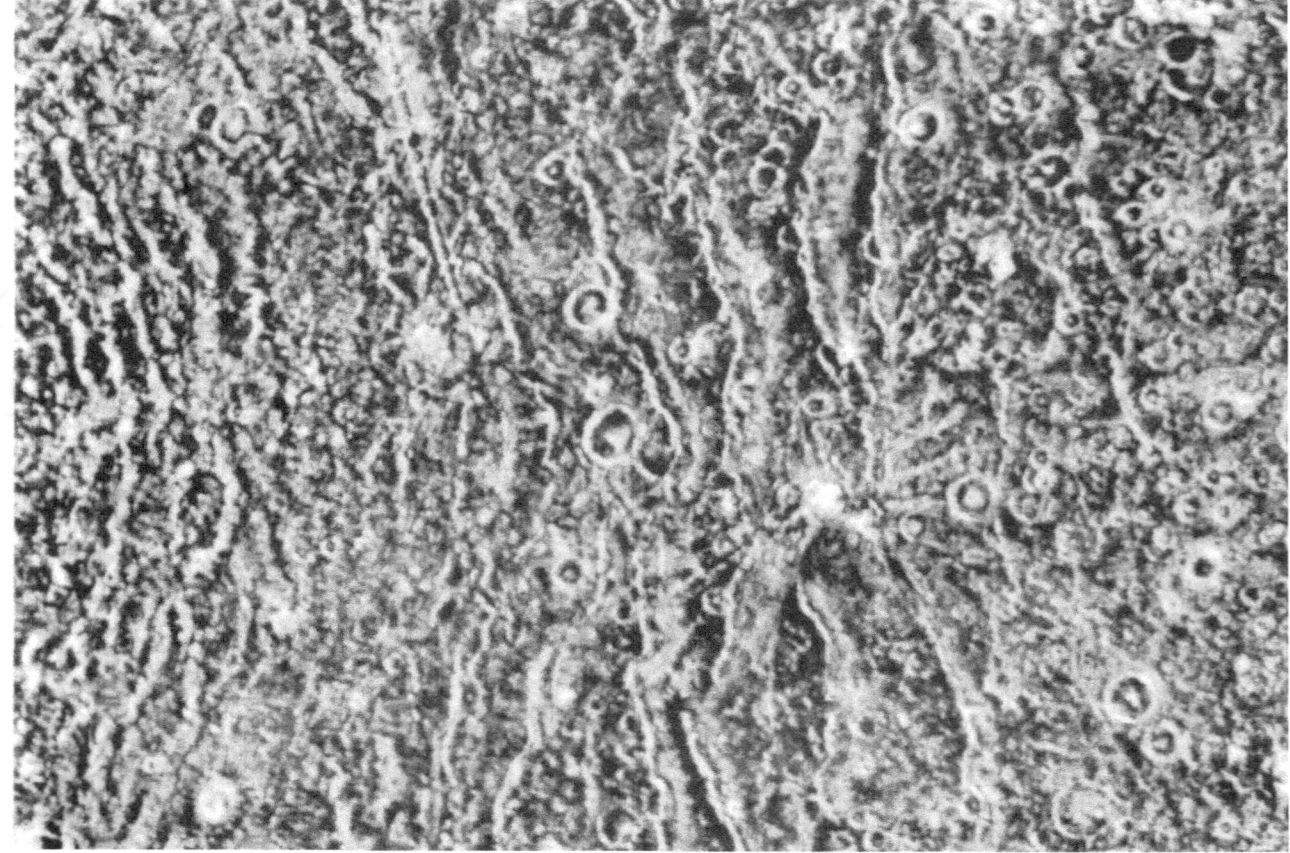

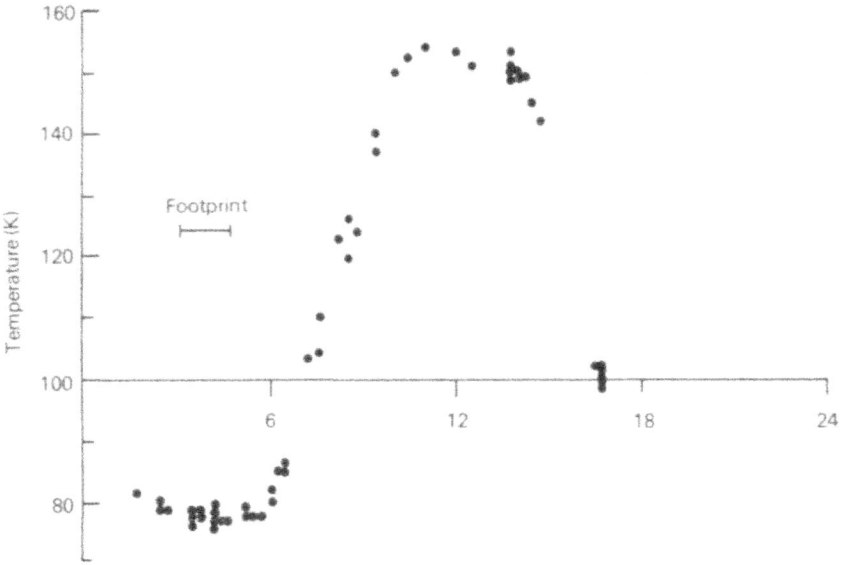

The IRIS instrument measured the temperature of spots on the surface of Callisto as each Voyager sped past. The measurements shown here were all made at equatorial latitudes (between −10° and 25°). Shown are very low predawn temperatures (−190° C) followed by an increase to a noon-time maximum of about −120° C, and then a drop again as the Sun sets. [260-735]

Apparently the ice-rock composition of Callisto alters the ability of the crust to retain large craters. Geologists speculate that the ice flows over many millions of years, filling in crater floors and gradually obliterating the largest craters. There is also a conspicuous absence of mountains on Callisto, again suggestive of a weak, icy crust.

The most prominent features in the Voyager pictures are the ghost remnants of what must have been immense impact basins. The largest of these, the "bullseye" of the Voyager 1 images, has been named Valhalla, for the home of the Norse gods. These ghost basins have lost nearly all their vertical relief. What remains is a central, light-colored zone (probably the location of the original crater), surrounded by numerous concentric rings of subdued, bright ridges. Such features had never been seen before on any planet, and they appear to be the characteristic geologic feature of an ice-rock planet.

Little is known about the composition of Callisto's surface, the material from which sunlight is reflected. It appears to be primarily dark rock or soil, but it lacks diagnostic spectral features, except for one infrared band due to water molecules bound in the soil. The many lighter spots and arcs that outline craters in the high-resolution pictures may be regions in which the ice is showing through, but these cover only a very small fraction of the exposed surface. (It should be noted that, although Callisto is the darkest of the Galilean satellites, the term "dark" is relative, for even Callisto is brighter than Earth's moon.)

The daytime surface temperature of Callisto, observed both from the ground and by Voyager, is about −118° C. The Voyager infrared interferometer spectrometer also determined the minimum temperature, reached just before dawn, of −193° C. No atmosphere is expected at these cold temperatures, and none was seen.

Analysis of Voyager images provided an improved diameter for Callisto of 4840 kilometers, yielding an average density of 1.8 grams per cubic centimeter. As noted previously, it is this low density that leads to the conclusion that ice or water is an important component of the interior of Callisto. The ice has never been detected directly, but the peculiar nature of the craters seen by Voyager adds strong circumstantial support to this conclusion.

Callisto, with its heavy cratering, is the most familiar-looking of the Galilean satellites; if all of them had turned out to be as geologically dead as Callisto, planetary geologists would certainly have been disappointed. However, each satellite, progressing in toward Jupiter, presents increasing evidence of internal activity.

Ganymede

The largest of the Galilean satellites (5270 kilometers in diameter), Ganymede was expected to be similar to Callisto in many ways. Both have low densities (for Ganymede, 1.9 grams per cubic centimeter), indicating a bulk composition of about half rocky materials and half water. In addition, their diameters differ by only eight percent, and both are far enough from Jupiter to escape the severe pounding Io receives from magnetospheric charged particles. Thus it was with great interest that Voyager scientists looked at the differences that emerged between these two satellites.

The surface of Ganymede as revealed by the Voyager cameras is one of great diversity, indicating differing periods of geologic activity. At one extreme there are numerous dark areas that resemble the surface of Callisto in both albedo (reflectivity) and crater density. The largest of these, Regio Galileo, stretches from the equator to latitude +45° and is 4000 kilometers across, nearly as large as the continental United States. This ancient terrain even preserves the remnants of a Callisto-type impact basin in the form of a system of parallel, curving, subdued ridges about 10 kilometers wide, 100 meters high, and spaced about 50 kilometers apart. The central part of this ghost basin is missing, however; it was presumably destroyed by subsequent geologic activity.

Other regions of the surface of Ganymede are clearly the product of intense internal geologic activity. Generally, these regions are of higher albedo and consist of many straight parallel lines of mountains and valleys. The Voyager geologists call these the grooved terrain because of their appearance from a great distance. Typically, these mountain ridges are 10–15 kilometers across and about 1000 meters high, similar in scale to some sections of the Appalachian Mountains in the Eastern United States. No higher relief exists, presumably for the same reason it is absent on Callisto. In many places the grooved terrain forms between areas of the older, darker surface, giving the appearance of mountains extruded between separating plates of ancient crust. In other areas the relationships are much more complex, with curved systems of grooves and ridges overlying each other, displaying intricate crosscutting relationships. Apparently Ganymede has experienced a series of mountain-building events.

The grooved terrains show a substantial range in age, as indicated by the crater densities. The oldest have nearly the same density as the ancient, dark plains, suggesting that formation of the grooved terrain began early, per-

Shaded relief map of Ganymede. [260-673]

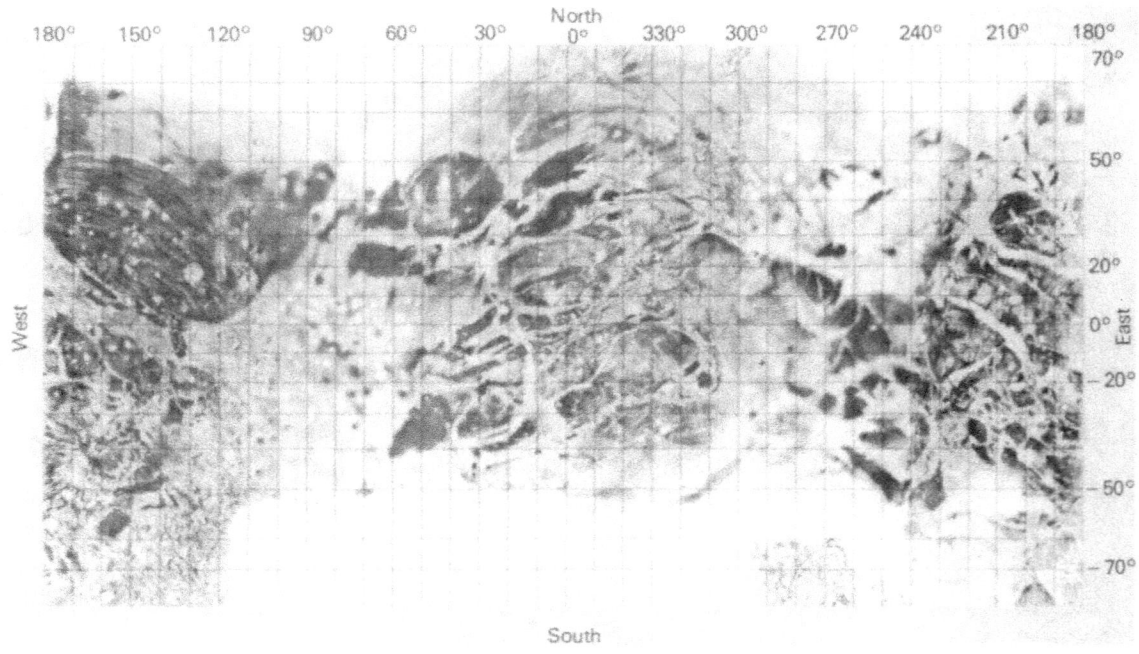

146

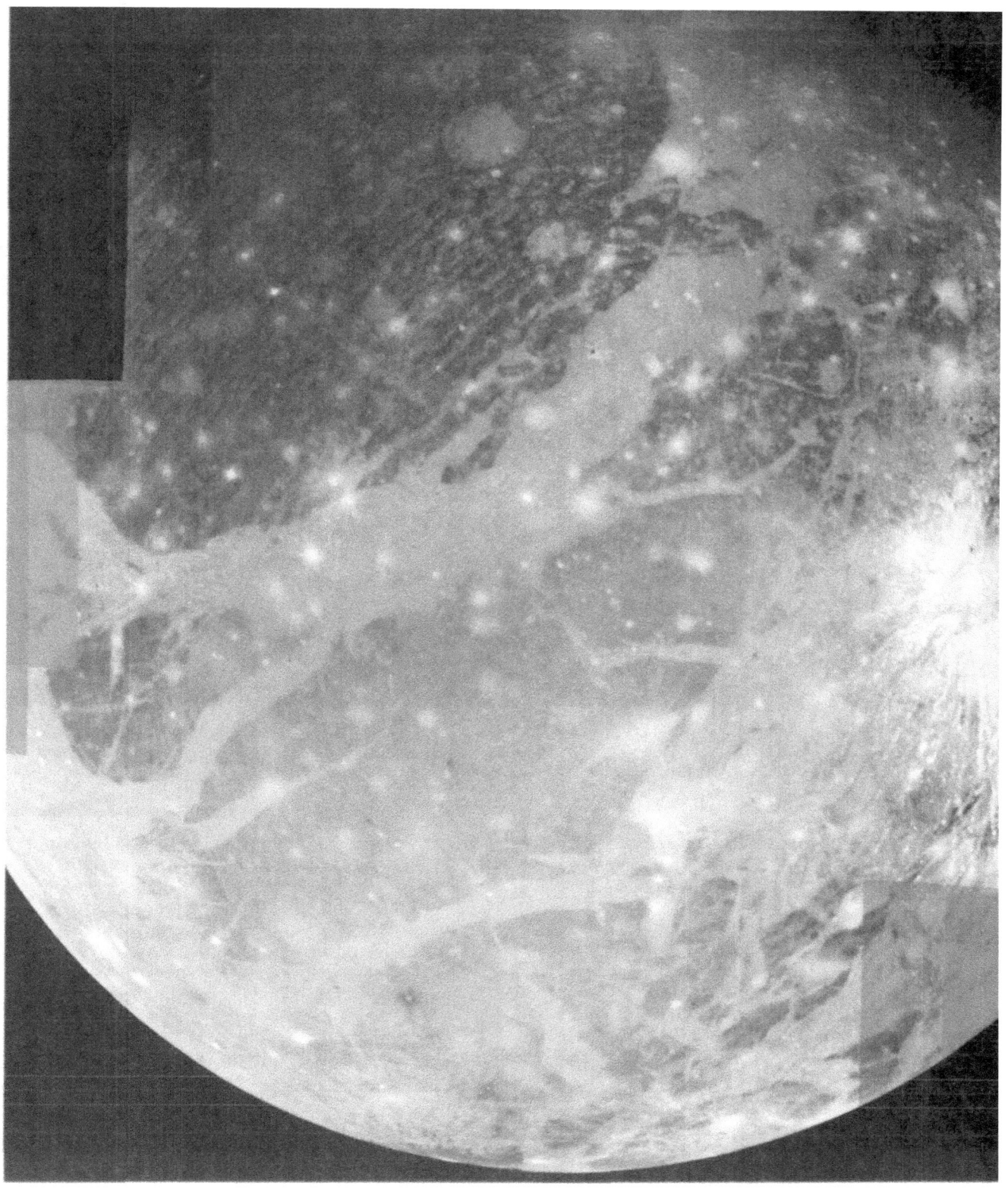

The hemisphere of Ganymede that faces away from the Sun displays a great variety of terrain. In this Voyager 2 mosaic, photographed at a range of 300 000 kilometers, the ancient dark area of Regio Galileo lies at the upper left. Below it, the lighter grooved terrain forms bands of varying width, separating older surface units. On the right edge, a prominent crater ray system is probably caused by water-ice, splashed out in a relatively recent impact. [260-671]

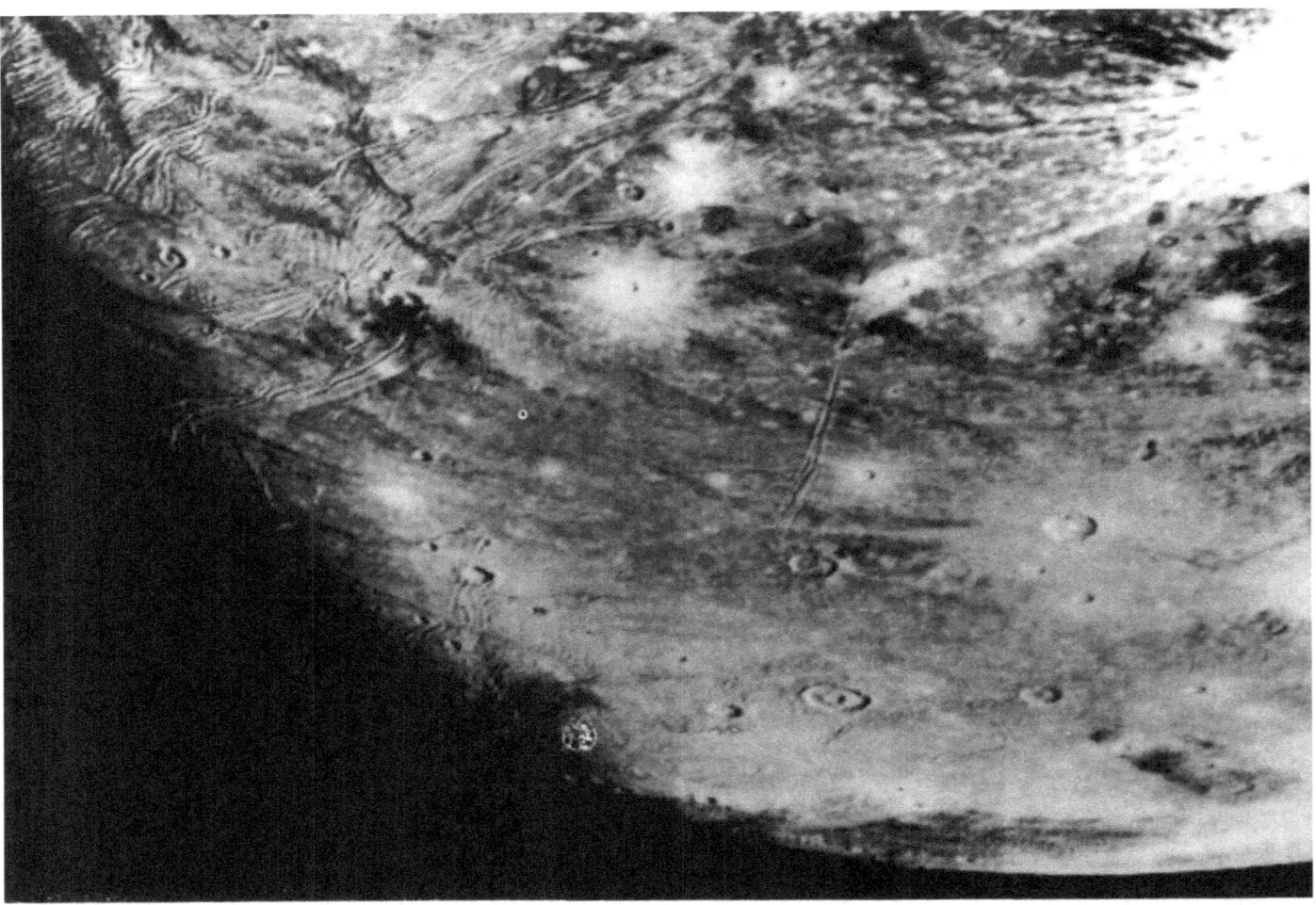

The sinuous nature of some of the narrower Ganymede groove systems can be seen in this oblique view, obtained on March 5 by Voyager 1. The area shown is about the size of California, with features visible as small as 5 kilometers across. The ridges appear to be the result of deformation of the crust of Ganymede. [P-21235]

haps 4 billion years ago. The youngest grooved terrain has only about one-tenth as many craters, but this is still as many as are seen in the 3.5-billion-year-old lunar plains. The Voyager geologists believe that even in these areas geologic activity ceased billions of years ago.

Other types of surfaces are seen on Ganymede. Some regions are lightly cratered and smooth, with no indication of mountain building. In one place, there is a rough mountainous area that looks more like the jumbled lunar mountains than the long ridges and valleys of the rest of Ganymede. Many of the larger craters are distinguished by brilliant white halos and rays that suggest that impacts may have splashed large quantities of water or ice over the surface.

Many of the geologic features seen on Ganymede appear to have been caused by breaking, faulting, or spreading of the crust. In

a few cases, there even seem to be indications of transverse, or sideways, motion along faults. This evidence is extremely exciting to geologists, since similar crustal motion on Earth is associated with the drift of continental plates, drawn by convection currents deep in the mantle. Such activity has never been seen before on another planet.

Astronomers on Earth had known since 1971 that about half the surface of Ganymede was covered with exposed water ice and about half with darker rock. An examination of the albedo variations in the Voyager pictures suggests that the ice is exposed near large craters and, to a lesser extent, in the grooved terrain, but no direct measurements were made by Voyager of the composition of different parts of the surface.

The presence of ice on the surface suggested to many astronomers that Ganymede

The complex patterns of the grooved terrain on Ganymede are apparent in high-resolution images. This picture, taken by Voyager 1 on March 5, has a resolution of about 3 kilometers and shows a region about the size of the state of Pennsylvania. The mountain ridges are spaced about 10 to 15 kilometers apart and rise about 1000 meters, similar to many of the mountains of Pennsylvania. The transections of different mountain systems indicate that they formed at different times. A degraded crater near the left center of the picture is crossed by ridges, indicating that it predated the period of crustal deformation and mountain building. [P-21279]

Ray systems of exposed water-ice are visible in this high-resolution mosaic of Ganymede, obtained by Voyager 2 on July 9 at a range of about 100 000 kilometers. The rough mountainous terrain at lower right is the outer portion of a large fresh impact basin that postdates most of the other terrain. At the bottom, portions of grooved terrain transect other portions, indicating an age sequence. The dark patches of heavily cratered terrain (right center) are probably ancient icy material formed prior to the grooved terrain. The large rayed crater at upper center is about 150 kilometers in diameter. [P-21770B/W]

might have a very tenuous atmosphere of water vapor or oxygen, which might be released by the breakdown of water vapor by sunlight. During the Voyager 1 flyby, a sensitive test for an atmosphere was made by the ultraviolet instrument from observations of the star Kappa Centauri as it was occulted by Ganymede. No dimming of the starlight was seen, yielding an upper limit for the surface pressure of the gases oxygen, water vapor, or carbon dioxide of 10^{-11} bar, or one hundred-billionth the atmospheric pressure at Earth.

The differences between the geologic histories of Ganymede and Callisto are surprisingly large. No one knows the reason. Perhaps only a small increase in internal temperature is necessary to initiate geologic activity in an icy planet, and for some reason Ganymede crossed this threshold for a part of its history, whereas Callisto did not.

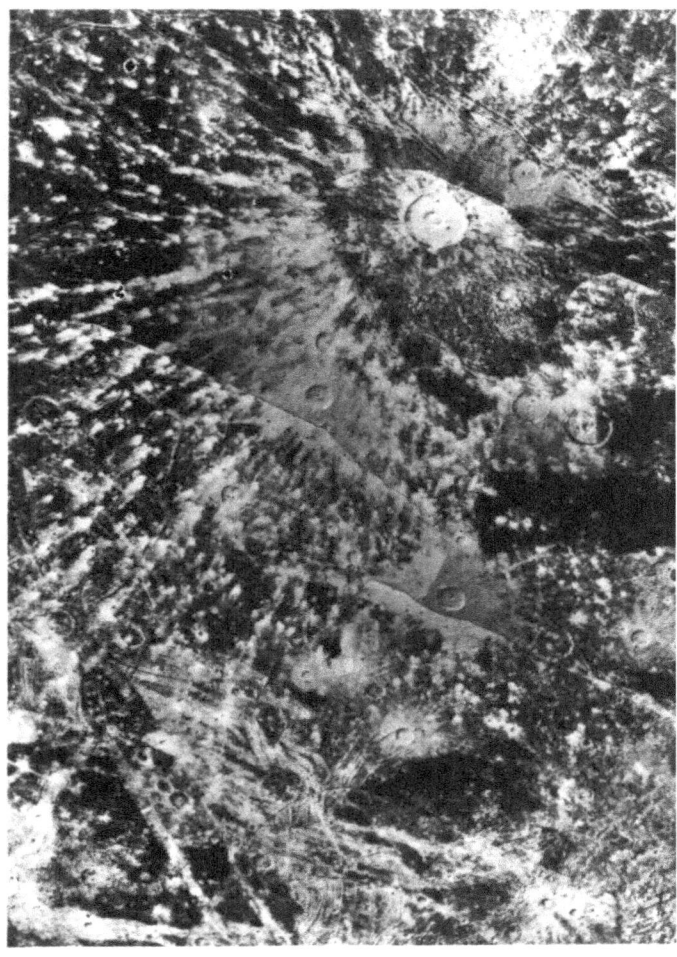

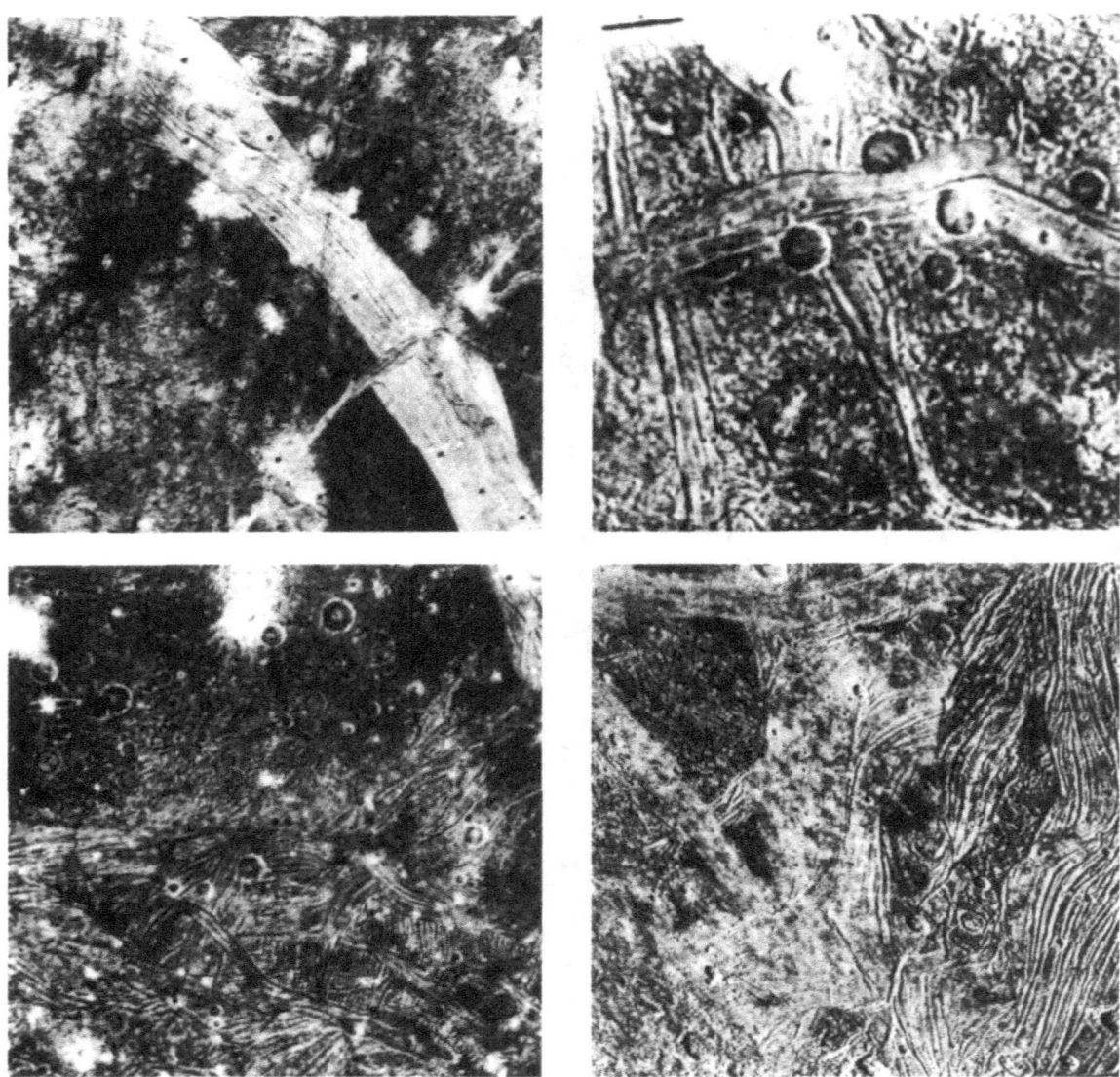

The many variants of smooth and grooved terrain on Ganymede suggest a complex geologic history for this satellite. Four high-resolution views by Voyager 2 are grouped together. (Top left) A band of low mountain ridges has apparently been cut and offset by a fault. (Top right) Multiple sets of mountain ridges transect at nearly right angles. Some impact craters were formed before, and some after, the grooved terrain. (Bottom left) Many short parallel ridges butt into each other, making a crazy-quilt pattern. (Bottom right) In the center of this frame is an unusual smooth area, perhaps the result of flooding of the surface by material that filled in the grooves. [260-678A]

Europa

As one proceeds inward through the Galilean satellites, these worlds become less and less familiar to the planetary geologist. This was an unexpected effect. Callisto and Ganymede were expected to have unusual properties as a result of their large percentage of ice. The densities of Europa and Io are more normal for the smaller, terrestrial-type planets, and before Voyager many scientists expected these two satellites might look much like the Moon, which they resemble in size.

Europa, with a diameter of 3130 kilometers, is about 15 percent smaller than the Moon. Its density is 3.0 grams per cubic centimeter, indicating a basically rocky composition. However, cosmic mixtures of rocky and metallic materials are often a bit denser than this, leaving room for a substantial component of ice or water. Calculations indicate that if all

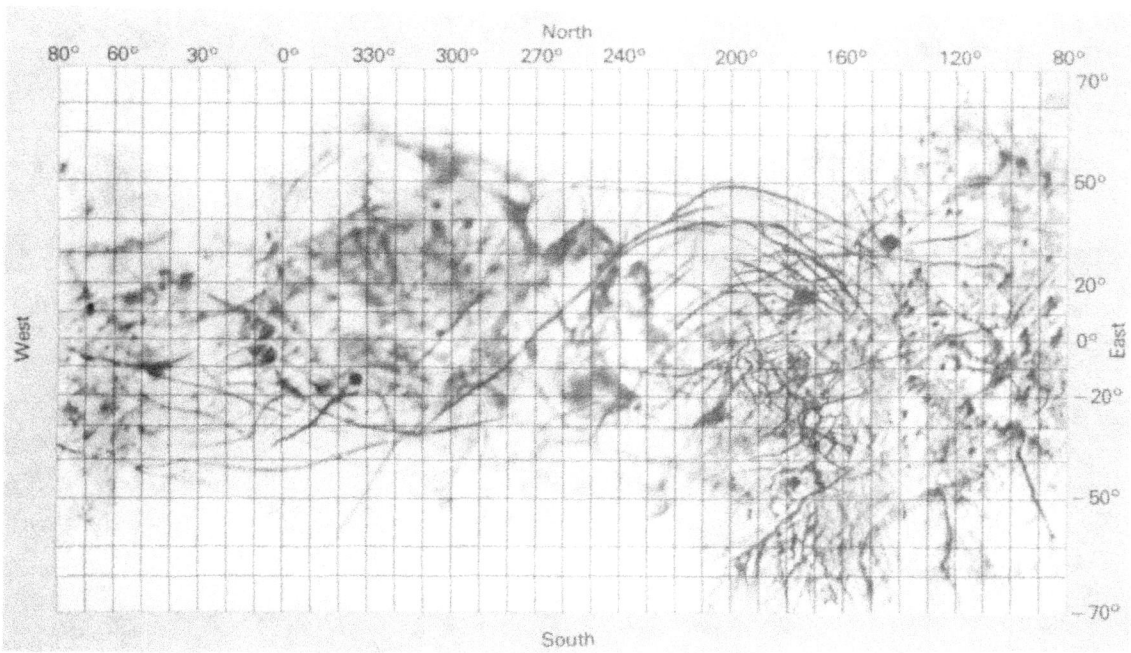

North

80° 60° 30° 0° 330° 300° 270° 240° 200° 160° 120° 80°

70°
50°
20°
0°
−20°
−50°
−70°

West

East

South

Shaded relief map of Europa. [260-659]

the ice were at the surface, it might form a crust up to 100 kilometers thick.

Telescopic observations of Europa demonstrated long before Voyager that this satellite is almost completely covered with ice. It is a white, highly reflecting body, looking, from a great distance, like a giant snowball. In the early Voyager pictures, Europa always showed a bland, white disk, in striking contrast to the spottiness of Ganymede or the brilliant colors of Io.

Voyager 1 never got closer to Europa than 734 000 kilometers, and at that distance it remained a nearly featureless planet, with no obvious impact craters or other familiar geologic structures. What did show in the Voyager 1 pictures, however, were numerous thin, straight dark lines crisscrossing the surface, some extending up to 3000 kilometers in length. To the members of the Imaging Team, these features were "strongly suggestive of global-scale tectonic processes, induced either externally (as by tidal despinning) or internally (as by convection)." It was with the greatest interest that the Voyager 2 images, taken from about four times closer, were anticipated.

The spectacular pictures obtained of the satellite in July were perhaps more confusing than clarifying. Europa is entirely covered with dark streaks that vary in width from several

kilometers to approximately 70 kilometers and in length from several hundred to several thousand kilometers. Most streaks are straight, but others are curved or irregular. The streaks lie on otherwise smooth, bright terrain, featureless except for numerous random dark spots, most less than 10 kilometers in diameter.

Voyager 2 photos showed, in addition to the smooth terrain with its dark streaks, regions of darker, mottled terrain. This mottled terrain appears rough on a small scale, and it may contain small craters just on the limit of resolution (about 4 kilometers). Only three definite impact craters have been identified, each about 20 kilometers across. This small number of craters suggests either that the surface is relatively young or that craters are not preserved for long in the icy crust.

Although the dark streaks give Europa a cracked appearance, the streaks themselves are not obviously cracks. They are not depressed below their surroundings; in fact, they have no topographic structure whatever. Europa is extraordinarily smooth, and the dark streaks look rather like marks made with a felt-tipped pen on a white billiard ball. The streaks are not even very dark; the contrast with adjacent smooth terrain is only about 10 percent.

One of the most remarkable geologic phenomena discovered by Voyager is the light streaks that appear on Europa. These are smaller than the dark streaks, only about 10 kilometers in width, but much more uniform. Seen at low Sun angle, they also show vertical relief of less than a few hundred meters. These light ridges are seen best at low Sun and tend to be invisible at higher illumination angles.

The most amazing thing about the light ridges is their form. Instead of being straight, they form scallops or cusps, with smooth curves that repeat regularly on a scale of 100 to a few hundred kilometers. In some of the low-Sun-angle pictures, the surface of Europa seems to be covered with a beautiful network of these regular curving lines. The impression is so bizzare that one tends not to believe the reality of what is seen. Nothing remotely like it has ever been seen on any other planet.

At present the geology of Europa remains beyond our understanding. Presumably there is a thick ice crust, perhaps floating on a liquid water ocean. Presumably there is sufficient heat coming from the interior to have produced cracking or motion in the ice crust, and the light and dark streaks preserve a pattern in some way related to this internal activity. However, the actual mechanisms for producing the observed features so lightly traced on this smooth white world remain for scientists to decipher.

Io

The most spectacular of the Galilean satellites is Io. Even in low-resolution images, its brilliant colors of red, orange, yellow, and white set it apart from any other planet. The dramatic scale of its volcanic activity confirms that Io is in a class by itself as the most geologically active planetary body in the solar system.

The diameter of Io is 3640 kilometers, and its density is 3.5 grams per cubic centimeter. Both values are nearly identical to those of our Moon. Were it not for Io's proximity to Jupiter, it would probably be a dead, rocky world much like Earth's satellite.

Careful examination of all the Voyager images of Io, some of which have resolutions as good as 1 kilometer, has failed to reveal a single impact crater. Yet the flux of crater-producing impacts at Io must be even greater than for the other Galilean satellites, because of the focusing effect of Jupiter's gravity. The absence of craters alone would indicate that Io has an extremely young and dynamic planetary surface, even without the observation of active volcanoes. Calculations indicate that craters on Io

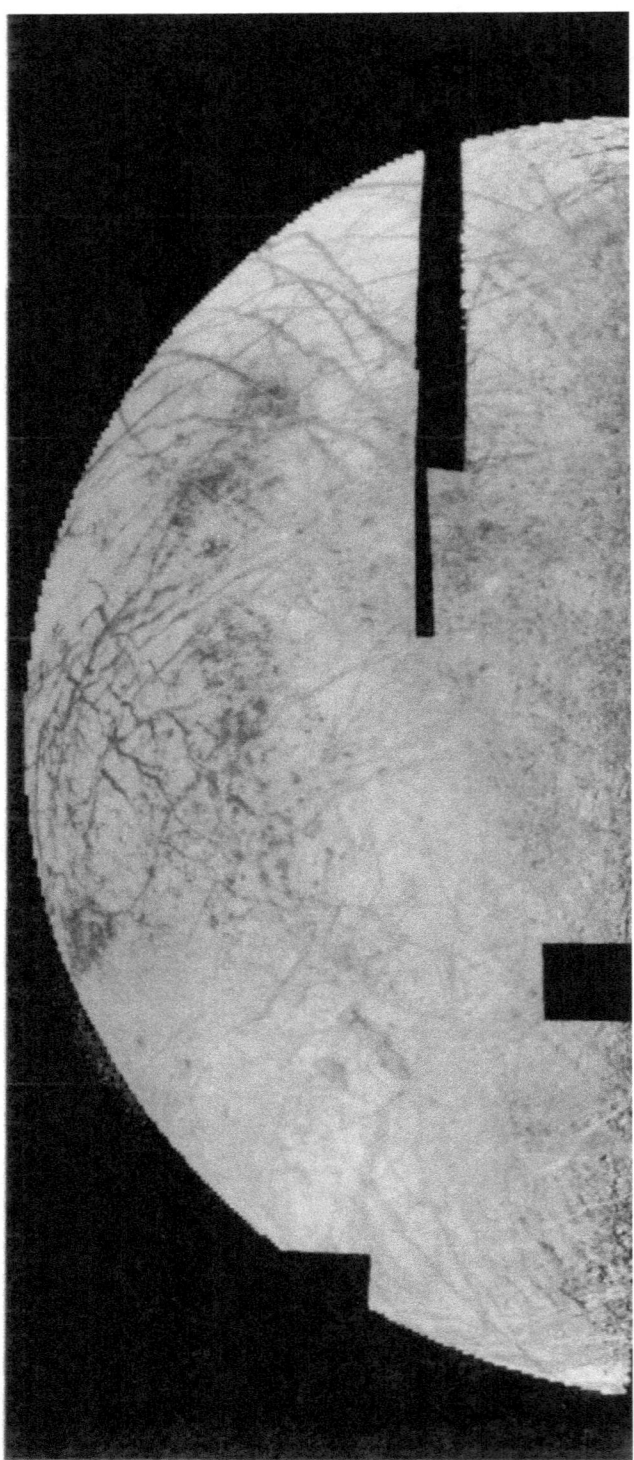

Europa looks like a cracked egg in this computer mosaic of the best Voyager 2 images. In this presentation, the variation of surface brightness due to the angle of the Sun has been removed by computer processing, so that surface features can be seen equally well at all places. The many broad dark streaks show up well, but this presentation does not bring out the much fainter and more enigmatic light streaks. These pictures were taken from a distance of about 250 000 kilometers and show features as small as 5 kilometers across. [260-686]

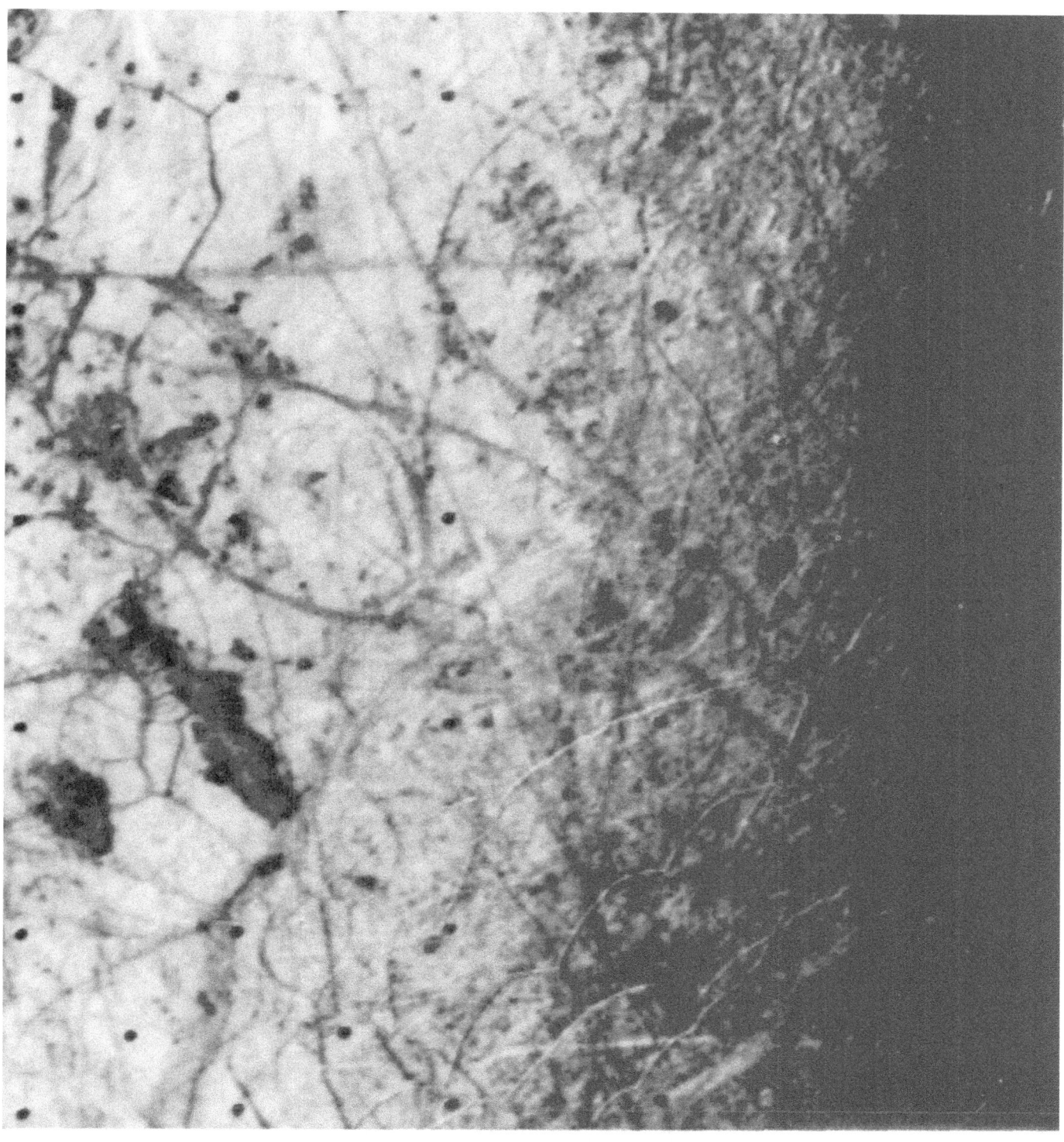

One of the most remarkable of all the Voyager discoveries was the arcuate white ridges on Europa. Visible only at very low Sun angle, these curved bright streaks are 5 to 10 kilometers wide and rise at most a few hundred meters above the surface. Their graceful scalloped pattern is unique to this planet and has defied explanation. Also visible in this view, taken by Voyager 2 on July 9 at a range of 225 000 kilometers, are dark bands, more diffuse than the light ridges, typically 20 to 40 kilometers wide and hundreds to thousands of kilometers long. [P-21766]

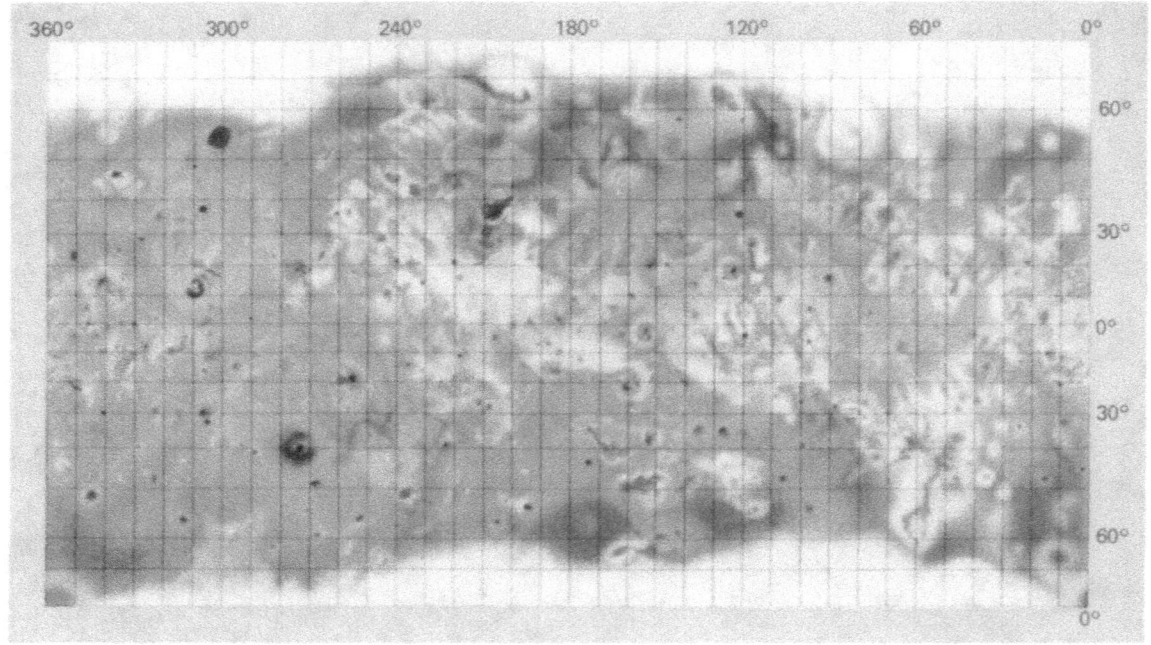

Shaded relief map of Io as it appeared in early March 1979. [260-634BC]

The great erupting volcanoes on Io produce distinctive surface markings. The top row shows three views of Prometheus (P_3). The bright ring on the surface rims the areas of fallout from the plume, and it is probably an area in which sulfur dioxide frost is being deposited on the surface. The bottom views are of Loki (P_2) as seen by Voyager 1. (Bottom left) The assymetric structure of the plume can be seen. (Bottom right) An ultraviolet image has been used to produce a false-color composite; the large ultraviolet halo above the visible-light plume may be due to scattering from sulfur dioxide gas rather than solid particles. [260-451]

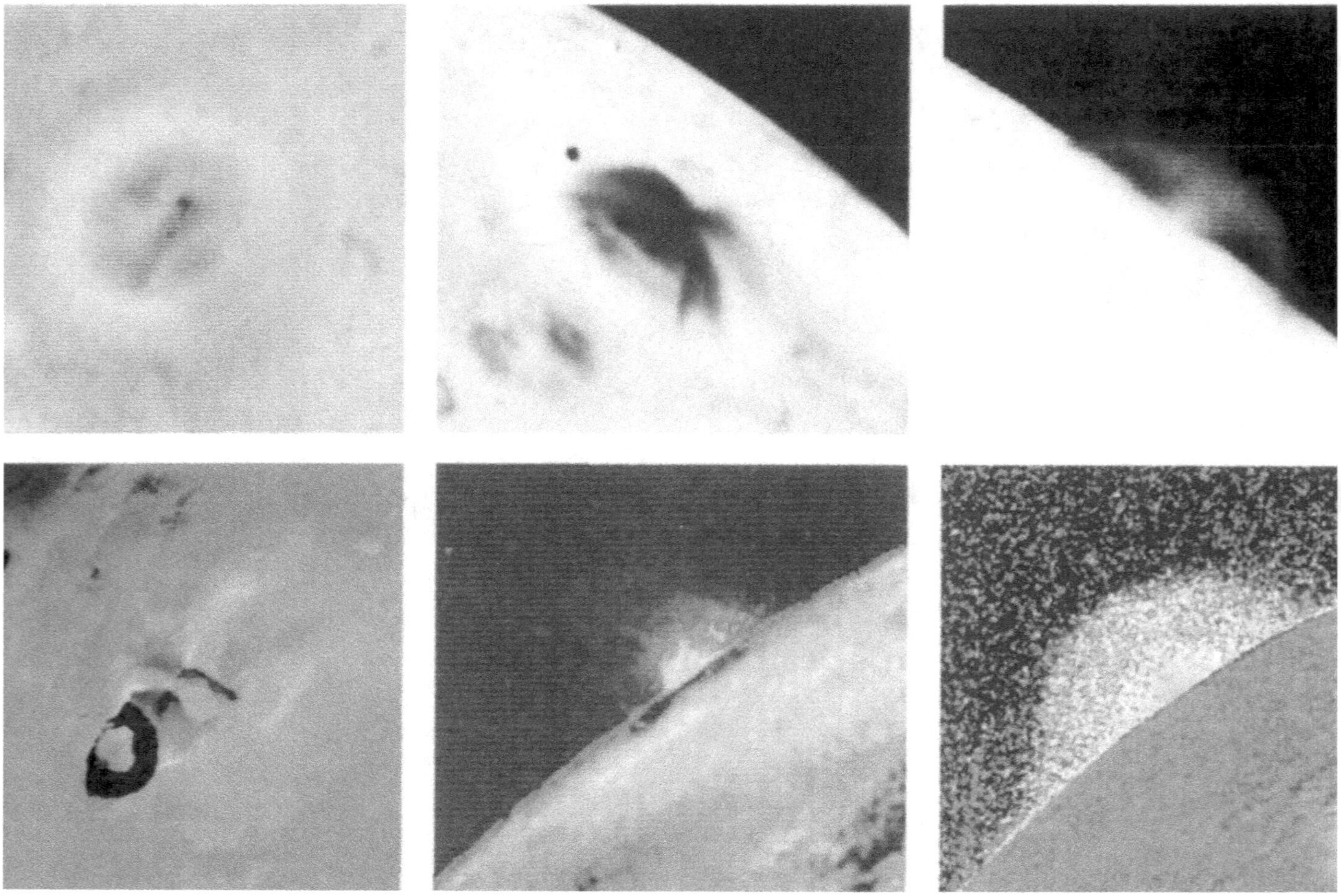

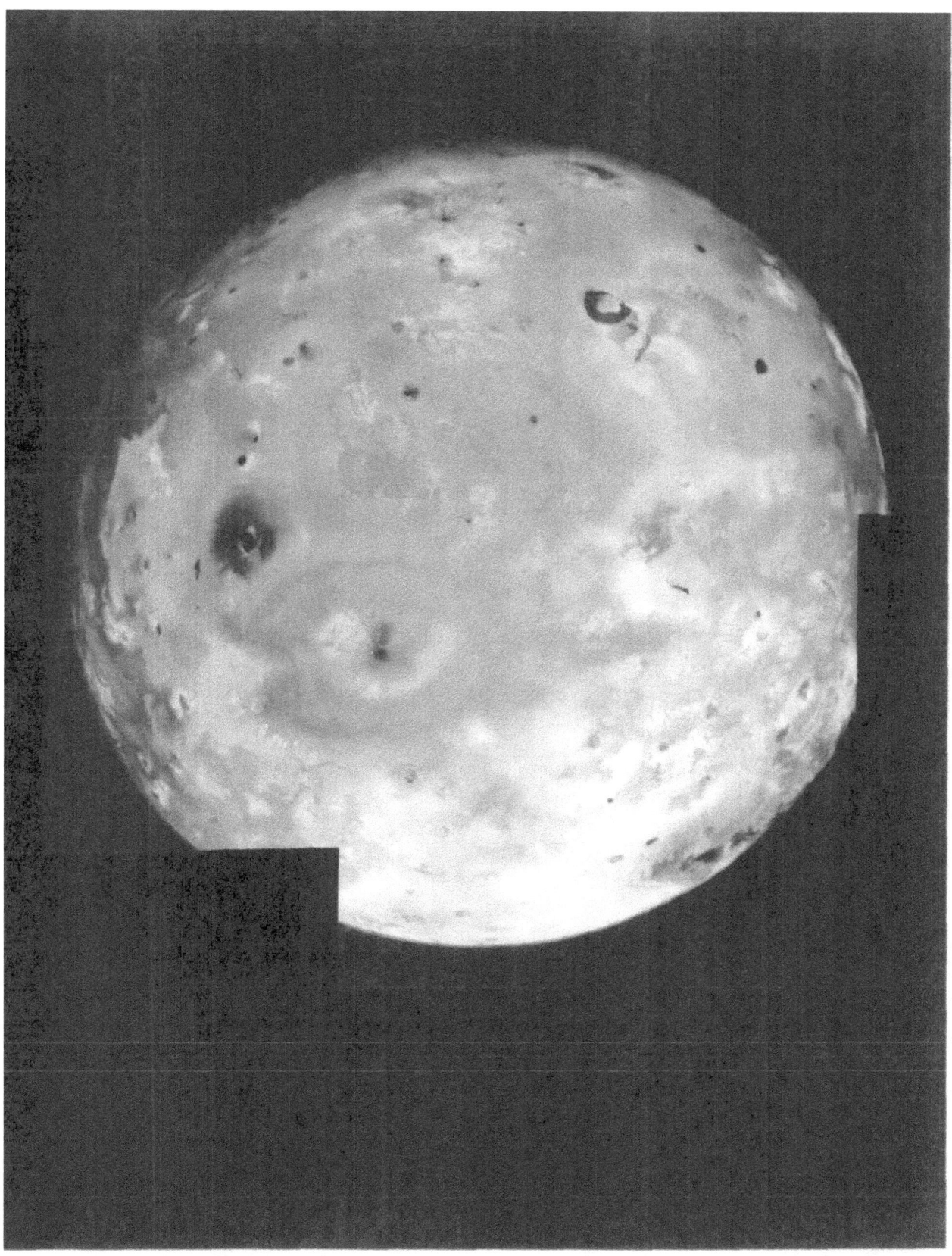

Perhaps the most spectacular of all the Voyager photos of Io is this mosaic obtained by Voyager 1 on March 5 at a range of 400 000 kilometers. A great variety of color and albedo is seen on the surface, now thought to be the result of surface deposits of various forms of sulfur and sulfur dioxide. The two great volcanoes Pele and Loki (upper left) are prominent. [260-464]

must be filled in or otherwise obliterated at a rate corresponding to the deposition of at least 100 meters per million years, and quite probably a factor of ten greater, or 1 meter every thousand years.

In place of impact craters, the surface of Io has a great many volcanic centers, which generally take the form of black spots a few tens of kilometers across. In a few cases, high-resolution pictures show the characteristic shapes associated with volcanic calderas on Earth and Mars, and, if the other volcanic centers are similar, about 5 percent of the entire surface of Io is occupied by calderas. These are extremely black, reflecting less than 5 percent of the sunlight; often they are surrounded by irregular, diffuse halos nearly as black as the central spot. The calderas seem more like the Valles caldera in New Mexico, which is associated with vents that produced large quantities of ash, than with those of Hawaiian-type shield volcanic mountains.

There is evidence in many of the Voyager photos of extensive surface flows on Io. These originate in dark volcanic centers and either spread to fan shapes, typically 100 kilometers across, or else snake out in long, twisting tentacles. Some of the flows are lighter than the background and some are darker. Most are red or orange in color, often outlined by fringes of contrasting albedo.

The equatorial regions of Io are quite flat, with no vertical relief greater than about 1 kilometer high; indeed, many of the volcanic centers do not appear to correspond to mountains or domes at all. There are, however, a number of long, curvilinear cliffs or scarps and narrow, straight-walled valleys a few hundred meters deep. These appear to be places in which the crust has broken under tension, somewhat similar to terrestrial faults and the valleys called graben. A few rugged mountains of uncertain origin are visible in low Sun elevation pictures.

Near the poles of Io the terrain is more irregular. There are few volcanic centers, but more mountains, some with heights of several kilometers. In addition, there are regions that appear to be made of stacked layers of material. These so-called layered terrains are revealed when erosion cuts into them, exposing the layers along the cliff or scarp. The largest such plateau or mesa has an area of about 100 000 square kilometers. The scarps sometimes intersect each other, suggesting a complex history of deposition, faulting, and erosion. Voyager geologists believe that these scarps

may be areas in which the release of liquid sulfur or sulfur dioxide has undercut cliffs, analogous to internal sapping by groundwater at similar scarps on Earth.

Perhaps the most distinctive surface features on Io are the circular or oval albedo markings that surround the great volcanoes. The first of these to be seen was the 300-kilometer-wide white donut of Prometheus, on the equator at longitude 150°. Much more spectacular is the hoofprint of Pele, about 700 by 1000 kilometers. These symmetric rings mark the locations of the kinds of eruptions that generate large fountains or plumes, and may be produced by condensible sulfur or sulfur dioxide raining down from the volcanic fountain. At least one new ring appeared during the four months between the Voyager encounters, centered at longitude 330°, latitude +20°, but by the time Voyager 2 photographed this area, no plume remained active.

During the Voyager 1 flyby, temperature scans of the surface of Io were made with the infrared interferometer spectrometer (IRIS). A number of localized warm regions were found, the most dramatic being just south of the volcano Loki. Here the images showed a strange, U-shaped black feature about 200 kilometers across. The IRIS team interpreted its data to indicate a temperature for the black feature of 17° C (or room temperature), in contrast to the surrounding surface at −146° C. Perhaps the dark feature was some sort of lava lake, either of molten rock or molten sulfur. The melting point of sulfur is 112° C. If there were a scum of solidifying sulfur on top of the "lake," this interpretation might well be the correct one.

The brilliant reds and yellows of the surface of Io immediately suggest the presence of sulfur. When heated to different temperatures and suddenly cooled, sulfur can assume many colors, ranging from black through various shades of red to its normal light-yellow appearance.

Even before Voyager, laboratory studies had shown that sulfur matches the overall properties of the spectrum of Io, including the low albedo in the ultraviolet and the high reflectivity throughout the infrared. Contemporary with the Voyager flybys, additional telescopic observations and laboratory studies by Fraser Fanale at JPL and Dale Cruikshank at the University of Hawaii identified another component on Io, sulfur dioxide. Sulfur dioxide is an acrid gas released from terrestrial volcanoes, where it combines with water in the Earth's atmosphere

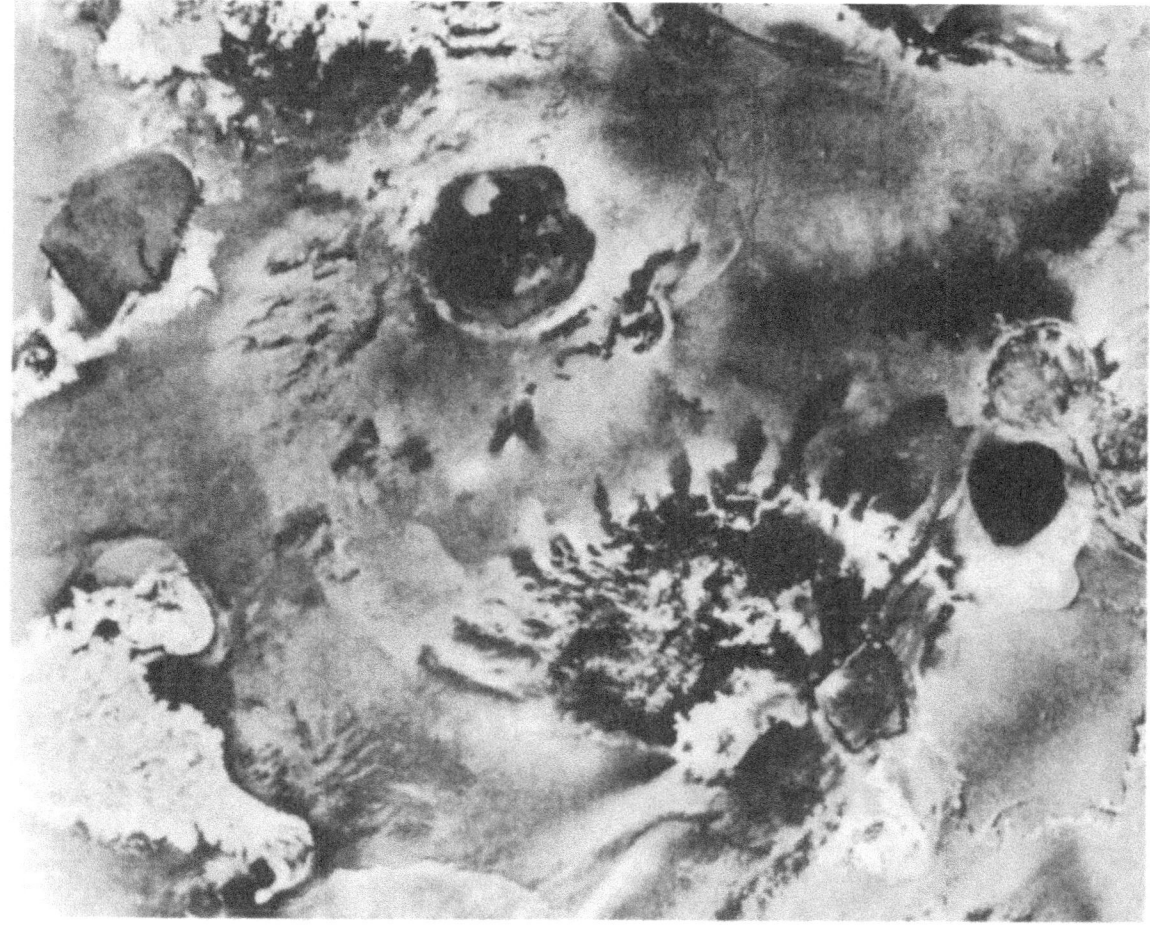

Close-ups of Io reveal a wide variety of volcanic phenomena. This Voyager 1 view of an equatorial region near longitude 300° shows several large surface flows that originate in volcanic craters or calderas. At the right edge is a light flow about 250 kilometers long. Another dark, lobate flow with bright edges is just left of center, with an exceedingly dark caldera to its left. [260-468A]

to produce sulfuric acid. At the temperature of the surface of Io, sulfur dioxide is a white solid. Researchers guessed that the extensive bright white areas in the Voyager pictures of Io might be covered with sulfur dioxide frost or snow. The presence of this material on Io was confirmed when the infrared IRIS instrument obtained a spectrum of sulfur dioxide gas over the erupting volcano Loki during the Voyager 1 encounter.

The discovery of the ongoing eruptions on Io, made shortly after the Voyager 1 flyby, did much to clarify the confused evidence pouring in concerning the apparent youth of Io's surface. Here, under the very eyes of Voyager, eruptions were taking place on a scale that dwarfed anything ever seen before. The discovery picture alone, taken from a distance of 4 million kilometers, showed two eruptions (Pele and Loki), each of which was much larger than the most violent volcanic eruption ever recorded on Earth.

Voyager 1 found eight giant eruptions, with fountains or plumes rising to heights of between 70 and 280 kilometers. To reach these altitudes, the material must have been ejected from the vents at speeds of between 300 and 1000 meters per second, several times greater than the highest ejection velocities from terrestrial volcanoes. Although widely spaced in longitude, these volcanoes were concentrated toward the equator; seven of the eight were at latitudes between +30° and −30°, and the eighth at −44°.

When Voyager 2 arrived four months later, it was able to reobserve seven of the eight volcanoes. (To be identified reliably, the volcanic plumes must be silhouetted against dark space at the edge of the disk.) Six of these were still erupting; one, Pele, the largest plume seen by Voyager 1, had ceased activity. The plume associated with Loki had also changed markedly, increasing in height from 100 to 210 kilometers in visible light. (All the plumes ap-

157

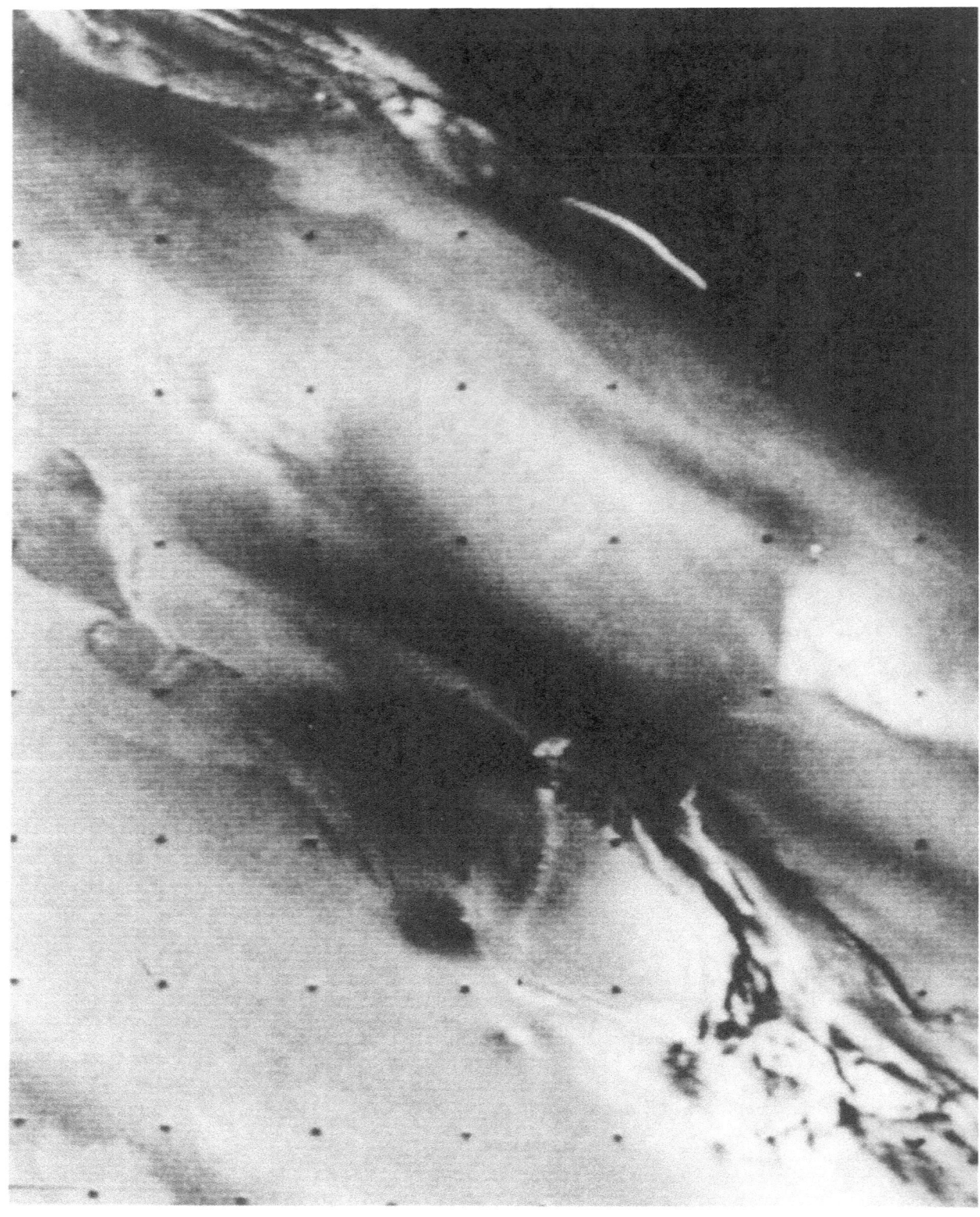

Differences in surface elevation can clearly be seen in a few of the Io close-ups from Voyager 1. This remarkable picture is of the center of the great volcano Pele, at latitude 15°S and longitude 224°. A low mountain with flow features can be seen. In the background, there are several large irregular depressions with flat floors that appear to be the result of collapse. The diffuse dark features in the center are probably the ejecta plumes being erupted from the Pele vent. [P-21220B/W]

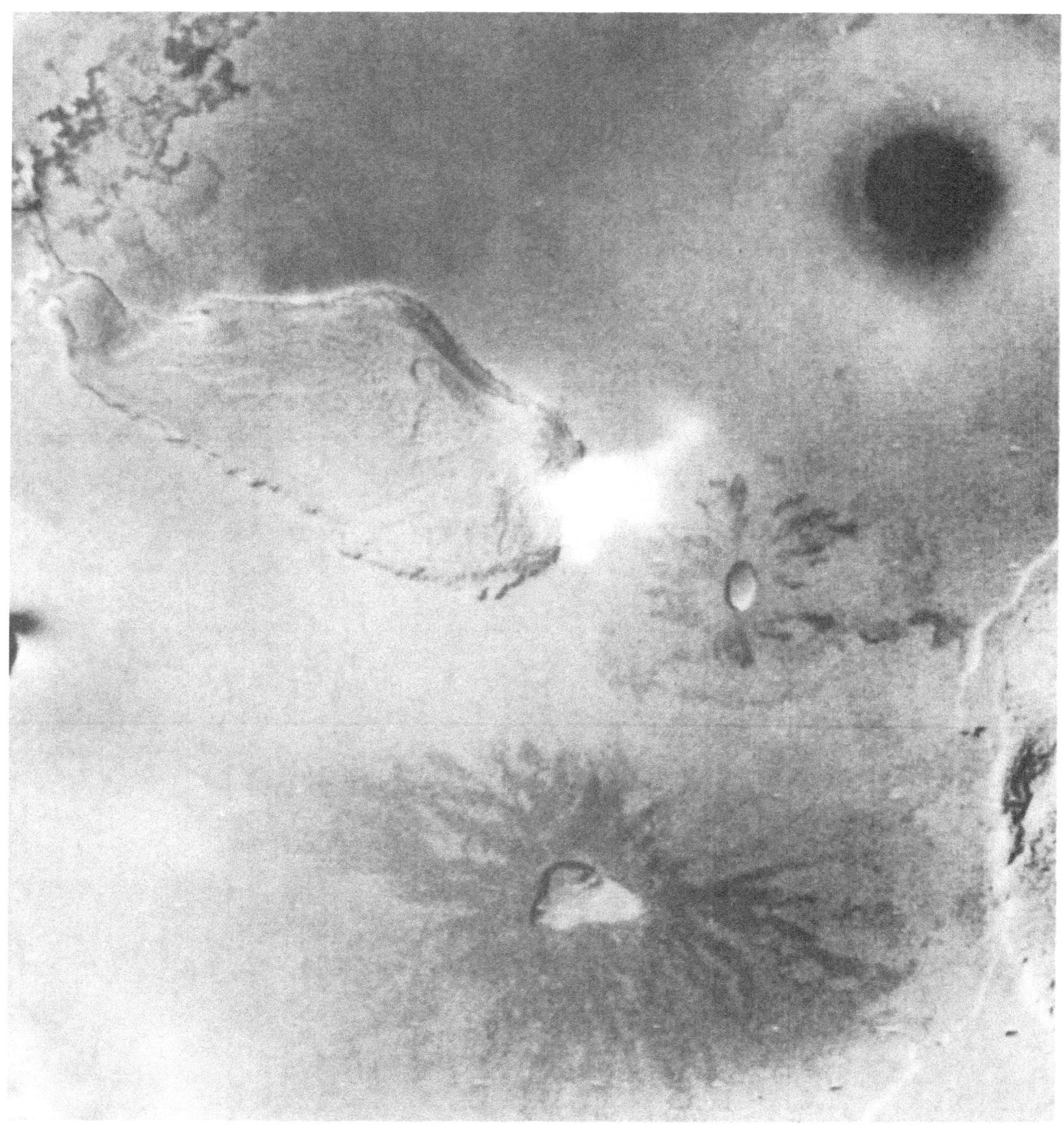

At the highest resolution obtained by the Voyager cameras, Io revealed some landscapes that looked familiar to terrestrial geologists. This picture, taken by Voyager 1 at a range of only 31 000 kilometers, shows a region about the size of the state of Maryland at a resolution of 300 meters. Clearly seen is a volcano not too different from some of those on the Earth or Mars. At the center is an irregular composite crater or caldera about 50 kilometers in diameter with dark flows radiating from its rim. The style of volcanism illustrated here is quite different from the explosive plumes or fountains with their associated rings of bright material deposited on the surface. This volcano is located at about longitude 330°, latitude 70°S. [260-502]

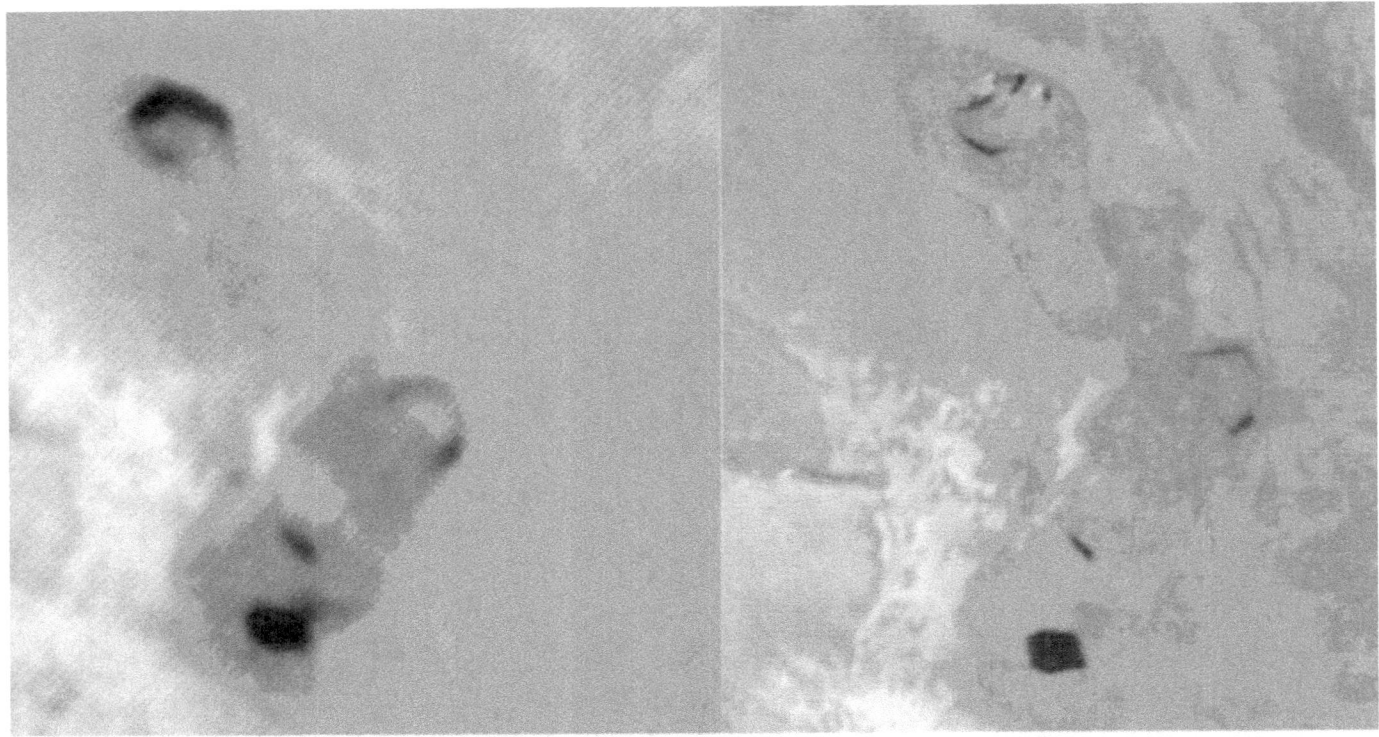

In addition to its giant volcanic plumes or fountains, Io possesses other indications of current volcanic activity. One of these takes the form of intermittent blue-white patches that may be caused by gas venting from the interior. In this pair of photographs, the same region of the surface is shown about six hours apart. On the right, there is an arcuate bright gas cloud; on the left the same region is black. It is believed that the venting gas is sulfur dioxide, and that the condensation of this gas produces fine particles of "snow" that look blue. [260-508]

Some of the most dramatic changes in the surface of Io between March and July took place in the vicinity of the volcano Loki, at longitude 310° and latitude 15°N. On the left is a Voyager 1 view; on the right, one from Voyager 2. The "lava lake" associated with Loki has become less distinct, apparently as a result of deposits that fell on the northern part of the dark U-shaped feature. Perhaps the surface had also cooled between these photos. In the upper left center, a new dark volcanic caldera with bright spots near it and a large, faint bright ring had appeared by July, although it was not active at the time Voyager 2 flew by. [260-687AC]

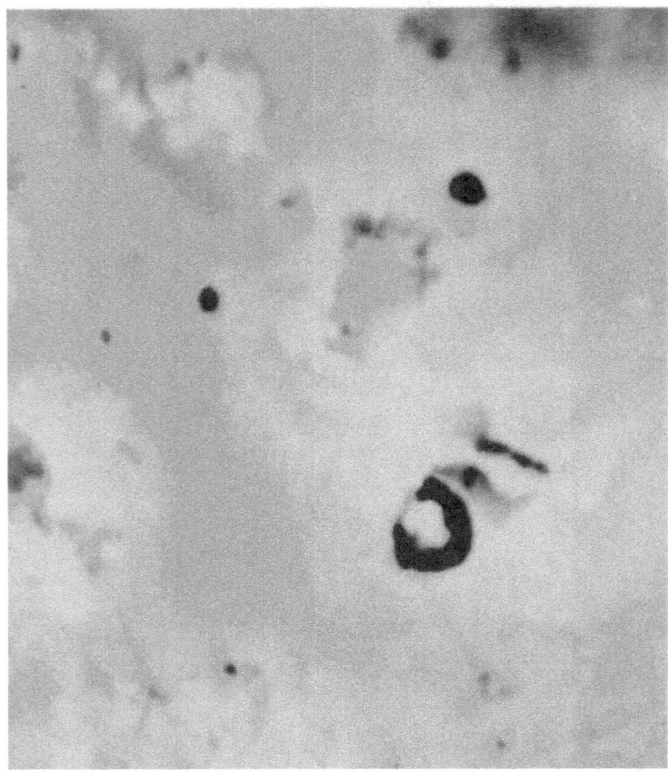
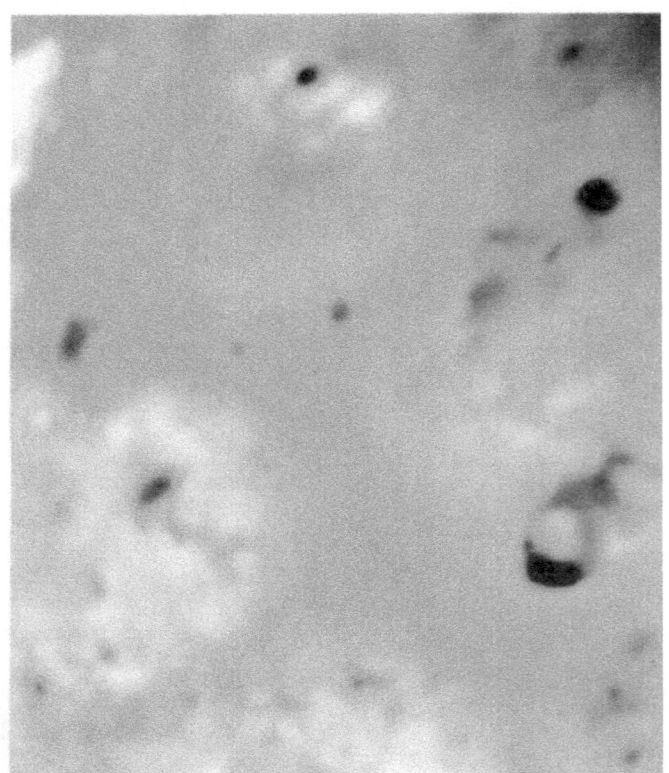

pear larger when viewed in the ultraviolet.) Loki had developed a more complex structure; in March it appeared to originate near the south end of a 250-kilometer-long dark feature, but in July there was a double plume, with activity at both ends of the dark feature.

No new eruptions were seen by Voyager 2; between them, the two spacecraft effectively surveyed the whole surface of Io for plumes down to 40 kilometers height. Interestingly, the smallest plume seen was 70 kilometers high; there appeared to be a real absence of smaller eruptions.

From the number and size of the observed eruptions, it is possible to calculate the resurfacing rate for Io due to these plumes. The result is that each plume is erupting about 10 000 tons of material per second, or more than 100 billion tons per year. This quantity corresponds to about 10 meters of deposition over the whole surface in a million years. When additional note is taken of surface flows, the deposition rate could easily be ten times higher, or 100 meters per million years, in agreement with the rate estimated from the absence of impact craters.

Energy for the Io Volcanoes

Clearly, something extraordinary is happening to Io to generate the observed level of volcanism. The primary heat source for the interiors of the terrestrial planets is the decay of the long-lived radioactive elements thorium and uranium. But Io would have to be supplied with a hundred times its quota of these elements to explain the observed activity.

A way out of this difficulty was provided by a theoretical investigation carried out by Stanton Peale of the University of California at Santa Barbara and Pat Cassen and Ray Reynolds at the NASA Ames Research Laboratory. Working in the months before the first Voyager flyby, they calculated that the tidal effects of Jupiter on Io could generate large-scale heating of the satellite. Io is about the same distance from Jupiter as the Moon from Earth, but the much greater mass of Jupiter raises enormous tides in its satellite. These tides distort its shape, but no other effect would be present if Io remained at a constant distance from Jupiter. What Peale, Cassen, and Reynolds realized was that the distance of Io from Jupiter varies as the result of small gravitational perturbations from the other Galilean satellites. Therefore the tidal distortions also vary, in effect squeezing and

unsqueezing Io each orbit. Such flexing pumps energy into the interior of Io in the form of heat; theorists calculated that the heat supplied could be as high as 10^{13} watts. They predicted, in a paper published just three days before the Voyager flyby of Io, that "widespread and recurrent surface volcanism might occur," and that "consequences of a largely molten interior may be evident in pictures of Io's surface returned by Voyager."

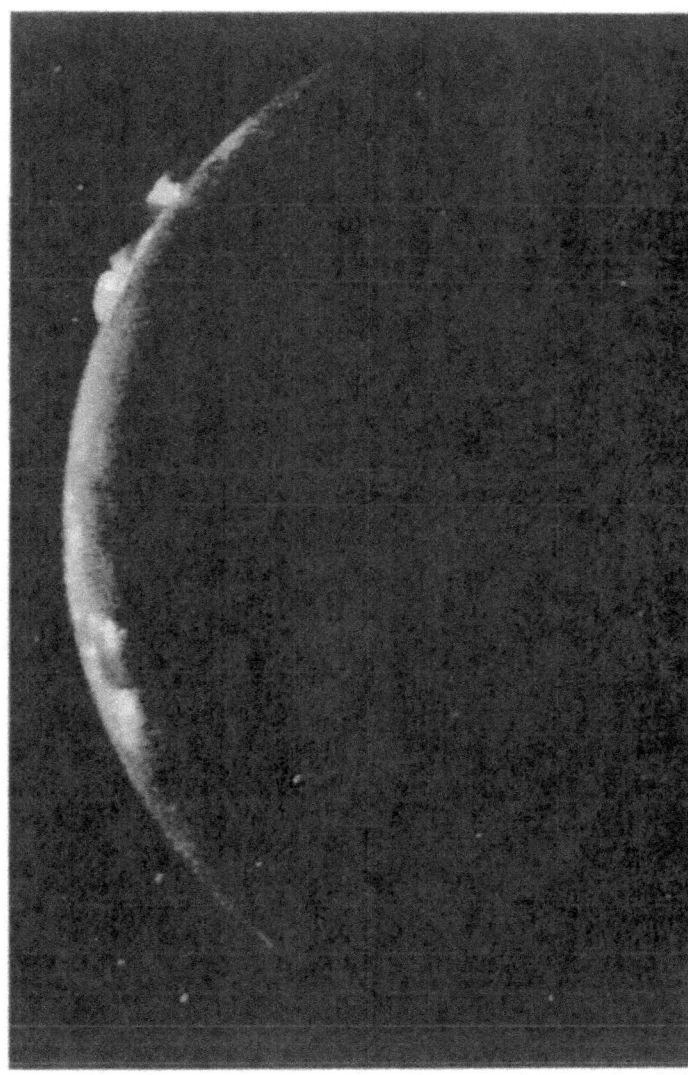

Voyager 2 obtained beautiful views of the volcanic eruptions during its ten-hour Io volcano watch on July 9. On the edge of the crescent image are the volcanoes Amirani (P_c) and below it Maui (P_6), each sending up fountains about 100 kilometers above the surface. The blue color is probably the result of sunlight scattered by tiny particles of sulfur dioxide snow condensing in the erupting plume. [P-21780]

Plume Number	Name	Location (latitude/longitude)	Height During Voyager 1 Flyby (kilometers)	Activity During Voyager 2 Flyby
1	Pele	−20°/255°	280	ceased
2	Loki	+20°/300°	100	increased
3	Prometheus	−5°/155°	70	increased
4	Volund	+20°/175°	95	no data
5	Amirani	+25°/120°	80	similar
6	Maui	+20°/120°	80	similar
7	Marduk	−25°/210°	120	similar
8	Masubi	−40°/50°	70	similar

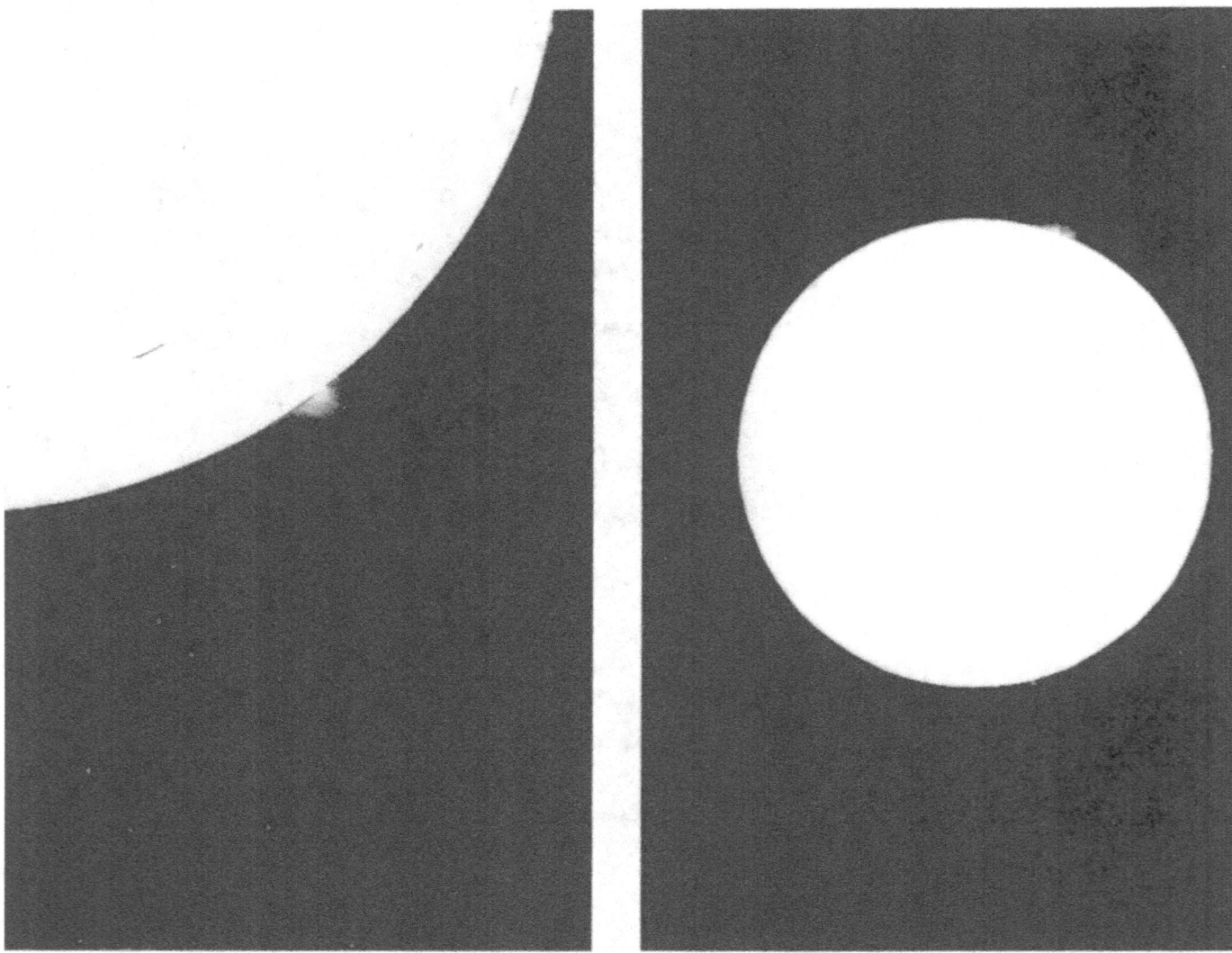

Between the two encounters, the volcanic eruption at Loki (P$_2$) changed character. The single plume emanating from the western end of an apparent dark fissure seen in March was joined by a second fountain of similar size about 100 kilometers to the east. The plume also increased in height, from about 120 kilometers in the Voyager 1 image (left) to 175 kilometers in the Voyager 2 image (right). [260-662A and B]

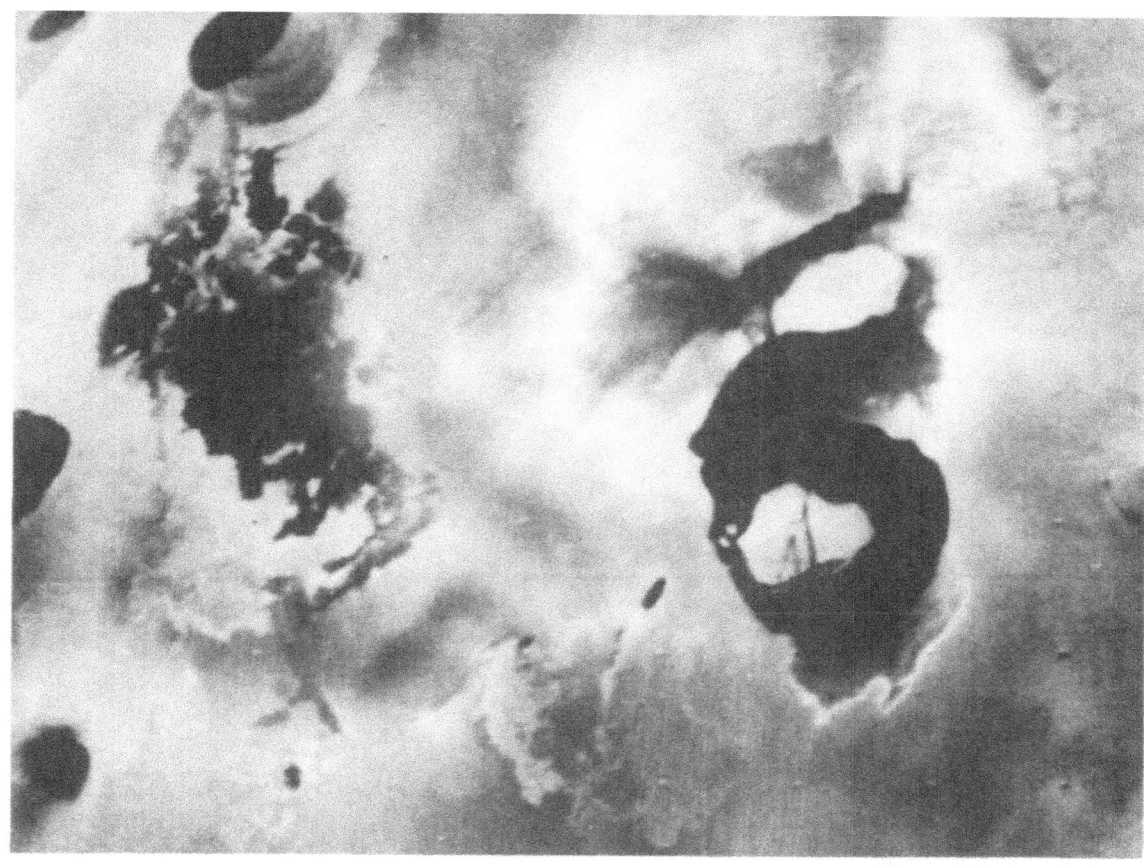

The detailed structure near the volcano Loki is like nothing seen elsewhere on Io. (Top) When this Voyager 1 picture was taken, the main eruptive activity (P_2) came from the lower left of the dark linear feature (perhaps a rift) in the center. Below it is the "lava lake," a U-shaped dark area about 200 kilometers across. In this specially processed image, detail can be seen in the dark surface of this feature, possibly due to "icebergs" of solid sulfur in a liquid sulfur lake. (Bottom) The IRIS on Voyager 1 found this "lava lake" to be the hottest region on Io, with a temperature about 150° C higher than that of the surrounding area. [260-642B]

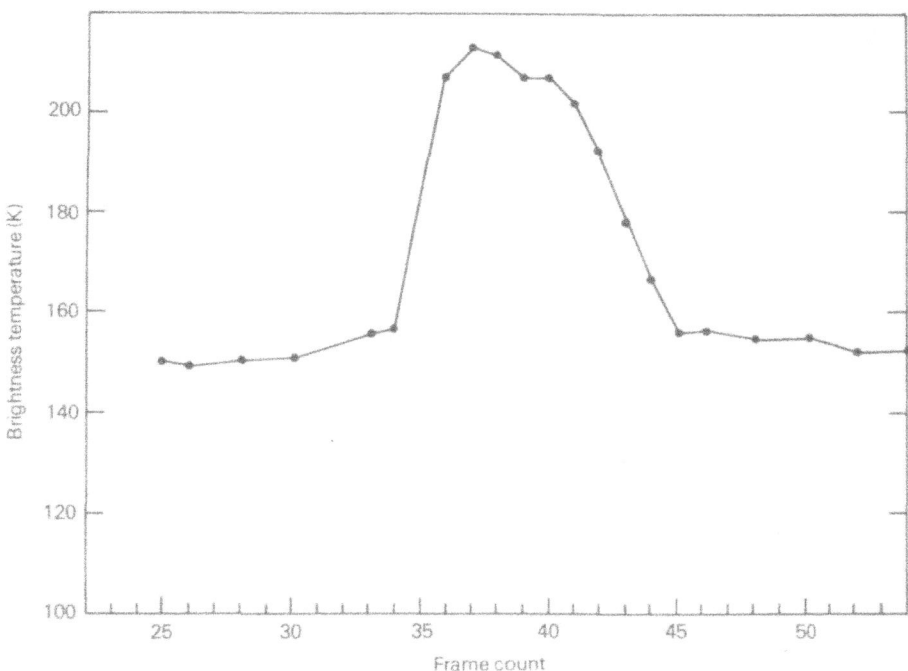

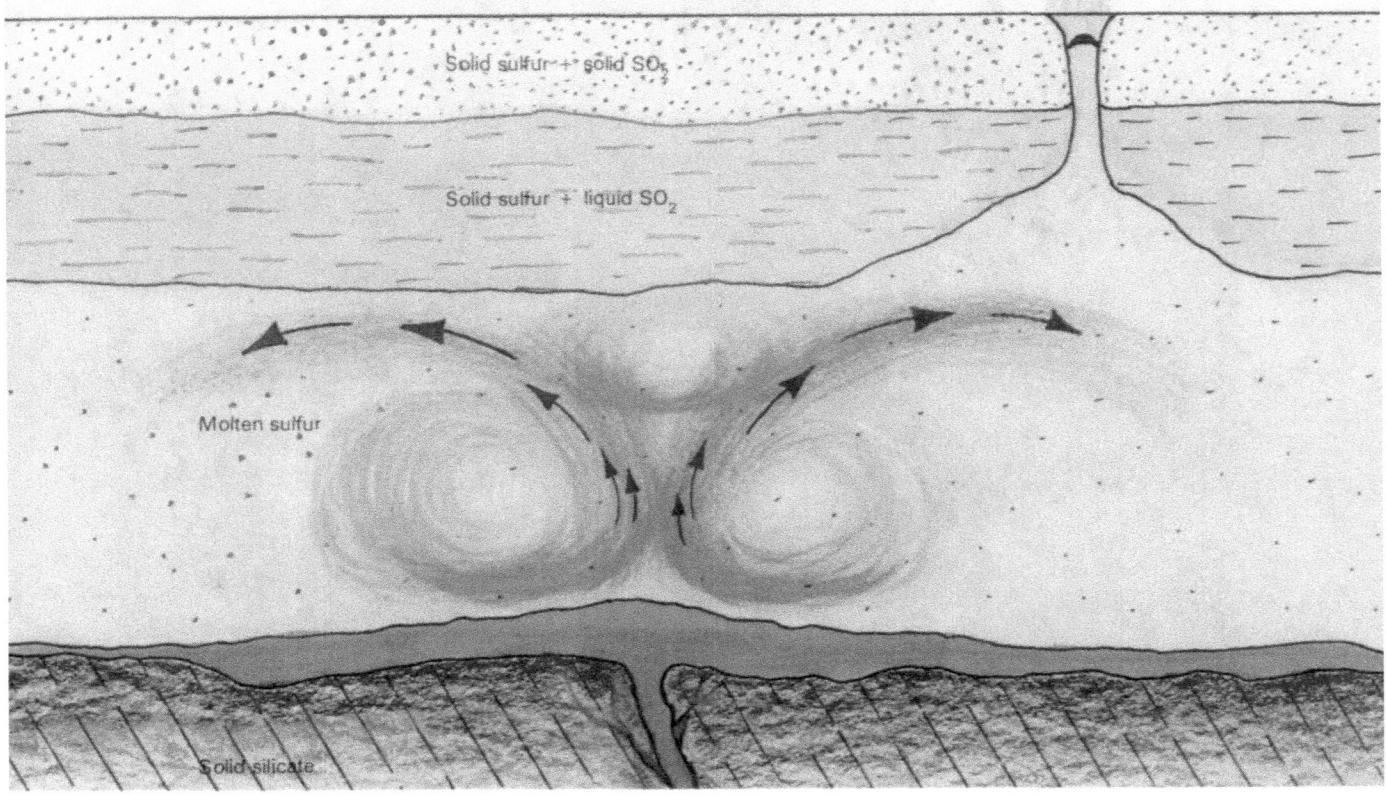

One model for the structure of Io indicates that an ocean of liquid sulfur with a solid sulfur crust covers most of the satellite. Heat escapes from the interior in the form of lava, which erupts beneath the sulfur ocean. Secondary eruptions in the sulfur ocean heat liquid sulfur dioxide, which is mixed with solid sulfur in the crust. The rapid expansion of sulfur dioxide gas then produces the great eruptive plumes, which consist of a mixture of solid sulfur, sulfur dioxide gas, and sulfur dioxide snow.

The Voyager observations appear to confirm the theoretical calculations. The tidal heat source has presumably been acting since Io was formed more than 4 billion years ago. With a totally molten interior and continuing large-scale volcanism, Io has had an opportunity to thoroughly sort out its composition. In the process it would have lost all the volatile gases such as water and carbon dioxide, explaining why Io now has no appreciable atmosphere in spite of the outpouring of material from the interior. In addition, most of the sulfur from the interior could have risen to the surface, where it would be constantly recycled through volcanic activity.

The presence of large amounts of sulfur on the surface may help explain the extraordinary nature of the Io volcanoes. One model considered for the satellite postulates that it is covered by a sea of liquid sulfur several kilometers deep, with a crust of solid sulfur and, below the surface, liquid sulfur dioxide. Calculations by Sue Kieffer of the U.S. Geological Survey and others indicate that the expansion of the sulfur dioxide in such a model can explain the observed eruption velocities of up to a kilometer per second.

The volcanic plumes on Io appear to be made primarily of sulfur and sulfur dioxide. Both are molten as they emerge from the vent, but they quickly cool as the plume rises 100 or more kilometers into the near vacuum of space. Unlike terrestrial volcanoes, there is almost no gas in the plumes. It requires about half an hour for the fine particles of solidified sulfur and sulfur dioxide snow to fall back to the surface, where they form the colorful rings that mark the major eruptive sites.

Almost all the roughly 100 000 tons of material erupted each second by the Io vol-

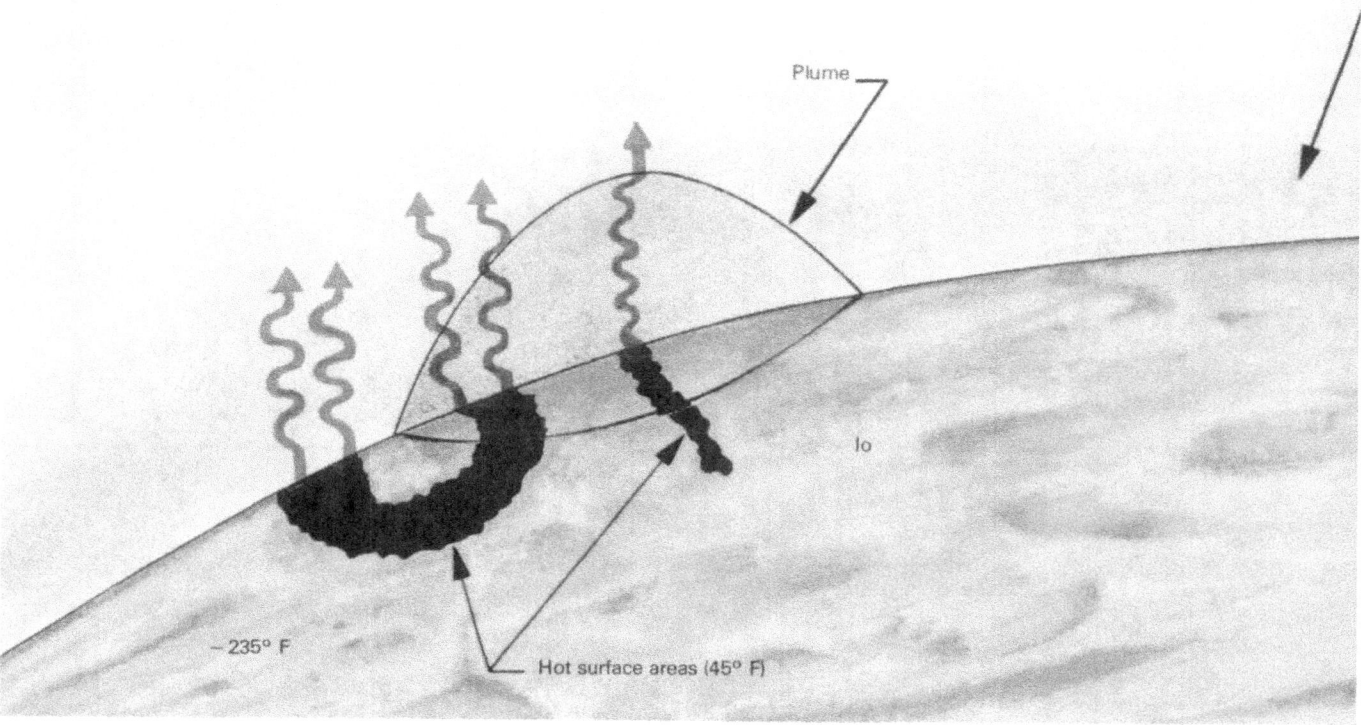

Sulfur dioxide gas cloud

Plume

Io

Hot surface areas (45° F)

−235° F

Direct evidence of an atmosphere on Io was obtained during the Voyager 1 flyby by the IRIS. In the region near the volcano Loki and its associated "lava lake," infrared spectra clearly showed the signature of sulfur dioxide gas. It is not known whether this gas was a temporary feature associated with the eruption of Loki or if it might be present on Io more generally. Other evidence, however, points to the sulfur dioxide atmosphere as a transient feature. A small amount of the sulfur dioxide escapes and is broken apart by sunlight to provide the oxygen and sulfur ions observed in the Io torus.

canoes snows back to the surface. But apparently a part—perhaps 10 tons per second—escapes from Io and is captured by the Jovian magnetosphere. Another part contributes to an ionosphere—a tenuous atmosphere of electrons and ions—that surrounds Io. The injection of several tons of particles each second into the magnetosphere has dramatic consequences that can be seen even from Earth.

The Io Torus

Surrounding Jupiter at the distance of Io is a donut-shaped volume, or torus, of plasma that originates at the satellite. At first, the atoms escaping from Io expand outward as a gas, but soon they are stripped of electrons and become electrically charged. Some of these gases, such as sulfur dioxide, apparently originate in the large volcanic eruptions; other, such

as the sodium cloud being studied with Earth-based telescopes, result from sputtering of surface materials by energetic particles in the magnetosphere. After they are ionized by the loss of one or more electrons, the atoms are caught by the spinning magnetic field of Jupiter and become a part of what is called a co-rotating plasma, spinning at 74 kilometers per second with the same ten-hour period as Jupiter itself.

The Io torus was easily detected on Voyager by the ultraviolet spectrometer, even from a distance of 150 million kilometers. The strongest ultraviolet radiation comes from twice-ionized sulfur (atoms that have lost two electrons) (S III), emitting a wavelength of 69 nanometers or about one-eighth the wavelength of visible light. The spectrometer also detected glows from atoms of triply ionized sulfur (S IV) and twice-ionized oxygen (O III).

165

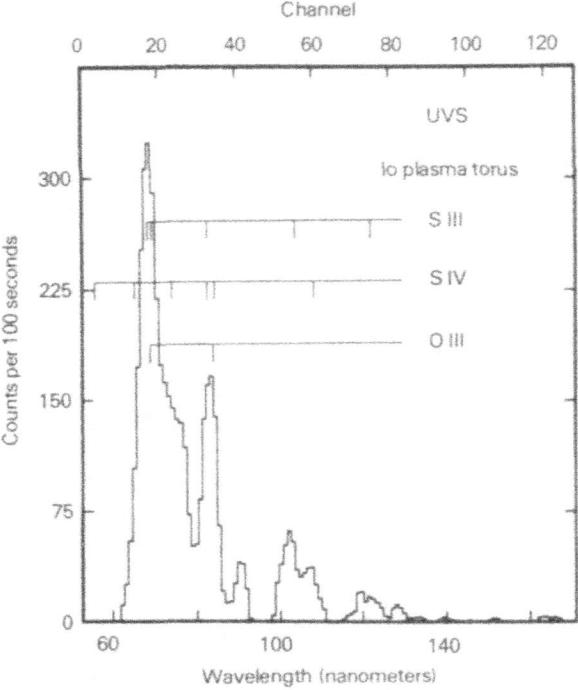

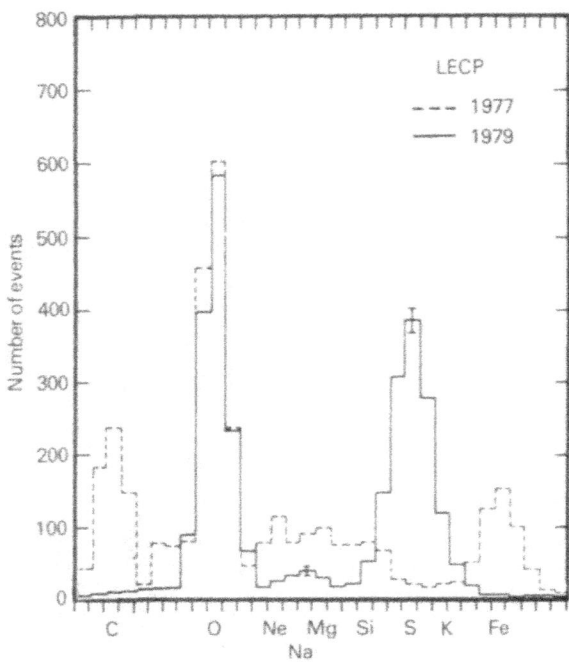

At the time of the Voyager 1 encounter, the most abundant heavy ions in the Jovian magnetosphere were sulfur and oxygen. Multiply ionized sulfur and oxygen both emit strongly in the ultraviolet, where they could be observed by the ultraviolet spectrometer. This spectrum of the Io torus registers the tremendous amount of ultraviolet energy (about a million million watts) being radiated. To emit so strongly, the temperatures in the torus must be near 100 000 K.

Direct measurement of the heavy ions associated with the Io torus were made by the Voyager 1 LECP instrument. Here the amounts of various elements are shown for two cases: the Jovian inner magnetosphere (solid line) and a typical solar event. Both the Jovian and the solar particles have been scaled to show similar amounts of oxygen, but the solar particles are also rich in carbon and iron, whereas Jupiter has a great deal of sulfur. Evidence such as this demonstrates that the sulfur does not come from the Sun; rather, the sulfur and most of the oxygen appear to be the product of the Io sulfur dioxide volcanic eruptions.

Scans across the torus showed that it had a thickness of 1.0 R_J and was centered at a distance of 5.9 R_J from Jupiter. The torus is centered on the magnetic, rather than the rotational, equator of the planet. To produce the intense glow observed, the electron temperatures in the torus must be 100 000 K, with an electron density of about 1000 per cubic centimeter. The brightness in the ultraviolet corresponds to a radiated power of more than a million million (10^{12}) watts. This enormous amount of energy must be continuously supplied by the magnetosphere.

The ultraviolet emissions from the Io torus seen by Voyager were dramatically different from those seen in 1973 and 1974 with the simpler ultraviolet instrument on board Pioneers 10 and 11. These changes correspond to more than a factor of 10 in brightness. As noted by the Voyager Team, "Because of the remarkable differences we conclude that the Jupiter-Io environment has changed significantly since

December 1973. The observed differences are so spectacularly large that this conclusion does not depend on a detailed comparison of the two instruments, their calibrations, or the observing geometry." The reason for this change, or the degree to which it reflects a large-scale variation in the volcanic activity of Io, is one of the major questions arising from the Voyager mission.

The Io torus can also be observed from the ground. In 1976, spectra showed the glow of singly ionized sulfur (S II), and in 1979 singly ionized oxygen (O II) was detected. The observations of S II and O II are particularly interesting because they provide a measure of the density and temperature of the plasma. In April 1979, between the two Voyager flybys, Carl Pilcher of the University of Hawaii succeeded in obtaining a telescopic image of the torus in the light of S II. He measured nearly the same ring diameter (5.3 – 5.7 R_J) as had Voyager; in-

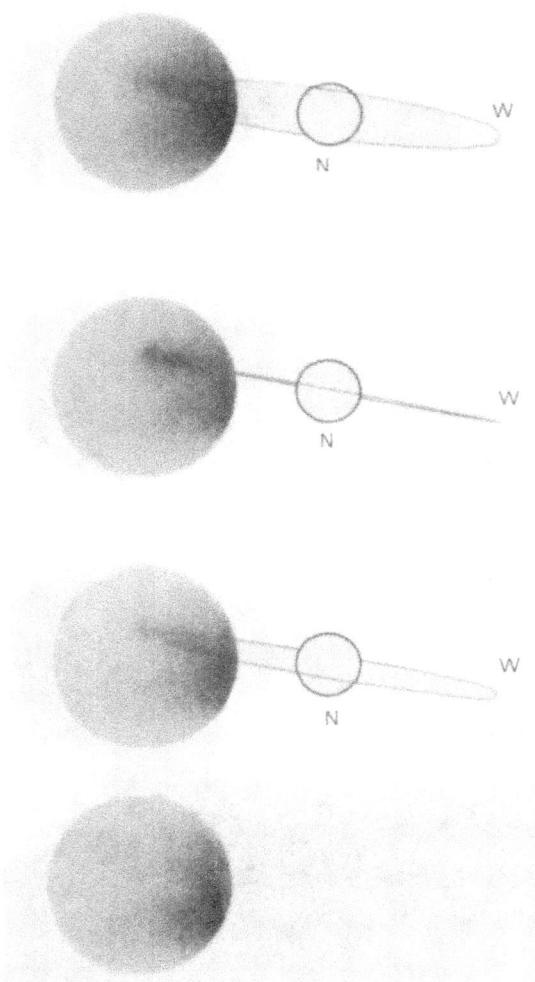

The emission of light from sulfur ions in the Io torus is so strong it can be measured from the Earth. In these pictures, University of Hawaii astronomer Carl Pilcher photographed the torus in the light of ionized sulfur on the night of April 9, 1979. As the planet rotates, the torus is seen first partly opened, then edge-on, and again opened in the opposite direction. The dark band on the right of each image is due to light from Jupiter scattered in the telescope, as shown in the bottom picture, which contains the scattered light only.

terestingly, both agree that the sulfur torus is centered slightly inside the orbit of Io (6.0 R_J).

Direct measurements of the torus were made from Voyager 1 as the spacecraft passed twice through this region, once inbound and once outbound. The low energy charged particle instrument and the cosmic ray instrument both determined that the composition of the ions in the Io torus was primarily sulfur and oxygen. Ionized sodium was also observed. For several years, ground-based telescope observations had revealed a cloud of neutral sodium around Io; the Voyager instruments picked up these atoms after they had each lost an electron and become trapped in the magnetosphere. These instruments also derived the electron and ion density (about 1000 per cubic centimeter) and confirmed that the ions were co-rotating with the inner magnetosphere.

Another Io-associated phenomenon searched for by Voyager was the Io flux tube. As a conductor moving through the Jovian magnetic field, Io generates an electric current, estimated to have a strength of about 10 million (10^7) amperes and a power of the order of a million million (10^{12}) watts. The region of space through which this current flows from the satellite to Jupiter is called the flux tube.

Voyager 1 was targeted to fly through the Io flux tube. This was an important decision, since this option precluded the possibility of obtaining occultations by either Io or Ganymede. The event was to take place on March 5, just after closest approach to Io. The effects of the flux tube were clearly observed by the magnetometer, the LECP instrument, and other particle and field instruments; however, subsequent analysis indicated that the spacecraft had not penetrated the region of maximum current flow; it probably missed the center of the flux tube between 5000 and 10 000 kilometers.

The flux tube is not the only connection between Io and Jupiter. Radio emissions from the atmosphere are triggered by Io's orbital position, and the aurorae that illuminate Jupiter's polar regions are the result of charged particles falling into the planet from the Io torus. Other charged particles can occasionally escape outward and be detected as far away as Earth.

Io is unquestionably a remarkable world. The only planetary body known to be geologically more active than the Earth, it provides many extreme examples to test the theories of geoscientists. Its intimate interconnections with the Jovian magnetosphere and the planet itself provide a unifying theme to the complex processes taking place in the inner parts of the Jovian system.

CHAPTER 10

RETURN TO JUPITER

A Successor to Voyager

The spectacular discoveries of the Voyagers did not exhaust our interest in the Jovian system. Both the giant planet and its system of satellites will almost certainly play a central role in any future program of solar system exploration and research. Thus, even as the two Voyager spacecraft directed their attention further outward toward Saturn, NASA had begun development of the next Jupiter mission, named Galileo.

Galileo is an ambitious, multiple-vehicle planetary mission. It has two major interlocking elements: a probe to be placed in the atmosphere of Jupiter and an orbiter to explore Jupiter, its satellites, and its magnetosphere. By using individual satellite flybys to alter its orbit, the Galileo spacecraft can carry out a satellite "tour" consisting of flybys of the Galilean satellites at different geometries and a deep penetration into the magnetosphere in the unexplored region of space behind Jupiter.

Both an atmospheric probe for the planet and a long-lived orbiter to study the satellites and magnetosphere are logical successors to Voyager. In 1974, three years before Voyager launch, the Space Science Board of the National Academy of Sciences was already emphasizing the scientific advantages of both of these approaches. In suggesting goals for 1975–1985, the Board wrote, "We recommend that a significant effort in the NASA planetary program over the next decade be devoted toward the outer solar system. Jupiter is the primary object of outer solar system exploration." Looking at specific mission goals, the Board recommended that "the primary objectives in the exploration of Jupiter and its satellites for the period 1975–1985 in order of importance are (1) determination of the chemical composition and physical state of its atmosphere, (2) the chemical composition and physical state of the satellites, and (3) the topology and behavior of the magnetic field and the energetic particle fluxes. In order to carry out this program, it will be necessary to utilize orbiting spacecraft and probe-delivering spacecraft."

In the same period NASA carried out studies of two possible orbiter and probe missions. Working through the Ames Research Center, a scientific panel chaired by James Van Allen explored the adaptation of the Pioneer 10 and 11 spinning spacecraft to carry a probe to Jupiter and to carry out an orbiter mission emphasizing magnetospheric studies. William B. Hubbard of the University of Arizona chaired a JPL-based panel investigating the use of a Mariner-class fully stabilized spacecraft similar to Voyager to carry out a satellite-oriented orbiter mission. In 1976 these concepts were combined in a study, again chaired by Dr. Van Allen, of a Voyager-type orbiter with probe-carrying capability. This mission concept was given the name JOP, for Jupiter Orbiter Probe, and lead responsibility was assigned by NASA to JPL, with Ames carrying out the design of the probe.

In 1977, as Voyager activity was building toward autumn launch, a struggle was under-

The Galileo Probe will make a fiery entry into the Jovian atmosphere, carrying a payload of scientific instruments for the first direct sampling of the atmosphere of a giant planet. Shown here is the moment, at a pressure level of about 0.1 bar, when the parachute is deployed and the still-glowing heatshield drops free from the Probe.

way in Washington to obtain approval for the new Jupiter orbiter and probe mission. Budgeting authority was requested in the President's Fiscal Year 1978 budget, but only after extensive testimony and several Congressional votes was the mission approved. The official new start for JOP, soon to be renamed Galileo, was set for July 1, 1977, and the scientific investigators and their instruments were selected in August.

At JPL, many members of the Voyager Team made a smooth transition to the Galileo Project. Much of the knowledge that had gone into the design of the Voyager spacecraft and its subsystems was now incorporated into Galileo. Similarly, at Ames the knowledge gained from the design of the Pioneer Venus probes, which were launched to Venus in 1978, a year after Voyager launch, was applied to design of a Jupiter probe. Among the individuals who brought their Voyager experience to Galileo were John Casani, who left the position of Voyager Project Manager to become Galileo Project Manager, and Torrence Johnson of the Voyager Imaging Team, who became Galileo Project Scientist.

The Scientific Capability of Galileo

The investigations of Jupiter and its system planned for the Galileo Project represented substantial advances over those carried out by Voyager. In part, this was the result of new spacecraft capabilities, particularly the atmospheric entry probe. It also represented increasing sophistication in scientific instrumentation over the seven-year interval between the selection of the payloads for the two missions.

The main emphasis in the study of Jupiter itself is on direct measurements with the Probe. For the first time it will be possible to examine directly the atmosphere of a giant planet. By measuring the temperature and pressure as it descends through the clouds, the Probe can determine the structure of the atmosphere with much higher precision than could ever be obtained from remote observations. The structure, in turn, provides information on dynamics—the circulation and heat balance of the Jovian atmosphere. In addition, the Probe can make direct measurements of the composition of the gases, with sensitivity in some cases to quantities as low as a few parts per billion. In addition to the elemental abundance, the amount of different isotopes can also be measured.

Direct studies of the clouds of Jupiter can be made from the Galileo Probe. With a device called a nephelometer (literally, cloud-meter), the sizes and compositions of individual aerosol particles will be determined. An infrared instrument will determine the temperatures of the cloud layers and measure the amounts of sunlight deposited in different regions of the atmosphere. Another instrument will search for lightning; it has the ability to detect both the flash of light and the radio static generated by each bolt.

Additional studies of the atmosphere, similar to those of Voyager, can be carried out from the Galileo Orbiter. Television pictures, ultraviolet and infrared spectra, and measurements of the polarization of reflected light will all be obtained with the same scan platform instruments that are used to study the surfaces of the satellites.

A full battery of fields and particles instruments is planned for the Galileo Orbiter. Many of these are direct descendants of Voyager instruments. In general, their capabilities have been improved, particularly their ability to determine the composition of charged particles. There is a steady progression from Pioneer to Voyager to Galileo: The early measurements were concentrated on particle energies, but more sophisticated instruments yield the composition of the ions and the details of their motion.

Many of the advances expected from Galileo in magnetospheric studies result from the Orbiter's ability to explore many parts of the environment of Jupiter. The Pioneer and Voyager spacecraft made single cuts through the magnetosphere, and often it was difficult to distinguish temporal from spatial effects. Galileo will repeatedly swing around Jupiter, sampling conditions at many distances from the planet over a time span of two years or more. In addition, it is planned to adjust the orbit of Galileo to swing out into the magnetotail, the turbulent region of the magnetosphere that stretches "downwind" from Jupiter for hundreds of Jupiter radii. No flybys can reach the magnetotail; an orbiting spacecraft is required.

The Galilean satellites naturally will be a primary focus of Galileo science, particularly after Voyager. It is planned to have as many as a dozen individual encounters, most of them at much closer range than the Voyager flybys. To take advantage of these opportunities, the

The Orbiter section of the Galileo spacecraft will carry both remote sensing and direct measuring instruments for the study of Jupiter, its satellites, and its magnetosphere. Several remote sensing instruments—an imagery system, a near infrared mapping spectrometer, an ultraviolet spectrometer, and a photopolarimeter/radiometer—will be mounted on a scan platform. The particles and fields instruments will be on a spinning section of the spacecraft. The Orbiter is expected to operate for at least two years around Jupiter, providing one close flyby of Io and several each of Europa, Ganymede, and Callisto. [P-20772]

Galileo scan platform will carry two new remote sensing systems.

Instead of the vidicon television camera on Voyager, Galileo imaging will be done with a new solid-state detector called a charged coupled device (CCD). The CCD has a wider spectral response and greater photometric accuracy. In addition, its increased sensitivity permits shorter exposures, so that even on very close flybys the pictures will not be blurred by spacecraft motion. Substantial coverage at a resolution of 100 meters should be possible,

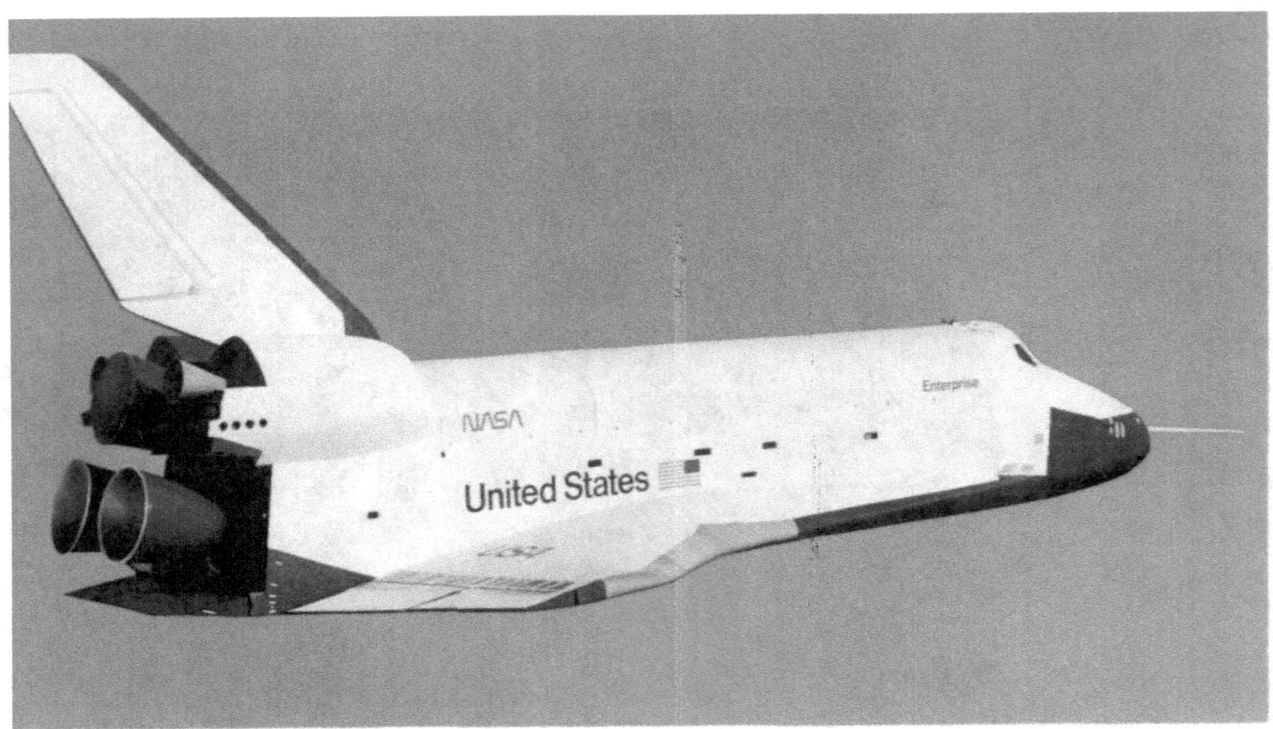

Galileo will be launched by the Space Shuttle, the central element of the new NASA Space Transportation System. Together with its upper stage launch rocket, Galileo will be placed into Earth's orbit in the large Shuttle bay, about 20 meters long and 5 meters in diameter. After releasing Galileo, the Shuttle will be piloted back to a landing at Cape Canaveral, to be used again for many future flights. [5-78-23599]

The Space Shuttle and the Inertial Upper Stage of Galileo as they will appear in Earth orbit. The upper stage has been released from the Shuttle bay and is being prepared for launch to Jupiter.

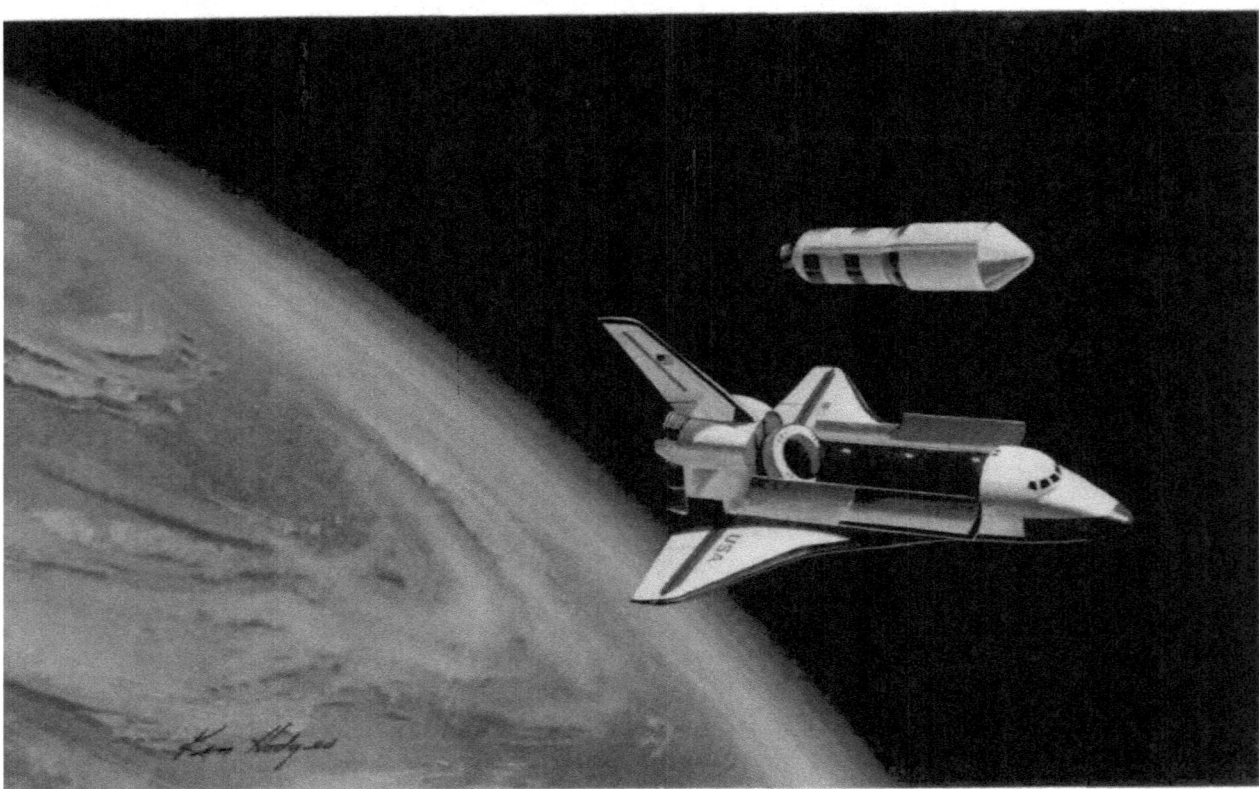

compared to Voyager's best resolution of 1 kilometer for Io and 4 kilometers for Europa.

To determine the composition of satellite surface materials, Galileo will also carry a near-infrared mapping spectrometer (NIMS). This instrument will obtain measurements over the visible and infrared spectra of areas as small as 10 kilometers across. With NIMS, it should be possible to investigate the composition of individual features as small as the volcanic calderas on Io or the ejecta blankets of Ganymede's craters.

Galileo Mission Design

The Galileo Orbiter and Probe are to be launched with NASA's new Space Shuttle and Inertial Upper Stage. To carry the maximum possible payload to Jupiter, a close flyby of Mars is planned en route. The gravitational field of Mars will give a boost to Galileo, just as that of Jupiter was used by Voyager to swing on to Saturn.

The exact launch date and trajectory for Galileo have not yet been specified, but if all goes well, the Orbiter spacecraft will approach Jupiter from the dawn side of the planet sometime in the mid-1980s. It will not be moving as fast as Voyager, since it must be placed into orbit around Jupiter rather than flashing past on its way to the outer solar system. On its initial trajectory, Galileo will probably come within 5 R$_J$ of Jupiter, slightly closer than Voyager 1. At this time it will fire its rocket engines (supplied by the Federal Republic of Germany in a cooperative program with NASA) to shed excess speed and let itself be captured by Jupiter's gravity. The first pass will also be the time for a close flyby of Io.

The most critical period of the Galileo flight will be the Probe entry at Jupiter. The Probe must strike the atmosphere at precisely the correct angle and speed to be slowed down without being destroyed. At a pressure level of about 0.1 bar the rapid deceleration period ends and the heat shield is released. A parachute is deployed to slow the descent further, and the Probe then has a period of nearly an hour to study the atmosphere and clouds of Jupiter. The Probe mission ends when its batteries run down or when it is crushed by the pressure of the Jovian atmosphere near the 20-bar level, whichever comes first. [SL78-545(3)]

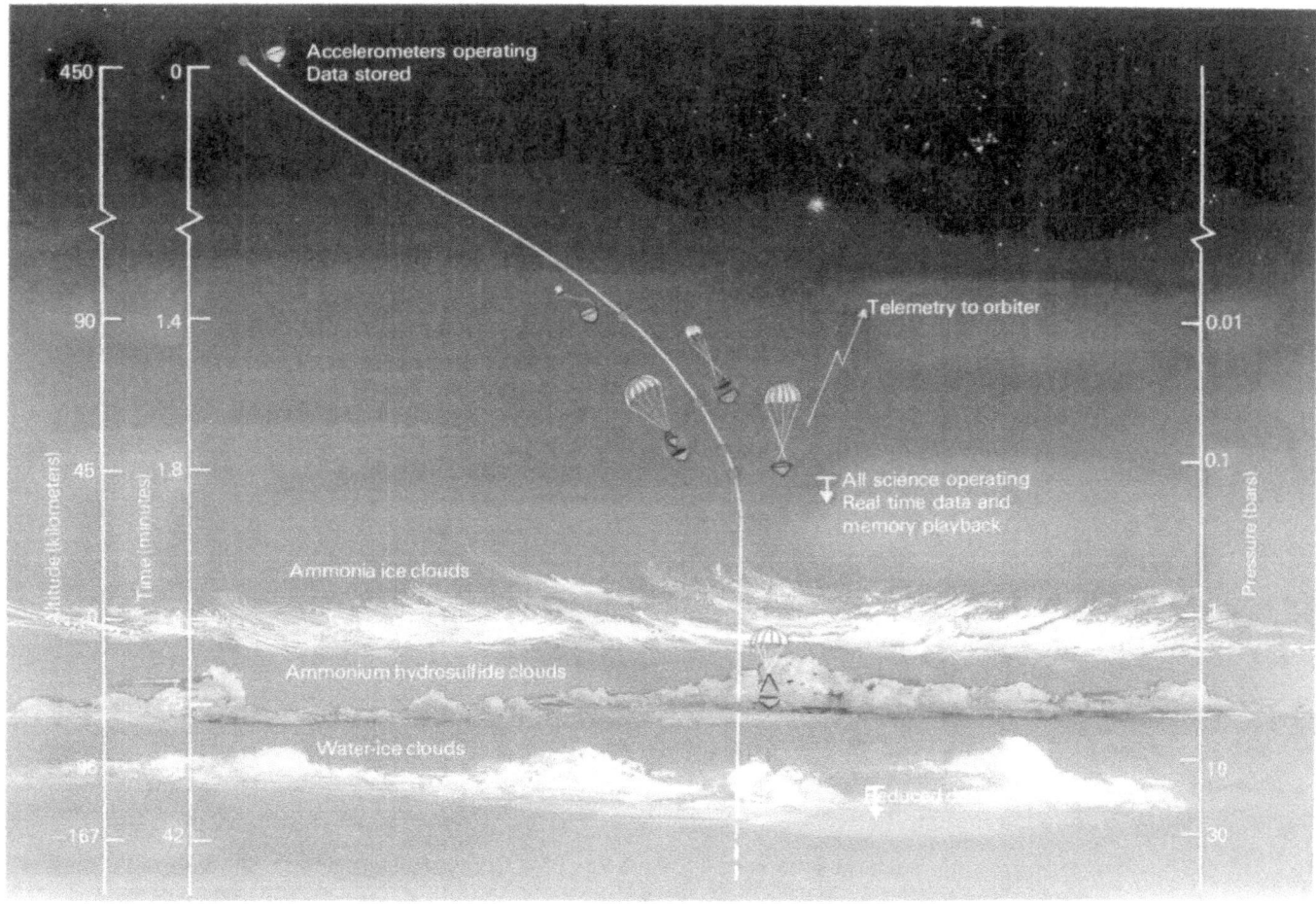

GALILEO PROBE SCIENCE INVESTIGATIONS
Probe Scientist: L. Collin, NASA Ames

Investigation	Principal Investigator	Primary Objectives
Atmospheric structure	A. Seiff, NASA Ames	Measure temperature, density, pressure, and molecular weight to determine the structure of Jupiter's atmosphere.
Neutral mass spectrometer	H. B. Neimann, NASA Goddard	Measure the composition of the gases in Jupiter's atmosphere and the variations at different levels in the atmosphere.
Helium abundance interferometer	U. von Zahn, Bonn U. (Germany)	Measure with high accuracy the ratio of hydrogen to helium in Jupiter's atmosphere.
Nephelometer	B. Ragent, NASA Ames	Determine the sizes of cloud particles and the location of cloud layers in Jupiter's atmosphere.
Net flux radiometer	R. W. Boese, NASA Ames	Measure energy being radiated from Jupiter and the Sun, at different levels in Jupiter's atmosphere.
Lightning and radio emission	L. J. Lanzerotti, Bell Labs	Measure lightning flashes in Jupiter's atmosphere, from the light and radio transmissions from those flashes.
Energetic particles	H. M. Fischer, U. Kiel (Germany)	Measure energetic electrons and protons in the inner regions of the Jovian radiation belts and determine their spatial distributions.

Because of the intense radiation environment, the Galileo Orbiter will not be able to spend much time in the inner magnetosphere, near the orbit of Io. To do so would risk damage to the spacecraft electronics and a premature end to the mission. Additional thruster firing during the first orbit can be used to raise the periapse to 10 R_J or greater. No more close passes by Io will be possible, but studies of this satellite can be made on each subsequent orbit with imaging resolutions of about 10 kilometers, sufficient to see details of the volcanic eruptions and monitor volcano-associated changes in the surface.

At each subsequent orbit, Galileo will be programmed for a close flyby of one of the other satellites. Several passes each of Callisto, Ganymede, and Europa should be possible. The satellite tour does not need to be fully planned in advance; by adjusting the spacecraft trajectory with small bursts of the thruster motors, navigation engineers can modify the orbit to permit adaptation to scientific needs. As the Orbiter mission progresses, the spacecraft will also sample many parts of the magnetosphere, including one long excursion, at least 150 R_J, into the magnetotail.

The total duration of the Orbiter mission is planned to be at least 20 months. Additions to the basic mission are possible if the spacecraft remains healthy and fuel reserves are adequate. In contrast, the Galileo Probe mission lasts only a few hours.

As the Probe approaches the atmosphere of Jupiter at the awesome speed of 26 kilometers per second, it will be traversing a region of space never before explored. An energetic particle detector will investigate the innermost magnetosphere before the entry begins. Then, within a period of just a few minutes, friction with the upper atmosphere must dissipate the Probe energy until it is falling gently in the Jovian air.

Jupiter, being the largest planet, presents the most challenging atmospheric entry mission ever undertaken by NASA. The design of the Galileo Probe calls for a massive heat shield to protect the instruments during the high-speed entry phase. After the Probe has slowed to subsonic velocities, a parachute will be deployed,

GALILEO ORBITER SCIENCE INVESTIGATIONS
Project Scientist: T. V. Johnson, JPL

Investigation	Principal Investigator	Primary Objectives
Solid state imaging	M. J. S. Belton, Kitt Peak Observatory (Team Leader)	Provide images of Jupiter's atmosphere and its satellites; study atmospheric structure and dynamics on Jupiter; investigate the composition and geology of the satellite surfaces; study the active volcanic processes on Io.
Ultraviolet spectrometer	C. W. Hord, U. Colorado	Study composition and structure of the upper atmospheres of Jupiter and its satellites.
Near-infrared mapping spectrometer (NIMS)	R. W. Carlson, JPL	Provide spectral images and reflected sunlight spectra of Jupiter's satellites, indicating the composition of their surfaces; measure reflected sunlight and thermal emission from Jupiter's atmosphere to study composition, cloud structure, and temperature profiles; monitor hot spots on Io.
Photopolarimeter/radiometer	J. E. Hansen, NASA Goddard	Measure temperature profiles and energy balance of Jupiter's atmosphere; measure Jupiter's cloud characteristics and composition.
Magnetometer	M. G. Kivelson, UC Los Angeles	Measure magnetic fields and the ways they change near Jupiter and its satellites; measure variations caused by the satellites interacting with Jupiter's field.
Plasma particles	L. A. Frank, U. Iowa	Provide information on low-energy particles and clouds of ionized gas in the magnetosphere.
Energetic particles	D. J. Williams, NOAA Space Environment Lab	Measure composition, distribution, and energy spectra of high-energy particles trapped in Jupiter's magnetosphere.
Plasma waves	D. A. Gurnett, U. Iowa	Investigate waves generated inside Jupiter's magnetosphere and waves radiated by possible lightning discharges in the atmosphere.
Dust detection	E. Grün, Max-Planck-Institut (Germany)	Determine size, speed, and charge of small particles such as micrometeorites near Jupiter and its satellites.
Celestial mechanics	J. D. Anderson, JPL (Team Leader)	Use the tracking data to measure the gravity fields of Jupiter and its satellites; search for gravity waves propagating through interstellar space.
Radio propagation	H. T. Howard, Stanford U. (Team Leader)	Use radio signals from the Orbiter and Probe to study the structure of the atmospheres and ionospheres of Jupiter and its satellites.

Interdisciplinary Scientists: F. P. Fanale (JPL), P. J. Gierasch (Cornell U.), D. M. Hunten (U. Arizona), H. Masursky (U.S. Geological Survey), M. B. McElroy (Harvard U.), D. Morrison (U. Hawaii), G. S. Orton (JPL), T. Owen (SU New York), J. B. Pollack (NASA Ames), C. T. Russell (UC Los Angeles), C. Sagan (Cornell U.), F. L. Scarf (TRW), G. Schubert (UC Los Angeles), C. P. Sonett (U. Arizona), J. A. Van Allen (U. Iowa).

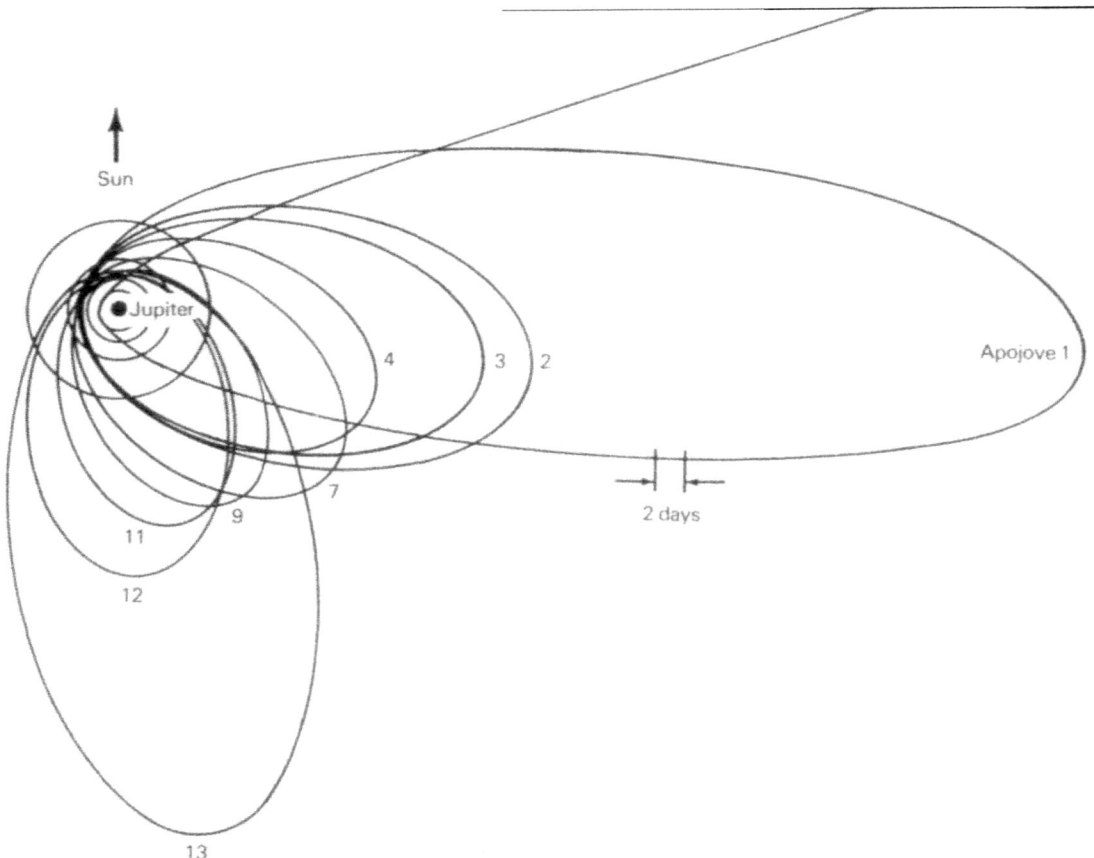

During its two years in orbit, the Galileo Orbiter will carry out many investigations of the planet, the Galilean satellites, and the Jovian magnetosphere. Repeated close flybys of the satellites are used to modify and shape the orbit to provide additional flybys at an optimum viewing geometry. Initially, the orbit is a long loop that extends in the general direction of the sunset side of Jupiter. The orbit is then contracted, and the encounters with the satellites rotate it behind the planet for a long excursion into the magnetotail late in the tour.

and the heat shield, having done its job, will be dropped free.

The Probe will spend nearly an hour descending from a pressure level of about 0.1 bar, where the heat shield is jettisoned, to a depth of 10–20 bars. During this time it will make most of its scientific measurements, relaying them back to Earth via the Probe carrier. Designers expect the Probe to sink through regions of ammonia clouds, ammonium hydrosulfide clouds, and ice and water clouds during this hour.

By the time it has descended below the water clouds, the increasing pressure will exceed the strength of some Probe components. Engineers expect the Probe to have completed its mission, exhausted its battery power, and been crushed by the atmospheric pressure before the 20-bar level is reached. Lifeless, it will then sink on into the thick, hot lower atmosphere of Jupiter.

Beyond Galileo

After Galileo, the future cannot be predicted. Perhaps there will no longer be a program of planetary exploration. But if humanity still has the vision to seek a future in the stars, there will surely be other Jupiter missions.

Perhaps the next mission will concentrate on Jupiter itself. Probes could be built to withstand pressures as high as several hundred bars, feeling their way deep into the murky depths of the planet. Or a hot-air balloon could be deployed from a probe to carry instruments for long-term studies of the atmosphere. A number of proposals have also been made for additional satellite missions, including orbiters or landers for Ganymede and Callisto. Or perhaps it will be desirable to land a vehicle on one of the satellites and collect a sample and return it to Earth for laboratory analysis.

Whatever the future holds, it is clear that the Pioneer and Voyager missions blazed the path to Jupiter and beyond. The little Pioneers proved that it could be done, and the Voyagers expanded their vision, exploring and discovering new worlds more remarkable and exciting than anyone could have imagined.

PICTORIAL MAPS OF THE GALILEAN SATELLITES

These maps were prepared for the Voyager Imaging Team by the U.S. Geological Survey in cooperation with the Jet Propulsion Laboratory, California Institute of Technology and the National Aeronautics and Space Administration. Copies are available from Branch of Distribution, U.S. Geological Survey, 1200 South Eads Street, Arlington, VA 22202, and Branch of Distribution, U.S. Geological Survey, Box 25286, Federal Center, Denver, CO 80225.

Preliminary Pictorial Map of Callisto

Atlas of Callisto
1:25,000,000 Topographic Series
Jc 25M 2RMN, 1979
I-1239

This map was compiled from Voyager 1 and 2 pictures of Callisto. Placement of features is based on predicted spacecraft trajectory data and is highly approximate. The linkage between Voyager 1 and Voyager 2 pictures is particularly tenuous. Placement errors as large as 10° are probably common throughout the map, and a few may be even larger. Feature names were approved by the International Astronomical Union in 1979. Airbrush representation is by P. M. Bridges.

Jc 25M 2RMN: Abbreviation for (Jupiter) Callisto, 1:25,000,000 series, second edition, shaded relief map, R, with surface markings, M, and feature names, N.

Scale 1:13 980 000 at 56° latitude
Polar stereographic projection

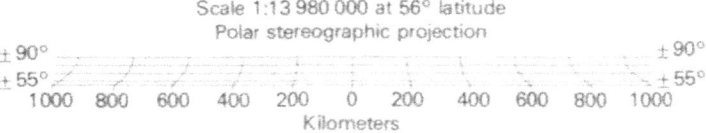

±90°
±55°
1000 800 600 400 200 0 200 400 600 800 1000
Kilometers

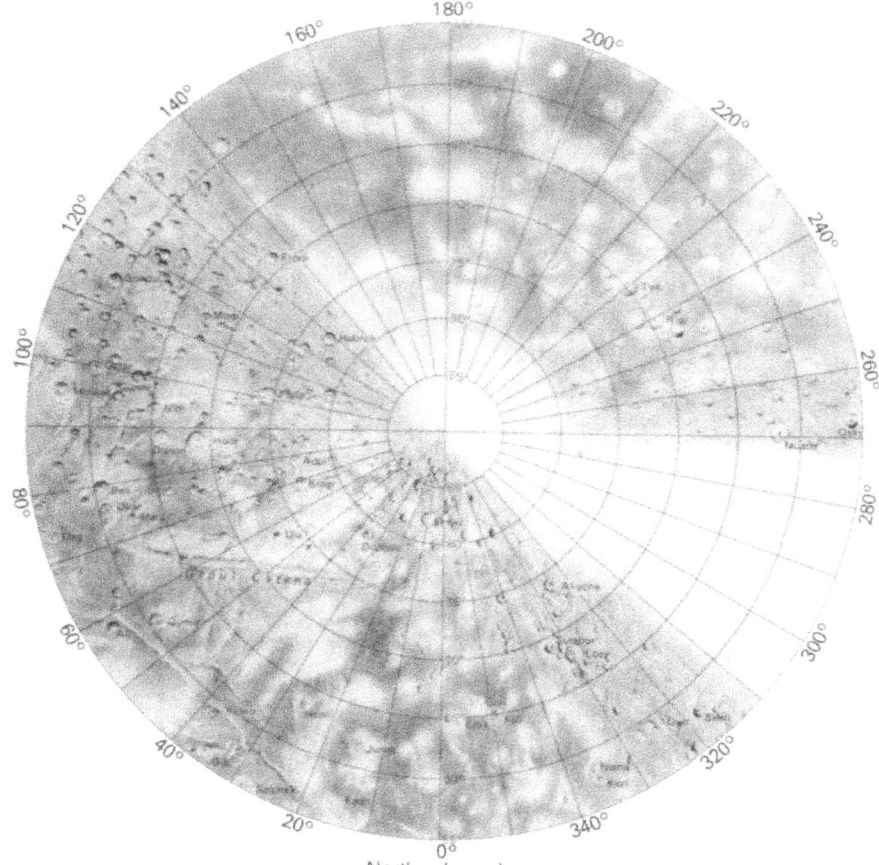

North polar region

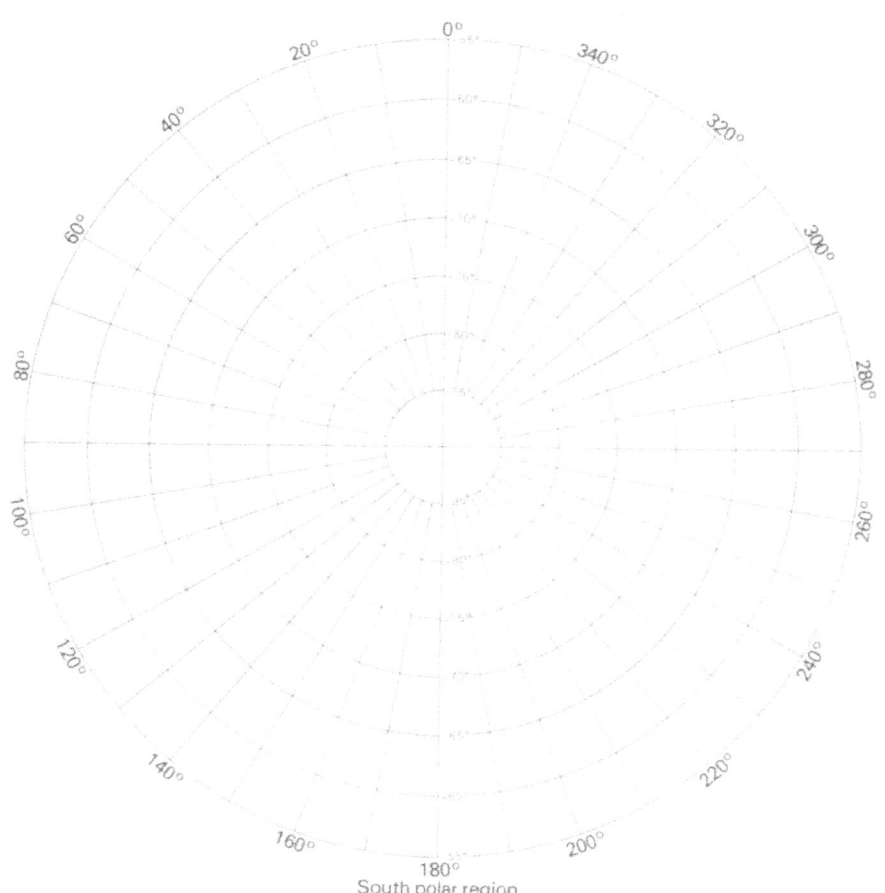

South polar region

Callisto

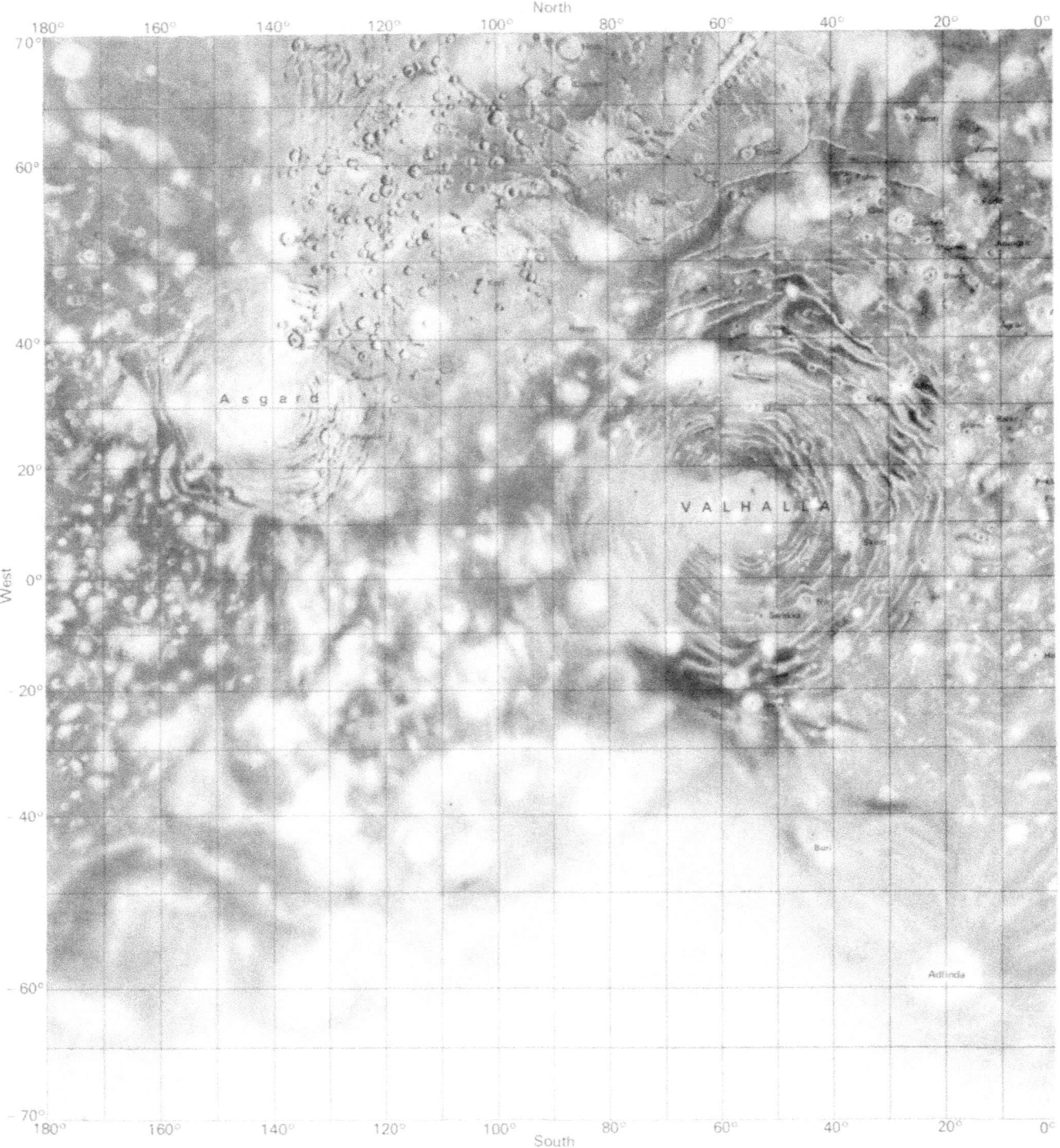

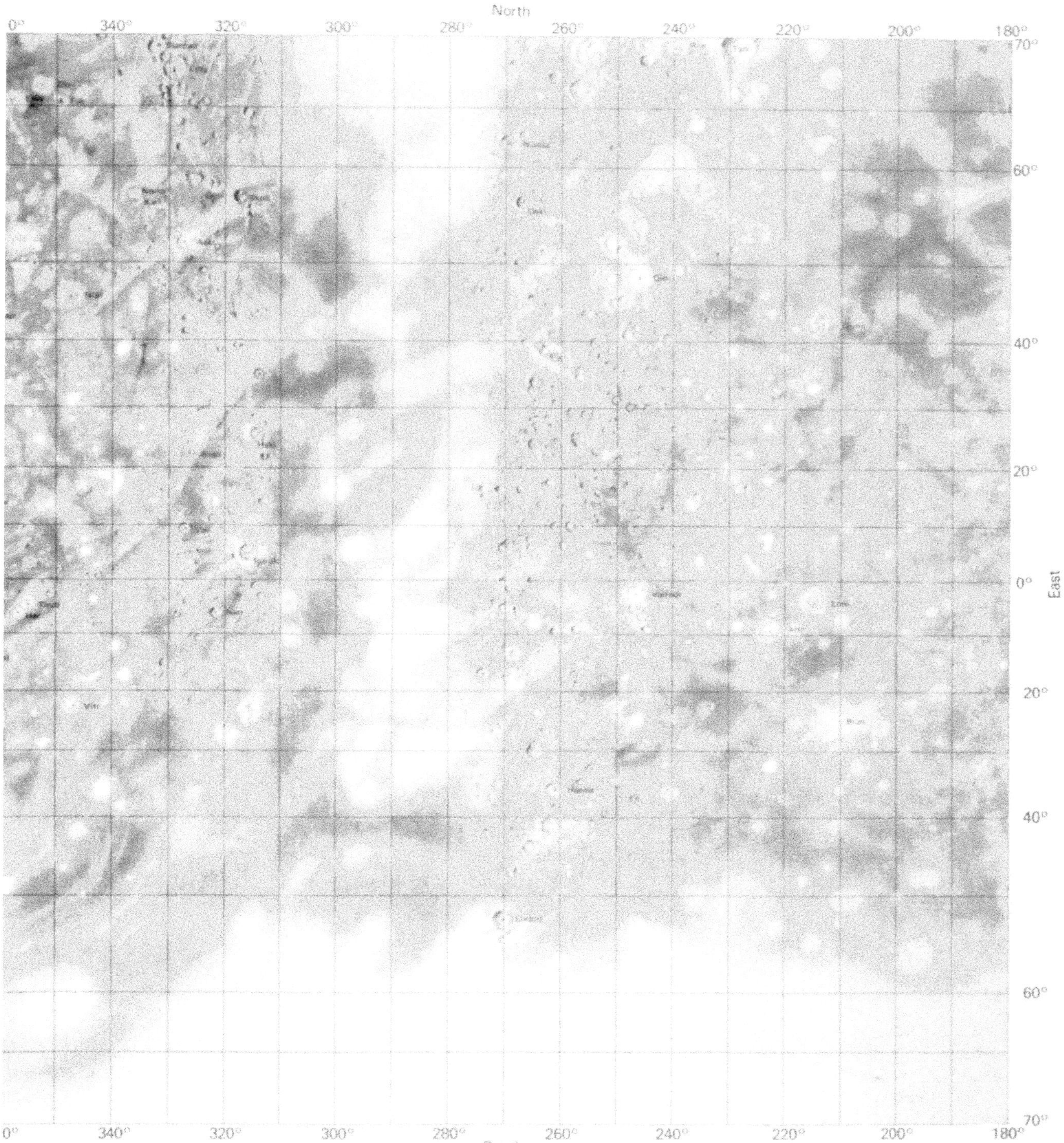

North

0° 340° 320° 300° 280° 260° 240° 220° 200° 180° 70°

60°

East
40°

20°

0°

20°

40°

60°

70°
0° 340° 320° 300° 280° 260° 240° 220° 200° 180°
South

Scale 1:25 000 000 at 0° latitude
Mercator projection

Kilometers
1000 800 600 400 200 0 200 400 600 800 1000
±70° ±70°
±50° ±50°
±30° ±30°
±10° ±10°

Preliminary Pictorial Map of Ganymede

Atlas of Ganymede
1:25,000,000 Topographic Series
Jg 25M 2RMN, 1979
I-1242

This map was compiled from Voyager 1 and 2 pictures of Ganymede. Placement of features is based on predicted spacecraft trajectory data and is highly approximate. The linkage between Voyager 1 and Voyager 2 pictures is particularly tenuous. Placement errors as large as 10° are probably common throughout the map, and a few may be even larger. A large unresolved discrepancy exists in the area bounded by the −45° and −55° parallels between 120° and 180° longitude. Relative placement of features is distorted in that area. Feature names were approved by the International Astronomical Union in 1979. Airbrush representation is by J. L. Inge.

Jg 25M 2RMN: Abbreviation for (Jupiter) Ganymede, 1:25,000,000 series, second edition, shaded relief map, R, with surface markings, M, and feature names, N.

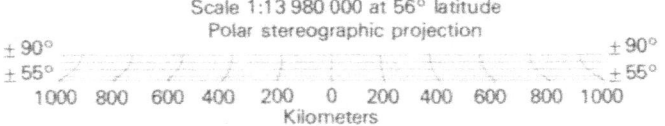

Scale 1:13 980 000 at 56° latitude
Polar stereographic projection

± 90° ± 90°
± 55° ± 55°

1000 800 600 400 200 0 200 400 600 800 1000
Kilometers

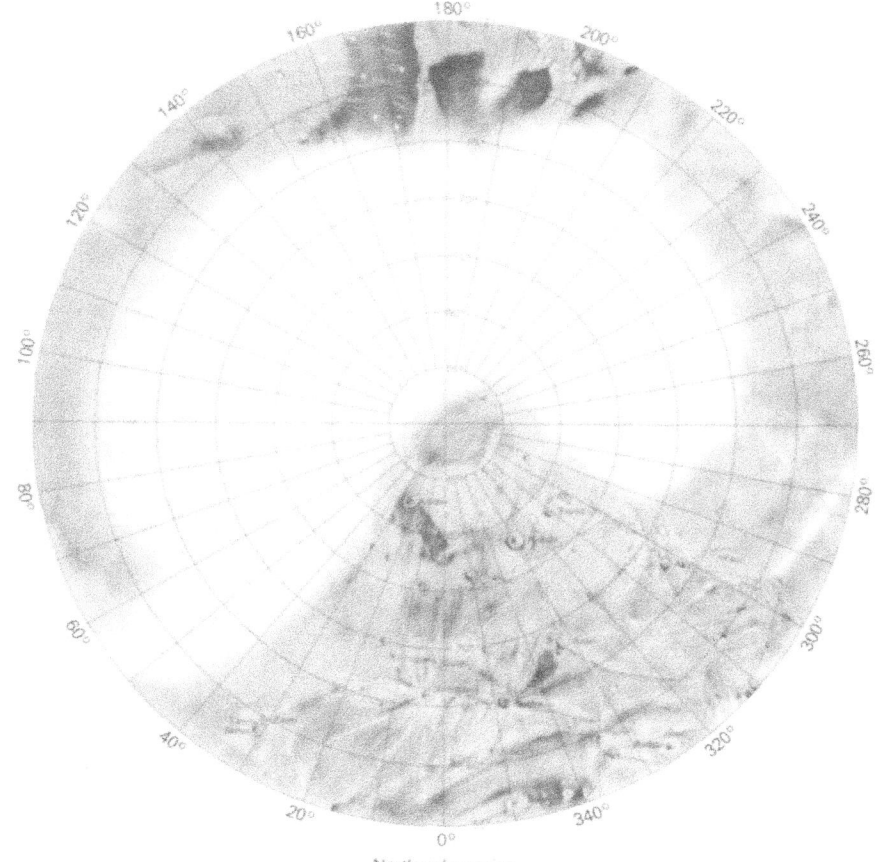

North polar region

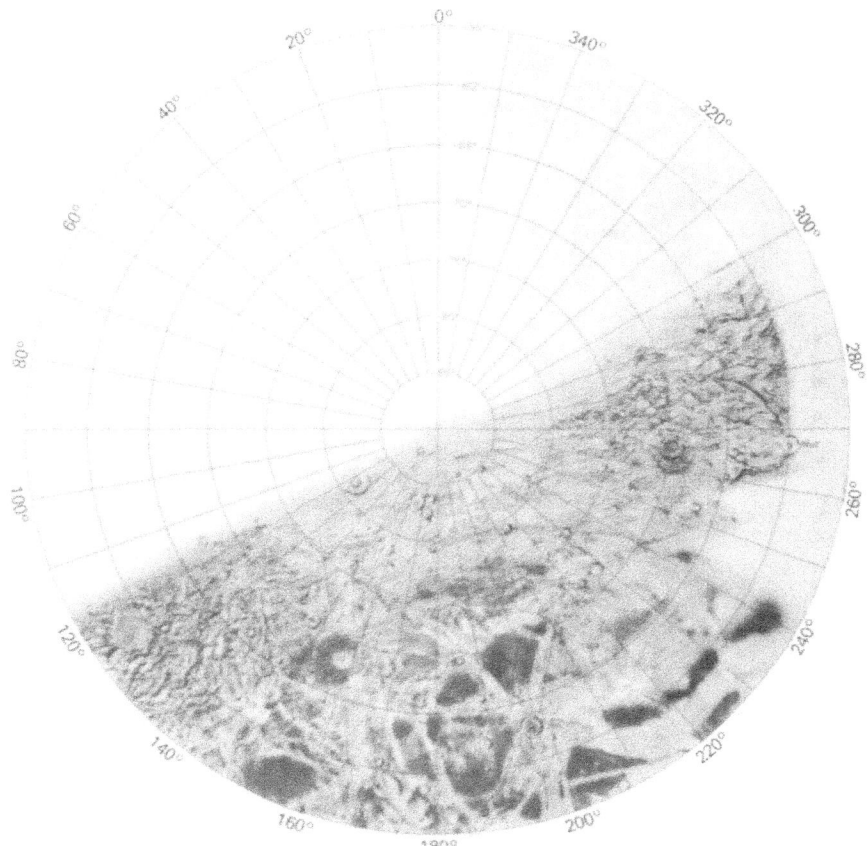

South polar region

Ganymede

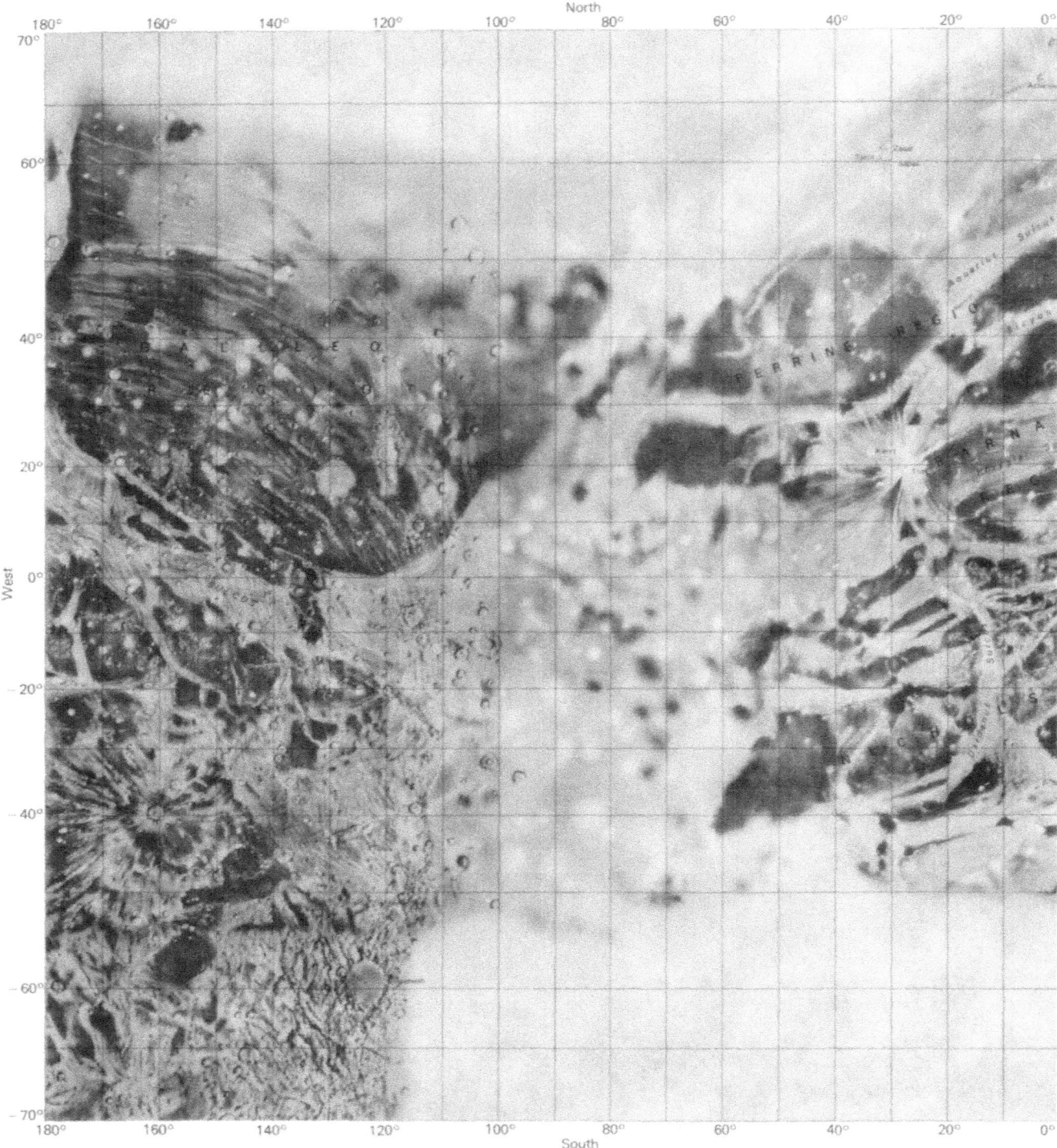

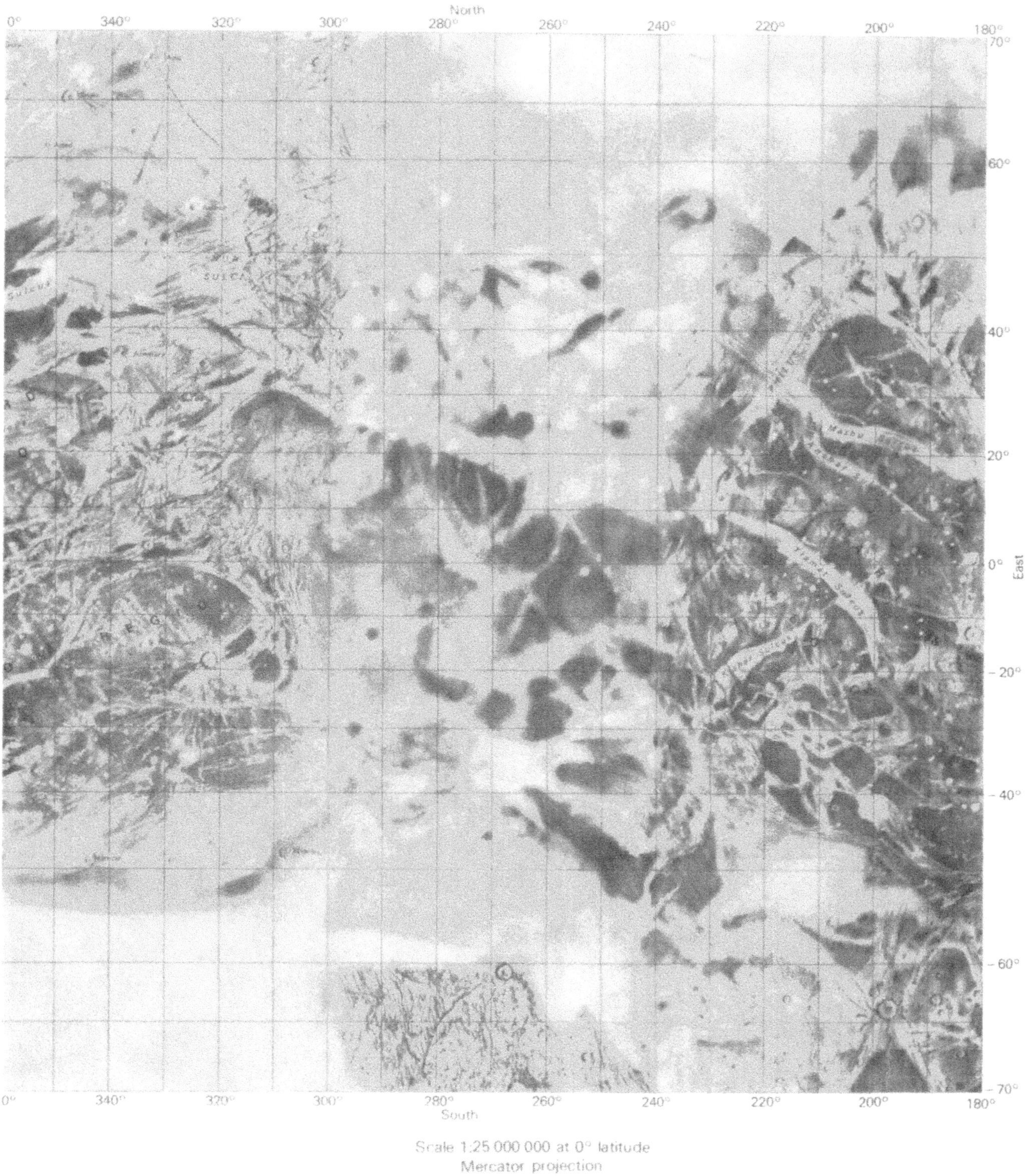

North

0° 340° 320° 300° 280° 260° 240° 220° 200° 180°

70°
60°
40°
20°
0° East
−20°
−40°
−60°
−70°

0° 340° 320° 300° 280° 260° 240° 220° 200° 180°

South

Scale 1:25 000 000 at 0° latitude
Mercator projection

Kilometers

1000 800 600 400 200 0 200 400 600 800 1000

±60°
±40°
±20°
0°

±60°
±40°
±20°
0°

185

Preliminary Pictorial Map of Europa

Atlas of Europa
1:25,000,000 Topographic Series
Je 25M 2RMN, 1979
I-1241

This map was compiled from Voyager 1 and 2 pictures of Europa. Placement of features is based on predicted spacecraft trajectory data and is highly approximate. Feature names were approved by the International Astronomical Union in 1979. Airbrush representation is by J. L. Inge.

Je 25M 2RMN: Abbreviation for (Jupiter) Europa, 1:25,000,000 series, second edition, shaded relief map, R, with surface markings, M, and feature names, N.

Scale 1:13 980 000 at 56° latitude
Polar stereographic projection

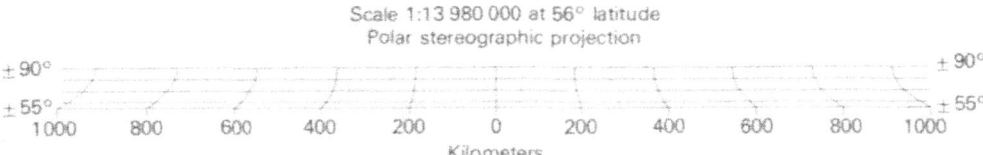

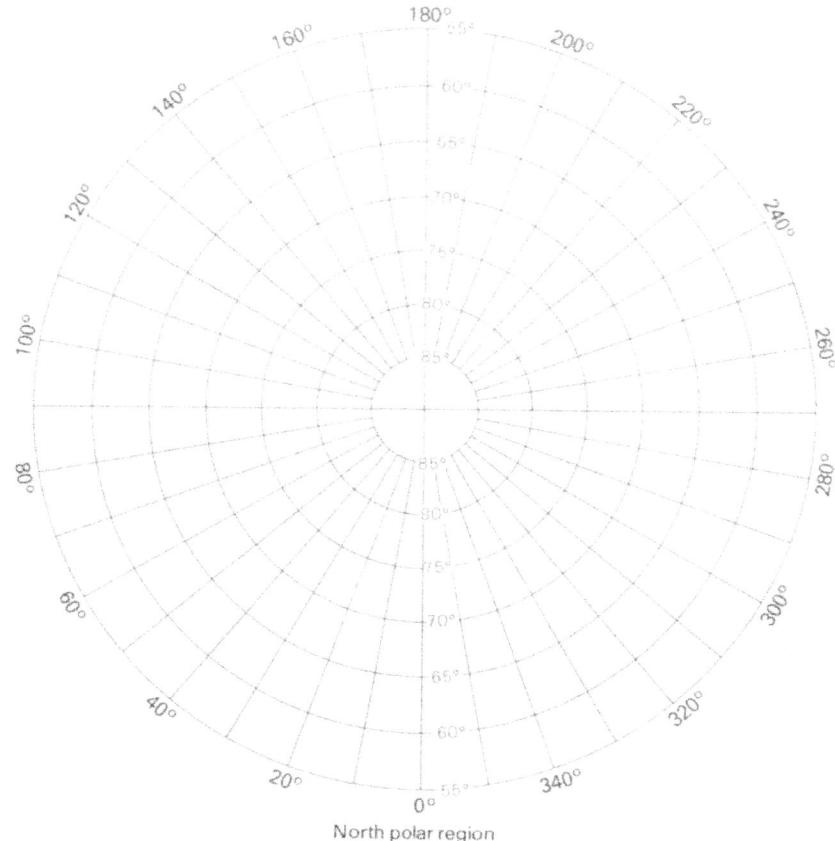

North polar region

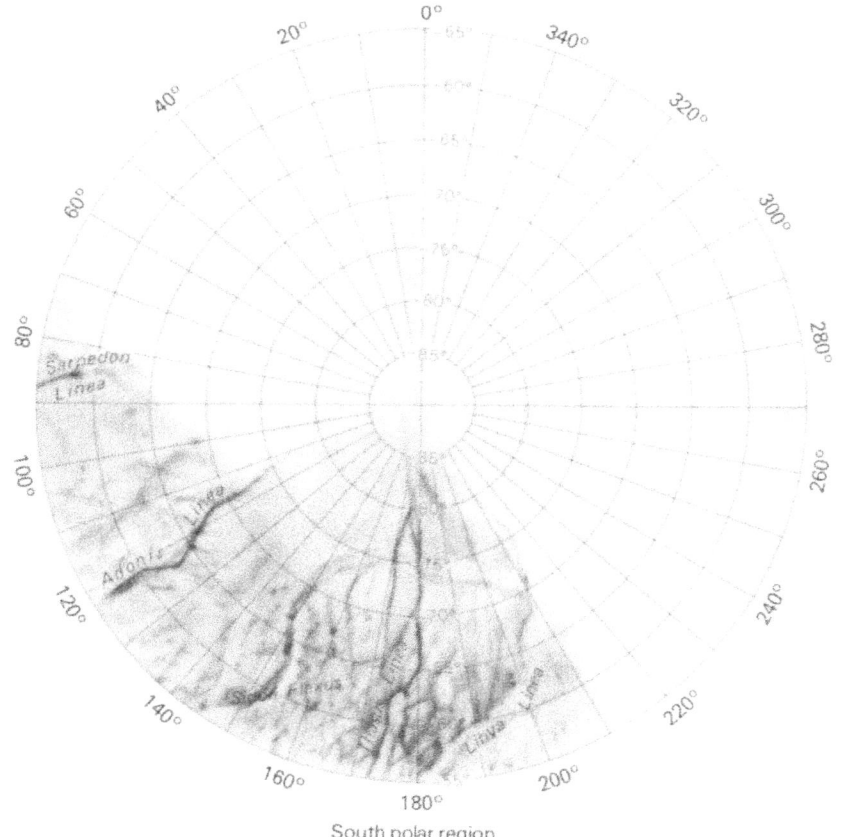

South polar region

Europa

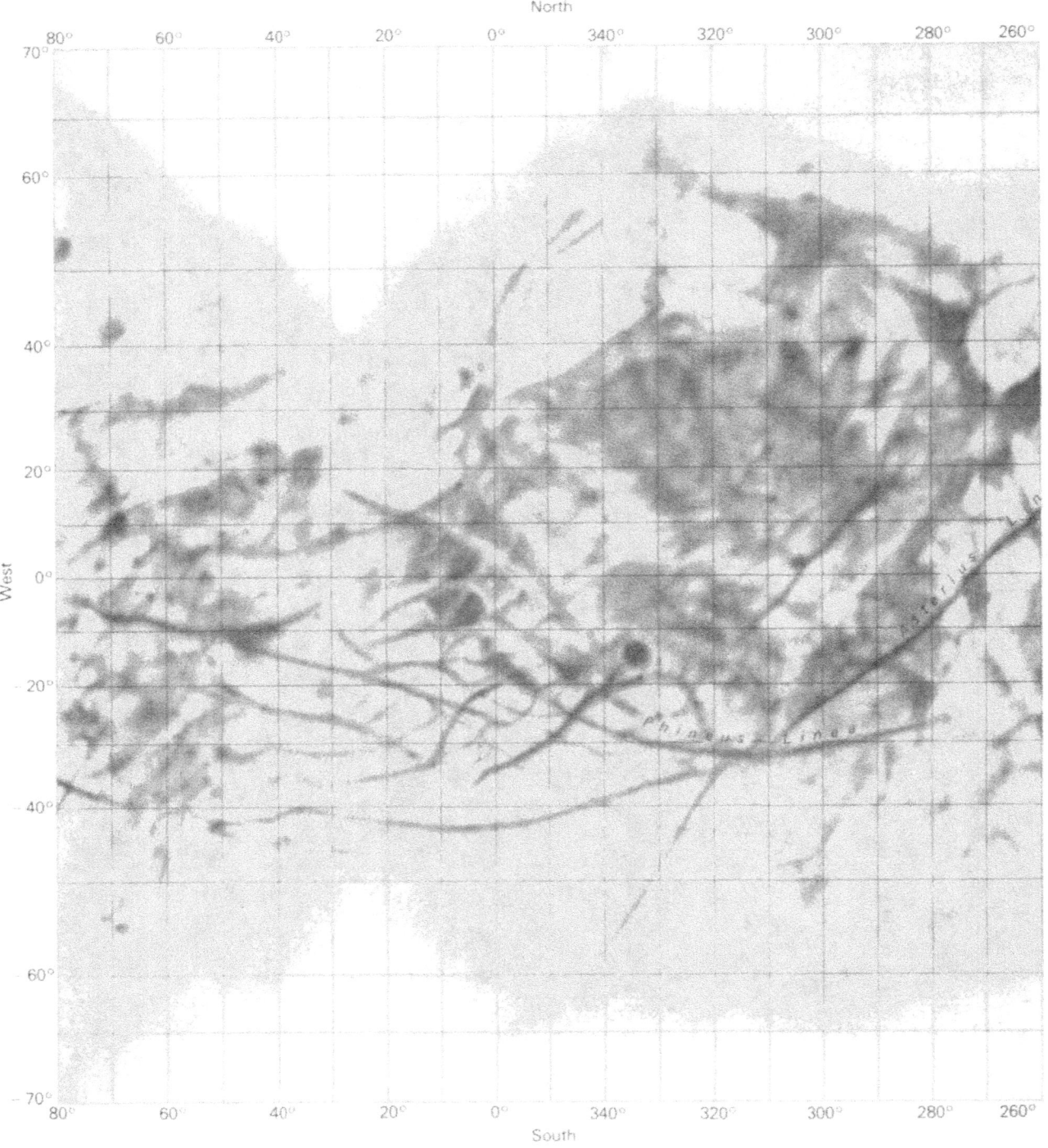

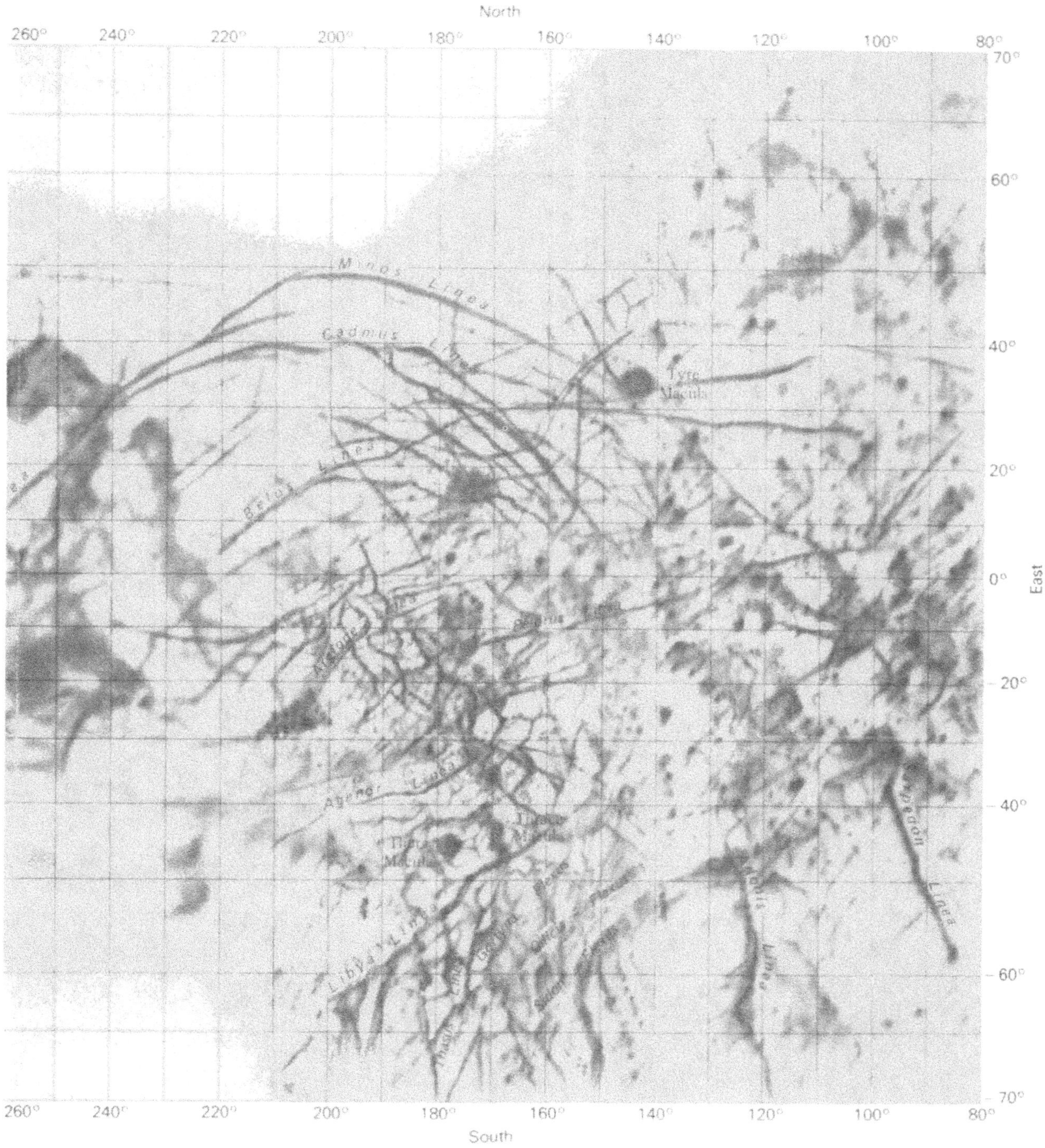

North

260° 240° 220° 200° 180° 160° 140° 120° 100° 80°

70°
60°
40°
20°
0°
20°
40°
60°
70°

East

Minos Linea

Cadmus Linea

Tyre Macula

South

260° 240° 220° 200° 180° 160° 140° 120° 100° 80°

Scale 1:25 000 000 at 0° latitude
Mercator projection

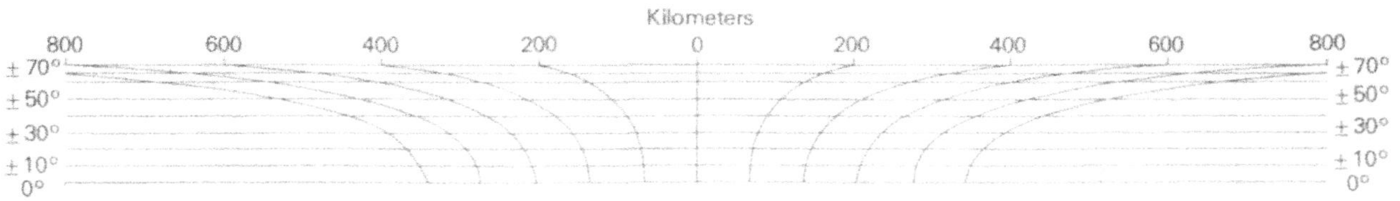

Kilometers

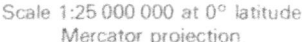

800 600 400 200 0 200 400 600 800

±70°
±50°
±30°
±10°
0°

±70°
±50°
±30°
±10°
0°

Preliminary Pictorial Map of Io

Atlas of Io
1:25,000,000 Topographic Series
Ji 25M 2RMN, 1979
I-1240

This map was compiled from Voyager 1 and 2 pictures of Io. Placement of features is based on preliminary control information provided by M. E. Davies of the Rand Corporation, Santa Monica, California, and is probably accurate within 50 to 100 km. Feature names were approved by the International Astronomical Union in 1979. Airbrush representation is by P. M. Bridges.

Ji 25M 2RMN: Abbreviation for (Jupiter) Io, 1:25,000,000 series, second edition, shaded relief map, R, with surface markings, M, and feature names, N.

Scale 1:13 980 000 at 56° latitude
Polar stereographic projection

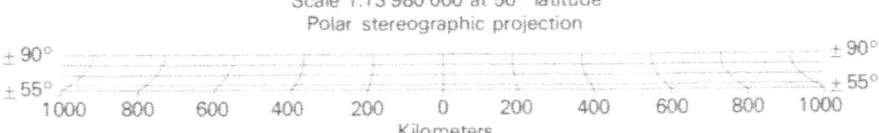

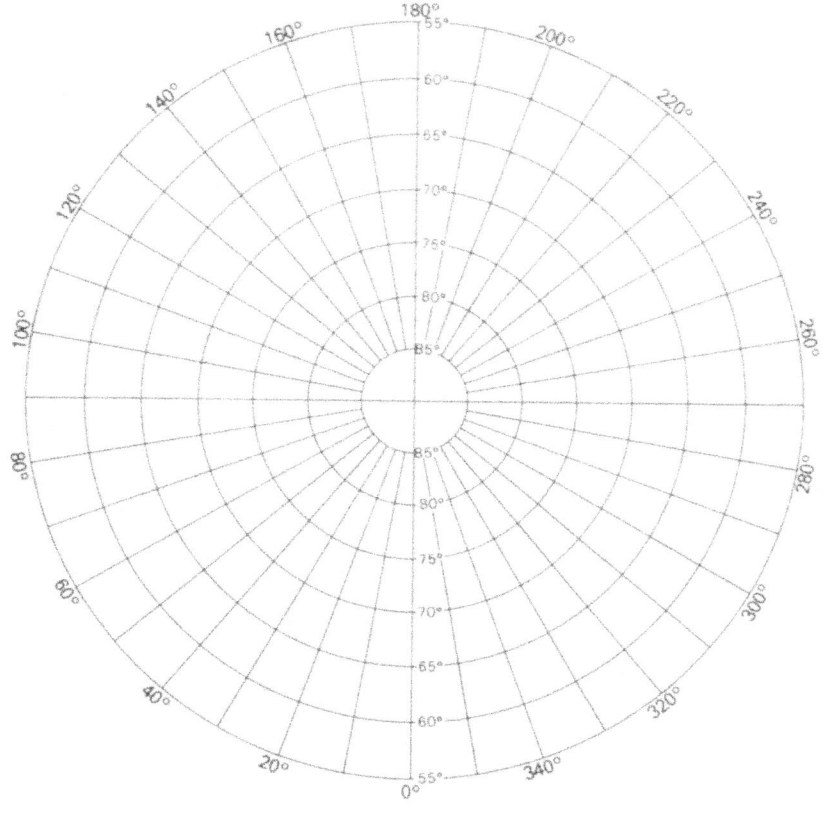

North polar region

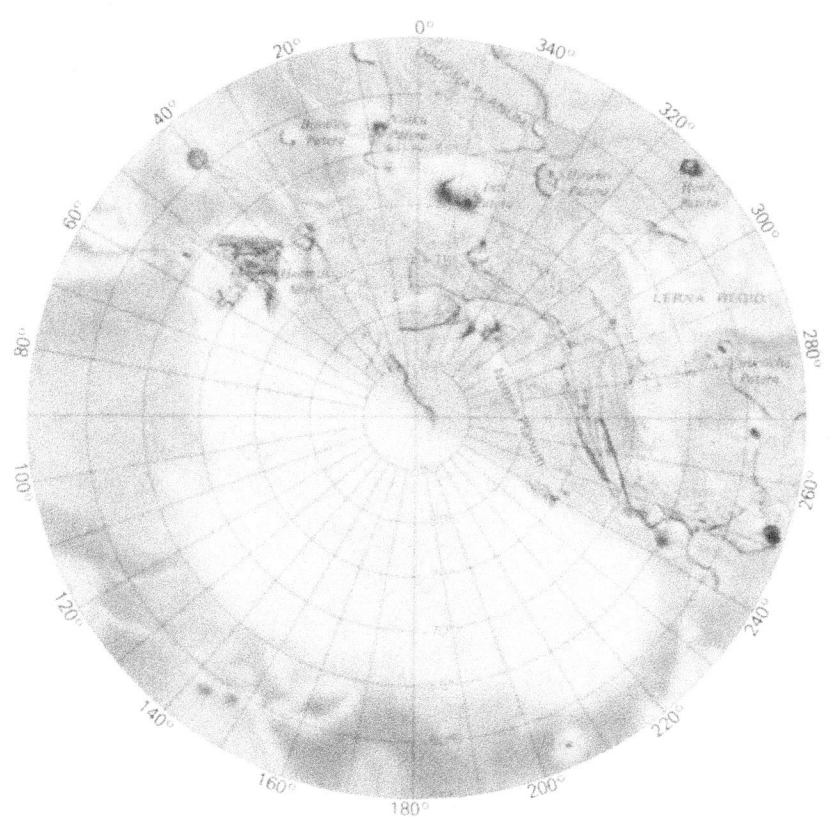

South polar region

Io

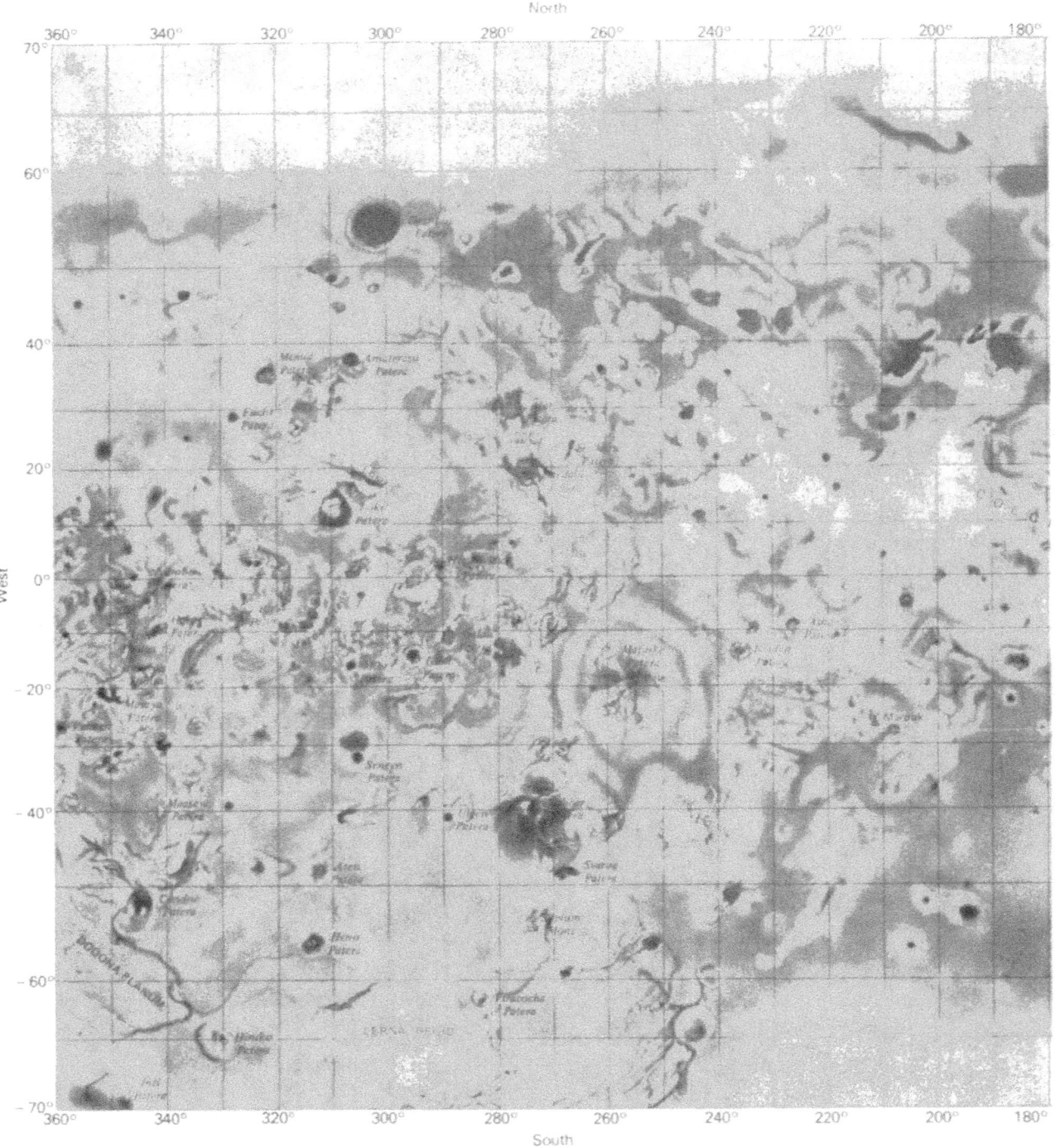

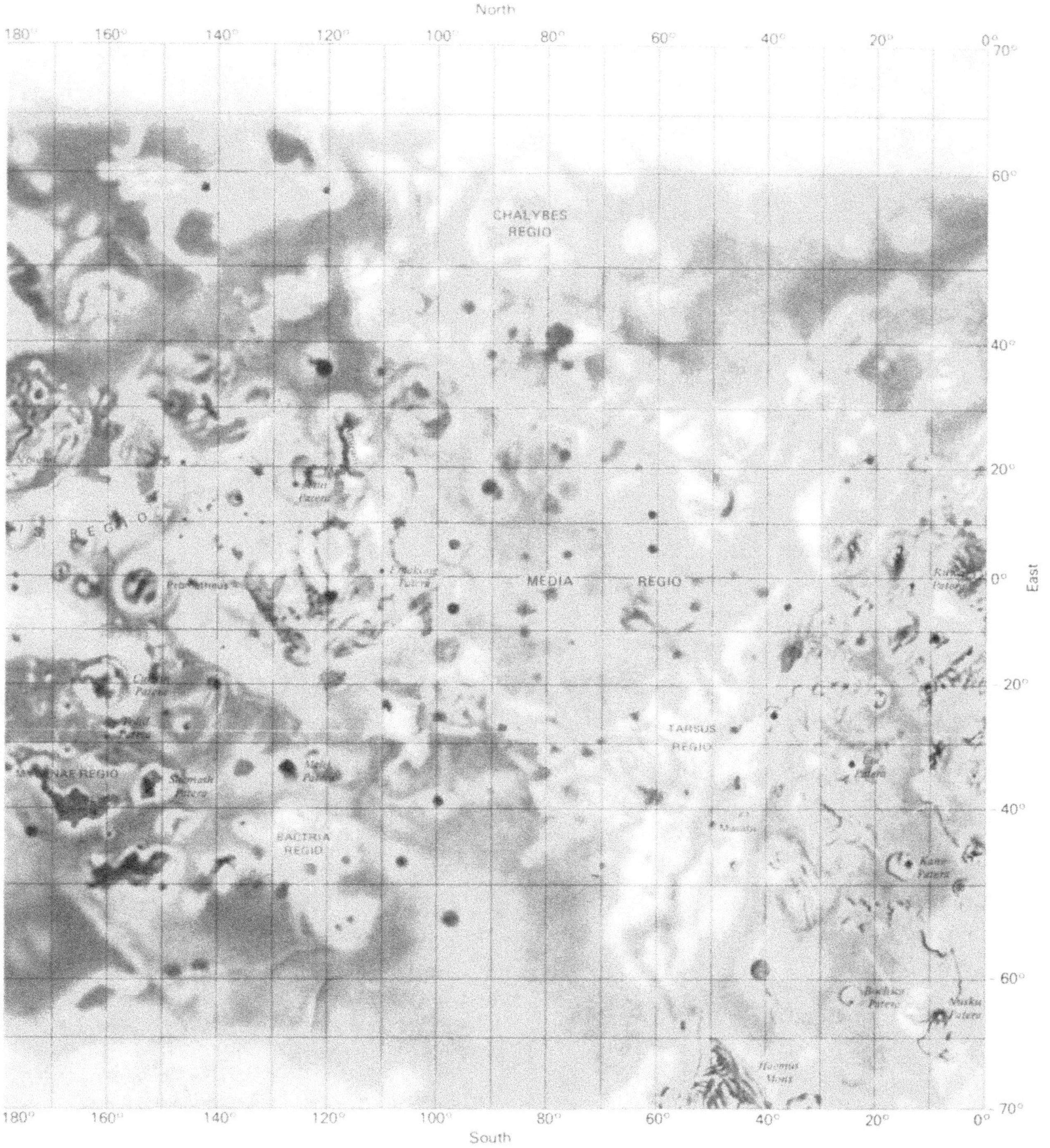

North

180° 160° 140° 120° 100° 80° 60° 40° 20° 0°
70°
60°
CHALYBES
REGIO
40°
20°
REGIO
Phoebus
Palus
MEDIA   REGIO   Kane
Palus
East
0°
20°
Oxia
Palus
TARSUS
REGIO
NAE REGIO
Shimash
...era
...lia
Palus
Masale
40°
BACTRIA
REGIO
Kane
Palus
60°
Buthies
Palus
Nuskis
Palus
Haemus
Mons
70°

180° 160° 140° 120° 100° 80° 60° 40° 20° 0°
South

Scale 1:25 000 000 at 0° latitude
Mercator projection

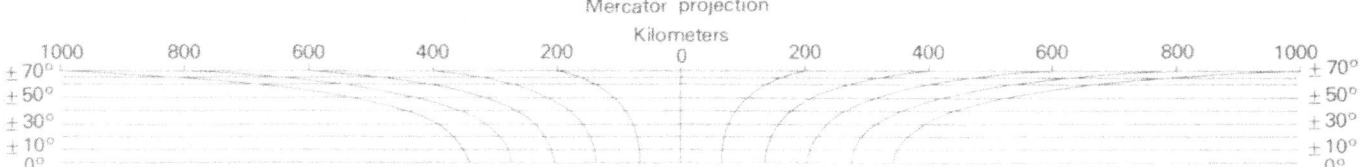

Kilometers

1000   800   600   400   200   0   200   400   600   800   1000
±70°   ±70°
±50°   ±50°
±30°   ±30°
±10°   ±10°
0°   0°

APPENDIX B

VOYAGER SCIENCE TEAMS

Imaging Science

Bradford A. Smith, University of Arizona, Team Leader
Geoffrey A. Briggs, NASA Headquarters
A. F. Cook, Smithsonian Institution
G. E. Danielson, Jr., California Institute of Technology
Merton Davies, Rand Corp.
G. E. Hunt, University College, London
Tobias Owen, State University of New York
Carl Sagan, Cornell University
Lawrence Soderblom, U.S. Geological Survey
V. E. Suomi, University of Wisconsin
Harold Masursky, U.S. Geological Survey

Radio Science

Von R. Eshleman, Stanford University, Team Leader
J. D. Anderson, Jet Propulsion Laboratory
T. A. Croft, Stanford Research Institute
Gunnar Lindal, Jet Propulsion Laboratory
G. S. Levy, Jet Propulsion Laboratory
G. L. Tyler, Stanford University
G. E. Wood, Jet Propulsion Laboratory

Plasma Wave

Frederick L. Scarf, TRW Systems, Principal Investigator
D. A. Gurnett, University of Iowa

Infrared Spectroscopy and Radiometry

Rudolph A. Hanel, Goddard Space Flight Center, Principal Investigator
B. J. Conrath, Goddard Space Flight Center

P. Gierasch, Cornell University
V. Kunde, Goddard Space Flight Center
P. D. Lowman, Goddard Space Flight Center
W. Maguire, Goddard Space Flight Center
J. Pearl, Goddard Space Flight Center
J. Pirraglia, Goddard Space Flight Center
R. Samuelson, Goddard Space Flight Center
Cyril Ponnamperuma, University of Maryland
D. Gautier, Meudon, France
S. Kuman, University of Southern California

Ultraviolet Spectroscopy

A. Lyle Broadfoot, Kitt Peak National Observatory, Principal Investigator
J. L. Bertaux, Service d'Aeronomie du CNRS, France
J. Blamont, Service d'Aeronomie du CNRS, France
T. M. Donahue, University of Michigan
R. M. Goody, Harvard University
A. Dalgarno, Harvard College Observatory
Michael B. McElroy, Harvard University
J. C. McConnell, York University, Canada
H. W. Moos, Johns Hopkins University
M. J. S. Belton, Kitt Peak National Observatory
D. F. Strobel, Naval Research Laboratory
Sushil Atreya, University of Michigan
William R. Sandel, University of Southern California
Donald Shemanski, University of Southern California

Photopolarimetry

Charles W. Hord, University of Colorado, Acting Principal Investigator

D. L. Coffeen, Goddard Institute for Space Studies

J. E. Hansen, Goddard Institute for Space Studies

K. Pang, Science Applications Inc.

Planetary Radio Astronomy

James W. Warwick, University of Colorado, Principal Investigator

Anthony Riddle, Radiophysics, Inc.

Jeffrey Pearce, Radiophysics, Inc.

J. K. Alexander, Goddard Space Flight Center

A. Boischot, Observatoire de Paris, France

W. E. Brown, Jet Propulsion Laboratory

T. D. Carr, University of Florida

Samuel Gulkis, Jet Propulsion Laboratory

F. T. Haddock, University of Michigan

C. C. Harvey, Observatoire de Paris, France

Y. LeBlanc, Observatoire de Paris, France

R. G. Peltzer, University of Colorado

R. J. Phillips, Jet Propulsion Laboratory

D. H. Staelin, Massachusetts Institute of Technology

Magnetic Fields

Norman F. Ness, Goddard Space Flight Center, Principal Investigator

Mario H. Acuna, Goddard Space Flight Center

K. W. Behannon, Goddard Space Flight Center

L. F. Burlaga, Goddard Space Flight Center

R. P. Lepping, Goddard Space Flight Center

F. M. Neubauer, Technische Universitat, F.R.G.

Plasma Science

Herbert S. Bridge, Massachusetts Institute of Technology, Principal Investigator

J. W. Belcher, Massachusetts Institute of Technology

J. H. Binsack, Massachusetts Institute of Technology

A. J. Lazarus, Massachusetts Institute of Technology

S. Olbert, Massachusetts Institute of Technology

V. M. Vasyliunas, Max Planck Institute, F.R.G.

L. F. Burlaga, Goddard Space Flight Center

R. E. Hartle, Goddard Space Flight Center

K. W. Ogilvie, Goddard Space Flight Center

G. L. Siscoe, University of California, Los Angeles

A. J. Hundhausen, High Altitude Observatory

Low-Energy Charged Particles

S. M. Krimigis, Johns Hopkins University, Principal Investigator

T. P. Armstrong, University of Kansas

W. I. Axford, Max Planck Institute, F.R.G.

C. O. Bostrom, Johns Hopkins University

C. Y. Fan, University of Arizona

G. Gloeckler, University of Maryland

L. J. Lanzerotti, Bell Telephone Laboratories

Cosmic Ray

R. E. Vogt, California Institute of Technology, Principal Investigator

J. R. Jokipii, University of Arizona

E. C. Stone, California Institute of Technology

F. B. McDonald, Goddard Space Flight Center

B. J. Teegarden, Goddard Space Flight Center

James H. Trainor, Goddard Space Flight Center

W. R. Webber, University of New Hampshire

APPENDIX C

VOYAGER MANAGEMENT TEAM

NASA Office of Space Science

Thomas A. Mutch, Associate Administrator
for Space Science
Andrew J. Stofan, Deputy Associate
Administrator
Adrienne F. Timothy, Assistant Associate
Administrator
Angelo Guastaferro, Director, Planetary
Division
Rodney A. Mills, Program Manager
Milton A. Mitz, Program Scientist
Walter Jakobowski, Viking and Flight
Support Manager

NASA Office of Space Tracking and Data Systems

William Schneider, Associate Administrator
of Space Tracking and Data Systems
Acquisition
Charles A. Taylor, Director, Network
Operations and Communication Programs
Frederick B. Bryant, Director, Network
System Development Programs
Maurice E. Binkley, Director, DSN Systems

NASA Office of Space Transportation Systems

John F. Yardley, Associate Administrator for
Space Transportation Systems
Joseph B. Mahon, Director, Expendable
Launch Vehicles
Joseph E. McGolrick, Chief, Small and
Medium Launch Vehicles
B. C. Lam, Titan III Manager

Jet Propulsion Laboratory, Pasadena, California

Bruce C. Murray, Laboratory Director
Gen. Charles H. Terhune, Jr., Deputy
Laboratory Director
Robert J. Parks, Assistant Laboratory
Director for Flight Projects
Raymond L. Heacock, Project Manager
Esker K. Davis, Deputy Project Manager
Peter T. Lyman, Deputy Project Manager
Richard P. Laeser, Mission Director
George P. Textor, Deputy Mission Director
Charles E. Kohlhase, Mission Planning Office
Manager
James E. Long, Science Directorate Manager
Charles H. Stembridge, Deputy
Arthur L. Lane, Assistant Project Scientist
for Jupiter
Francis M. Sturms, Sequence Design and
Integration Directorate Manager
Robert K. Wilson, Deputy
Michael J. Sander, Development, Integration
and Test Directorate Manager
Robert G. Polansky, Deputy
Michael W. Devirian, Space Flight Operations
Directorate Manager
Raymond J. Amorose, Deputy
Marvin R. Traxler, Tracking and Data System
Manager
Kurt Heftman, Mission Control and
Computing Center Manager

California Institute of Technology, Pasadena, California

Edward C. Stone, Project Scientist

ADDITIONAL READING

TECHNICAL

Jupiter, T. Gehrels, Ed., U. of Arizona Press, Tucson, 1254 pages (1976).

Space Science Reviews, special Voyager instrumentation issues, *Vol. 21*, No. 2, pgs. 75-232 (November 1977); *Vol. 21*, No. 3, pgs. 234-376 (December 1977).

"Melting of Io by Tidal Dissipation," by S.J. Peale, P. Cassen, and R.T. Reynolds, *Science, Vol. 203*, pgs. 892-894 (2 March 1979).

Science, special Voyager 1 issue, *Vol. 204*, pgs. 945-1008 (1 June 1979).

Nature, special Voyager 1 issue, *Vol. 280*, pgs. 725-806 (30 August 1979).

"Jupiter's Ring," by T. Owen et al., *Nature, Vol. 781*, pgs. 442-446 (11 October 1979).

Science, special Voyager 2 issue, *Vol. 206*, pgs. 925-996 (23 November 1979).

Geophysical Research Letters, special Voyager issue, *Vol. 7*, pgs. 1-68 (January 1980).

NONTECHNICAL

"The Solar System," special issue of *Scientific American, Vol. 223*, No. 3 (September 1975).

"The Galilean Satellites of Jupiter," by D.P. Cruikshank and D. Morrison, *Scientific American, Vol. 234*, No. 5, pgs. 108-116 (May 1976).

Pioneer Odyssey, by R.O. Fimmel, W. Swindell, and E. Burgess, NASA SP-396, 217 pages (1977).

Murmurs of the Earth: The Voyager Interstellar Record, by Carl Sagan et al., Random House, New York, 1978.

"Jupiter and Family," by J. Eberhart, *Science News, Vol. 115*, pgs. 164-173 (17 March 1979).

"The Far-Out Worlds of Voyager," by J.K. Beatty, *Sky and Telescope, Vol. 57*, pgs. 423-427 (May 1979) and pgs. 516-520 (June 1979).

"Return to Jupiter and Co.," by J. Eberhart, *Science News, Vol. 116*, pgs. 19-21 (14 July 1979).

"Voyage to the Giant Planet," by C. Sutton, *New Scientist, Vol. 83*, pgs. 217-220 (19 July 1979).

"Voyager Views Jupiter's Dazzling Realm," by R. Gore, *National Geographic, Vol. 157*, No. 1, pgs. 2-29 (January 1980).

"The Galilean Moons of Jupiter," by L.A. Soderblom, *Scientific American, Vol. 242*, No. 1, pgs. 88-100 (January 1980).

"The Great Red Spot," by D. Schwartzenburg, *Astronomy, Vol. 8*, pgs. 6-13 (July 1980).

"Four New Worlds," by D. Morrison, *Astronomy, Vol. 8* (September 1980).